Neuroprosthetic Supersystems Architecture

Considerations for the Design and Management of Neurocybernetically Augmented Organizations

Matthew E. Gladden

SYNTHYPNION
academic

Neuroprosthetic Supersystems Architecture: Considerations for the Design and
Management of Neurocybernetically Augmented Organizations (First Edition)

Published in the United States of America
by Synthypnion Academic, an imprint of Synthypnion Press LLC

Synthypnion Press LLC
Indianapolis, IN 46227
http://www.synthypnionpress.com

SYNTHYPNION academic
synthypnionpress.com

SYNTHYPNION
academic

ISBN 978-1-944373-07-8 (hardcover print edition)
ISBN 978-1-944373-16-0 (paperback print edition)
ISBN 978-1-944373-08-5 (ebook)
10 9 8 7 6 5 4 3 2
February 2017

For all those working to design new interfaces with technology
That respect human dignity and empower human creativity

Brief Table of Contents

Detailed Table of Contents

Preface

This volume emerged from a project investigating the implications of posthumanizing neuroprosthetics for organizational enterprise architecture that was undertaken at the Institute of Computer Science of the Polish Academy of Sciences in 2015 and 2016 under the title of "Enterprise Architecture for Neurocybernetically Augmented Organizational Systems: The Impact of Posthuman Neuroprosthetics on the Creation of Strategic, Structural, Functional, Technological, and Sociocultural Alignment." This text represents a significant expansion of that work; in particular, it explores in greater depth the converging characteristics of human agents and electronic information systems that result from the growing use of advanced neuroprostheses, the role of military organizations as early adopters of posthumanizing neuroprosthetic technologies, and the use of network topology as a conceptual tool for designing and analyzing the architectures of neuroprosthetic supersystems. I offer my heartfelt thanks to Serge Pukas, Paulina Krystosiak, Robert Pająk, Jacek Koronacki, and everyone affiliated with the Institute of Computer Science for their encouragement and support of my research.

This book also draws on a number of my other previously published texts and conference presentations relating to various aspects of advanced neuroprosthetics and the anticipated organizational and social impacts of such technologies. In particular, I have drawn material from earlier works such as *The Handbook of Information Security for Advanced Neuroprosthetics* (2015); "Utopias and Dystopias as Cybernetic Information Systems: Envisioning the Posthuman Neuropolity" (2015); "Cybershells, Shapeshifting, and Neuroprosthetics: Video Games as Tools for Posthuman 'Body Schema (Re)Engineering'" (2015); "Implantable Computers and Information Security: A Managerial Perspective" (2016); *Sapient Circuits and Digitalized Flesh: The Organization as Locus of Technological Posthumanization* (2016); and "Neural Implants as Gateways to Digital-Physical Ecosystems and Posthuman Socioeconomic Interaction" (2016). Details regarding these works and their specific contribution to this text can be found in this book's bibliography and footnotes.

I am thankful to the faculty, staff, and students of the universities and other research institutions at whose conferences such material was first presented, including those of the Jagiellonian University, the Warsaw University of Technology, the University of Silesia in Katowice, the Centrum Informacji Naukowej i Biblioteka Akademicka (CINiBA) in Katowice, the Faculty of Humanities of the AGH University of Science and Technology, the Facta Ficta Research Centre, the Digital Economy Lab of the University of Warsaw, and the EuroMed Research Business Institute. In particular, I offer my deep gratitude to Krzysztof Maj, Ksenia Olkusz, Magdalena Szczepocka, Natalia Juchniewicz, and Renata Włoch. I am especially grateful to those scholars exploring the social and organizational implications of posthumanizing technologies whom I have had the opportunity to hear present their research and with whom I have enjoyed valuable conversations about these topics – especially Bartosz Kłoda-Staniecko, Agata Kowalewska, Krzysztof Maj, and Magdalena Szczepocka. I am also thankful to Paweł Urbański for sharing his expertise in the field of enterprise architecture through a number of helpful conversations.

As always, I am indebted to my parents, my brother, and other relatives, friends, and colleagues for the support that they have provided me in my research, and I offer my heartfelt thanks to my wife for her insightful counsel, friendship, and continual encouragement. The individuals mentioned above – along with many other conference attendees and anonymous reviewers who have provided feedback on my texts – have contributed much to the development of whatever value this book might hold. Those flaws that the book still possesses after their helpful input are my responsibility alone.

Matthew E. Gladden
Pruszków, December 6, 2016

Introduction

The Nature of Neuroprosthetic Supersystems Architecture

From Neuroprosthetic Devices to Neuroprosthetic Supersystems

This volume is a resource for the design and analysis of *neuroprosthetic supersystems*, which can be defined as organizations – either small or large, simple or complex – whose human members have been neuroprosthetically augmented. Such supersystems must be distinguished from the neuroprostheses that make them possible. In itself, a neuroprosthesis is an artificial device that is integrated into the neural circuitry of a human being, typically in order to support or participate in the sensory, cognitive, or motor processes of its human host; however, it can also be used to gather data about its host's biological processes and transmit it to an external computer for medical, archival, or potentially even surveillance purposes.[1] It is possible to view a neuroprosthesis from the perspective of its existence as a self-contained technological device with its own internal structures and dynamics, independently of the way in which it acts on the organism of its human host. When understood as such, the design of neuroprostheses draws heavily on fields such as computer architecture, electronics engineering, and robotics and is in many ways comparable to the design of other complex, specialized computerized devices such as smartphones, communications satellites, or self-driving vehicles. Numerous excellent texts already address the engineering of neuroprostheses understood as this sort of information system that receives input, processes it, and generates output, and this topic is not the focus of the present volume.

---

[1] See Lebedev, "Brain-Machine Interfaces: An Overview" (2014); Gladden, "Enterprise Architecture for Neurocybernetically Augmented Organizational Systems" (2016); Lorence et al., "Transaction-Neutral Implanted Data Collection Interface as EMR Driver: A Model for Emerging Distributed Medical Technologies" (2009); Bonaci et al., "App Stores for the Brain" (2015), p. 35; Luber et al., "Non-invasive brain stimulation in the detection of deception: Scientific challenges and ethical consequences" (2009); and Gladden, *The Handbook of Information Security for Advanced Neuroprosthetics* (2015).

Similarly, in addition to constituting a computerized system in itself, a neuroprosthesis also serves as a constituent element of the larger biological-electronic system that it creates with the organism of its human host, insofar as the neuroprosthesis and its host's neural circuitry become integrated to form a new hybrid entity, or host-device system. The engineering of such 'cyborgs' and the brain-machine interfaces that constitute their distinguishing characteristic draws largely on fields such as biomedical engineering and neurocybernetics; it is also not the primary focus of this volume.[2] Rather, the subject that will be investigated in this text is the intentional creation of higher-order supersystems that allow multiple neuroprosthetically augmented human beings to interact with one another and with external information systems in order to accomplish some shared task. In essence, this can be understood as the work of designing and managing neuroprosthetically enhanced *organizations*.[3]

Unlike the design of neuroprosthetic devices or host-device systems, the design of neuroprosthetically enhanced organizations is not primarily a question of computer architecture, electronics engineering, biomedical engineering, or neurocybernetics; instead, it is largely a matter of organizational design, organizational architecture, enterprise architecture, and management cybernetics. It is about creating decision-making structures, processes of interaction, and technological systems that allow neuroprosthetically augmented individuals to collaborate in a manner that generates some strategic, operational, or tactical value for their organization, that is efficient and effective, and which satisfies relevant financial, legal, and ethical requirements.

---

[2] The growing number of ways in which even ordinary human beings are undergoing processes of 'cyborgization' to become functionally integrated with technological systems – and the broader forces of technologization and posthumanization of which the creation of cyborgs is an example – is discussed from various perspectives in, e.g., Haraway, *Simians, Cyborgs, and Women: The Reinvention of Nature* (1991); Tomas, "Feedback and Cybernetics: Reimaging the Body in the Age of the Cyborg" (1995); Hayles, *How We Became Posthuman: Virtual Bodies in Cybernetics, Literature, and Informatics* (1999); Anderson "Augmentation, symbiosis, transcendence: technology and the future(s) of human identity" (2003); Clark, *Natural-born cyborgs: Minds, Technologies, and the Future of Human Intelligence* (2004); Fleischmann, "Sociotechnical Interaction and Cyborg–Cyborg Interaction: Transforming the Scale and Convergence of HCI" (2009); and Gladden, *Sapient Circuits and Digitalized Flesh: The Organization as Locus of Technological Posthumanization* (2016).

[3] Lune defines an organization as "a group with some kind of name, purpose, and a defined membership" that possesses "a clear boundary between its inside and its outside" and which can take the form of either a formal organization with clearly defined roles and rules, an informal organization with no explicitly defined structures and processes, or a semi-formal organization that possesses nominal roles and guidelines that in practice are not always observed. See Lune, *Understanding Organizations* (2010), p. 2. Meanwhile, Daft et al. define organizations as "(1) social entities that (2) are goal-directed, (3) are designed as deliberately structured and coordinated activity systems, and (4) are linked to the external environment." See Daft et al., *Organization Theory and Design* (2010), p. 10.

While such neuroprosthetically enhanced organizations can correctly be understood as 'systems' in themselves, for the sake of clarity we will generally use the term 'neuroprosthetic device' to refer to a neuroprosthesis, 'host-device *system*' to refer to a single neuroprosthetically augmented human being, and 'neuroprosthetic *supersystem*' to refer to an organization comprising multiple neuroprosthetically augmented human beings.

The Organization of This Text

This book is divided into two parts. The three chapters of Part I provide an introduction to key building-blocks of neuroprosthetic supersystems by investigating the nature of neuroprosthetic devices and host-device systems and exploring the unique manner in which neuroprosthetically augmented individuals exist in and interact with their digital-physical environments. The four chapters of Part II then consider the reasons why organizations would seek to deploy neuroprosthetic supersystems, ways in which the discipline of enterprise architecture can be employed to manage such supersystems' implementation, the ways in which neuroprosthetic augmentation will impact an organization's use of enterprise architecture, and key principles for the development of effective architectures for neuroprosthetically augmented organizations. Below we describe each of these chapters in more detail.

Part I: Elements of Neuroprosthetic Supersystems

Chapter 1, "An Ontology of the Neuroprosthesis as Computing Device," formulates an ontology that can be employed to describe the fundamental characteristics of a neuroprosthesis in its role as a *computing device*. The ontology draws on existing neuroprosthetic device typologies and ontologies developed for other kinds of devices such as mobile devices and robotic systems. It describes four key aspects that shape the functioning of a neuroprosthesis as computing device: (1) the device's external context (including the human agents who participate in its development and use, factors impacting its availability, and its relationship to the body of its human host); (2) physical components of the neuroprosthesis (including the device's basic morphology, input and output mechanisms, and computational substrate); (3) processes utilized by the device (including computational processes and input and output modalities); and (4) the types of information generated or handled by the device (which may include data regarding the device's status and environment, data regarding the cognitive and biological processes of the device's human host, and procedural and declarative knowledge). The use of such an ontology allows the functionality of a neuroprosthesis as a computing device to be more easily analyzed or designed and facilitates interoperability between neuroprostheses, their human hosts and users, and external computer systems.

In Chapter 2, "Integrating Neuroprostheses into Human Sensory, Cognitive, and Motor Processes," a different sort of ontology is developed that envisions, captures, and describes the full range of ways in which a neuroprosthesis may participate in the sensory, cognitive, and motor processes of its human host. By considering anticipated future developments in neuroprosthetics and adopting a generic biocybernetic approach, this ontology is able to account for therapeutic neuroprostheses already in use as well as future types of neuroprostheses that are expected to be deployed for purposes of human enhancement.

The ontology encompasses three areas. First, a neuroprosthesis may participate in its host's processes of *sensation* by (a) detecting stimuli such as photons, sound waves, or chemicals; (b) fabricating sense data, as in the case of virtual reality systems; (c) storing sense data; (d) transmitting sense data within a neural pathway; (e) enabling its host to experience sense data through a sensory modality such as vision, hearing, taste, smell, touch, balance, heat, or pain; or (f) creating mappings of sensory routes – e.g., in order to allow sensory substitution. Second, a neuroprosthesis may participate in processes of *cognition* by (a) creating a basic interface between the device and the host's conscious awareness or affecting the host's (b) perception, (c) creativity, (d) memory and identity, or (e) reasoning and decision-making. Third, a neuroprosthesis may participate in processes of *motion* by (a) detecting motor instructions generated by its host's brain; (b) fabricating motor instructions, as in the case of a medical device controlled by software algorithms rather than its host's volitions; (c) storing motor instructions; (d) transmitting motor instructions as within a neural pathway; (e) effectuating physical action within effectors such as natural biological muscles and glands, synthetic muscles, robotic actuators, video screens, audio speakers, or wireless transmitters; (f) allowing the expression of volitions through motor modalities such as language, paralanguage, and locomotion; or (g) creating mappings of motor routes. The use of this type of ontology allows easier, more systematic, and more robust analysis of the biocybernetic role of a neuroprosthesis within its host-device system.

Yet another kind of ontology is developed in Chapter 3, "An Ontology of Neuroprostheses as Instruments of Cyborgization: Portals to the Experience of Posthumanized Digital-Physical Worlds." The incorporation of a neuroprosthetic device into one's being at the physical, cognitive, and social levels constitutes a form of 'cyborgization' that imposes new constraints on one's existence while simultaneously opening a path to new forms of experience. This chapter explores the boundaries of this qualitatively novel form of being by formulating an ontology of the neuroprosthesis as an instrument that shapes the way in which its human host experiences and acts within emerging posthumanized digital-physical ecosystems.

This ontology addresses four main roles that a neuroprosthetic device may play in this context. First, a neuroprosthesis may serve as a means of human augmentation by altering the cognitive and physical capacities possessed by its host. Second, it may manipulate the contents of information produced or utilized by its human host. Third, a neuroprosthesis may shape the manner in which its host inhabits a digital-physical body and external environment. And finally, a neuroprosthesis may regulate the autonomous agency possessed and experienced by its host. The use of this type of conceptual framework can allow researchers to better understand the psychological, social, and ethical ramifications of such technologies and can enable the architects of neuroprosthetic systems and the digital-physical ecosystems within which their human hosts operate to formulate principles of design and management that minimize the dangers and maximize benefits for the neuroprosthetically augmented inhabitants of such environments.

Part II: Enterprise Architecture for Neuroprosthetic Supersystems

Chapter 4, "The Organizational Deployment of Posthumanizing Neuro-prostheses," examines the types of organizations that are already working to intentionally deploy neuroprosthetic technologies for human enhancement among their workforce (or are expected to do so), factors that can motivate their adoption of such technologies, and the organizational roles that such neurotechnologies may play.

The current state of therapeutic neuroprosthetic device use is presented, along with an overview of posthumanizing neuroprostheses and the types of enhanced capacities that they offer human workers that may be relevant to organizations. A range of factors incentivizing or discouraging the organizational deployment of posthumanizing neuroprostheses is identified and discussed. The organizational roles of therapeutic and posthumanizing neuroprostheses are then analyzed. On the one hand, many organizations already unknowingly incorporate workers possessing therapeutic neuroprostheses. Meanwhile, two key paths for the organizational deployment of posthumanizing neuroprostheses are highlighted. First is the 'transitional augmentation' of human workers as a stopgap measure on the path to eventual full automation of business processes through the use of artificial intelligence. The second path involves retaining human workers in particular positions because exogenous factors (such as legal, ethical, or marketing requirements) mandate that human agents fill those roles, while augmenting the workers so that they can perform more competitively.

It is noted that military organizations play a key role among organizations likely to be early adopters of posthumanizing neuroprostheses. Known and hypothesized military programs for neuroprosthetic enhancement are discussed, along with characteristics of military organizations that remove obstacles that render the deployment of neuroprostheses impractical for most organizations. Other types of organizations are highlighted that share some traits as potential early adopters. Finally, enterprise architecture (EA) is discussed as a preferred management tool for many organizations that are likely to be early adopters. While EA does not directly address the serious ethical and legal questions raised by posthumanizing neuroprosthetics, it can facilitate the functional aspects of integrating neuroprosthetically augmented workers into an organization's personnel structures, business processes, and IT systems.

The potential uses of enterprise architecture as a tool for managing the organizational deployment of neuroprostheses is explored further in Chapter 5, "An Introduction to Enterprise Architecture in the Context of Technological Posthumanization." Enterprise architecture seeks to generate alignment between an organization's electronic information systems, human resources, business processes, workplace culture, mission and strategy, and external ecosystem in order to increase the organization's ability to manage complexity, resolve internal conflicts, and adapt proactively to environmental change. In this chapter, an introduction to the definition, history, organizational role, objectives, benefits, mechanics, and popular implementations of enterprise architecture is presented. The historical shift from IT-centric to business-centric definitions of EA is reviewed, along with the difference between 'hard' and 'soft' approaches to EA. The unique organizational role of EA is highlighted by comparing it with other management disciplines and practices.

The creation of alignment is then explored as the core mechanism by which EA achieves advantageous effects. Different kinds of alignment are defined, the history of EA as a generator of alignment is investigated, and EA's relative effectiveness at creating different types of alignment is candidly assessed. Attention is given to the key dynamic by which alignment yields deeper integration of an organization's structures, processes, and systems, which in turn grants the organization greater agility – which itself enhances the organization's ability to implement rapid and strategically directed change. The types of tasks undertaken by enterprise architects are discussed, and a number of popular enterprise architecture frameworks are highlighted. A generic EA framework is then presented as a means of discussing elements such as architecture domains, building blocks, views, and landscapes that form the core of many EA frameworks. The role of modelling languages in documenting EA plans is also addressed.

In light of enterprise architecture's strengths as a tool for managing the deployment of innovative forms of IT, it is suggested that by adopting EA initiatives of the sort described here, organizations may better position themselves to address the new social, economic, and operational realities presented by emerging posthumanizing technologies such as those relating to social robotics, nanorobotics, artificial life, genetic engineering, neuroprosthetic augmentation, and virtual reality.

Chapter 6, "The Deepening Fusion of Human Personnel and Electronic Information Systems," explores in more depth the implications of neuroprosthetic augmentation for enterprise architecture. When designing target architectures for organizations, EA has historically relied on a set of assumptions regarding the physical, cognitive, and social capacities of the human beings serving as organizational members. In this chapter we explore the fact that for those organizations that intentionally deploy posthumanizing neuroprosthetic technologies among their personnel, such traditional assumptions no longer hold true: the use of advanced neuroprostheses intensifies the ongoing structural, systemic, and procedural fusion of human personnel and electronic information systems in a way that provides workers with new capacities and limitations and transforms the roles available to them.

Such use of neuroprostheses has the potential to affect an organization's workers in three main areas. First, the use of neuroprostheses may affect workers' physical form, as reflected in the physical components of their bodies, the role of design in their physical form, their length of tenure as workers, the developmental cycles that they experience, their spatial extension and locality, the permanence of their physical substrates, and the nature of their personal identity. Second, neuroprostheses may affect the information processing and cognition of neurocybernetically augmented workers, as manifested in their degree of sapience, autonomy, and volitionality; their forms of knowledge acquisition; their locus of information processing and data storage; their emotionality and cognitive biases; and their fidelity of data storage, predictability of behavior, and information security vulnerabilities. Third, the deployment of neuroprostheses can affect workers' social engagement, as reflected in their degree of sociality; relationship to organizational culture; economic relationship with their employers; and rights, responsibilities, and legal status.

While ethical, legal, economic, and functional factors will prevent most organizations from deploying advanced neuroprostheses among their personnel for the foreseeable future, those select organizations that are already working to develop such technologies and implement them among their personnel will be forced to adapt their enterprise architectures to accommodate the new realities of human-computer integration brought about by posthumanizing neuroprosthetic technologies described in this chapter.

Finally, Chapter 7, "From Virtual Teams to Hive Minds," develops a model based on network topology that can be used to analyze or engineer the structures and dynamics of an organization in which neuroprosthetic technologies are employed to enhance the abilities of human personnel. It is argued that the expanded sensory, cognitive, and motor capacities provided by posthumanizing neuroprostheses may enable human beings possessing such technologies to collaborate using novel types of organizational structures that differ from the traditional structures that are possible for unaugmented human beings. The concept of network topology is then presented as a concrete approach to analyzing or engineering such neuroprosthetic supersystems. A number of common network topologies such as chain, linear bus, tree, ring, hub-and-spoke, partial mesh, and fully connected mesh topologies are discussed and their relative advantages and disadvantages noted.

Drawing on the notion of different architectural 'views' employed in enterprise architecture, we formulate a topological model that incorporates five views that are relevant for neuroprosthetic supersystems: the (1) physical and (2) logical topologies of the neuroprosthetic devices themselves; (3) the natural topology of social relations of the devices' human hosts; (4) the topology of the virtual environments, if any, created and accessed by means of the neuroprostheses; and (5) the topology of the brain-to-brain communication, if any, facilitated by the devices. Potential uses of the model are illustrated by applying it to four hypothetical types of neuroprosthetic supersystems: (1) an emergency medical alert system incorporating body sensor networks (BSNs); (2) an array of centrally hosted virtual worlds; (3) a 'hive mind' administered by a central hub; and (4) a distributed hive mind lacking a central hub. It is our hope that models such as the one formulated here will prove useful not only for engineering neuroprosthetic supersystems to meet functional requirements but also for analyzing the legal, ethical, and social aspects of potential or existing supersystems – to help ensure that the organizational deployment of neuroprosthetic technologies does not undermine the wellbeing of such devices' human users or of societies as a whole.

Part I

Elements of Neuroprosthetic Supersystems

Chapter One

An Ontology of the Neuroprosthesis as Computing Device

Abstract. The structure and behavior of a neuroprosthetic device can be analyzed from different perspectives. In this text, we formulate an ontology that can be employed to describe the fundamental characteristics of a neuroprosthesis in its role as a *computing device*. The ontology draws on existing neuroprosthetic device typologies and ontologies developed for other kinds of devices such as mobile devices and robotic systems. It describes four key aspects that shape the functioning of a neuroprosthesis as computing device: 1) the device's external context (including the human agents who participate in its development and use, factors impacting the device's availability, and the device's relationship to the body of its human host); 2) physical components of the neuroprosthesis (including the device's basic morphology, input and output mechanisms, and computational substrate); 3) processes utilized by the device (including computational processes and input and output modalities); and 4) the types of information generated or handled by the device (which may include data regarding the device's status and environment, data regarding the cognitive and biological processes of the device's human host, and procedural and declarative knowledge). The use of such an ontology allows the functionality of a neuroprosthesis as a computing device to be more easily analyzed or designed and facilitates interoperability between neuroprostheses, their human hosts and users, and external computer systems.

Introduction

A neuroprosthesis may be defined as *an artificial device that is integrated into the neural circuitry of a human being*, thereby creating a neurocybernetic host-device system that possesses both human and computerized elements.[1] In this chapter, we develop an ontology of the neuroprosthesis in its nature as a computing device. Subsequent chapters will consider the neuroprosthesis from other perspectives, formulating complementary ontologies of such tools as biocybernetic instruments that become integrated into the neural

---

[1] See Lebedev, "Brain-Machine Interfaces: An Overview," *Translational Neuroscience* 5, no. 1 (2014), and Gladden, "Enterprise Architecture for Neurocybernetically Augmented Organizational Systems" (2016).

circuitry of a human organism to participate in processes of sensation, cognition, and motor action and as instruments of 'cyborgization' that shape how such individuals experience posthumanized digital-physical ecosystems.[2]

The Need for a Neuroprosthetic Device Ontology

Before effective enterprise architectures can be developed for organizations deploying posthuman neuroprostheses among their personnel, it is necessary to possess an appropriate ontology for such pieces of equipment in their role as computing devices. Such an ontology should define the range of possible values for relevant characteristics that are critical for enterprise architecture and whose values can vary across the universe of neuroprosthetic devices. The ontology should include both the physical characteristics, processes, and contexts and relationships possessed by such devices.

Existing Neuroprosthetic Device Typologies

A number of approaches for classifying neuroprostheses exist in the form of device typologies developed by researchers in the field of neuroprosthetics. Such existing typologies are usually too abstract to be applied directly to the development of enterprise architectures. They are typically functional in nature: for example, a neuroprosthetic device might be classified based on the nature of its interface with the brain's neural circuitry (i.e., as sensory, motor, bidirectional sensorimotor, or cognitive[3]), its purpose (i.e., for restoration, diagnosis, identification, or enhancement[4]), or its location relative to the brain and body (i.e., as non-invasive, partially invasive, or invasive[5]). By analyzing and classifying neuroprostheses primarily according to a single aspect of its operation, such typologies lack the comprehensive approach and level of detail required for an effective device ontology.

Existing Device Ontologies for Other Advanced Technologies

Researchers and device manufacturers have, for example, made efforts to create generic ontologies for robotic systems[6] and mobile devices;[7] however, there does not yet exist a widely accepted ontology that focuses on the char-

---

[2] For a biomedical engineering perspective on the use of neuroprosthetic devices to create human beings who are, in effect, 'cyborgs,' see Lebedev (2014). Regarding potential implications of processes of cyborgization for organizational management, see Berner, *Management in 20XX: What Will Be Important in the Future – A Holistic View* (2004). For a discussion of processes of cyborgization from the perspective of critical and cultural posthumanism, see Haraway, *Simians, Cyborgs, and Women: The Reinvention of Nature* (1991); Hayles, *How We Became Posthuman: Virtual Bodies in Cybernetics, Literature, and Informatics* (1999); and Clark, *Natural-born cyborgs: Minds, Technologies, and the Future of Human Intelligence* (2004).

[3] Lebedev (2014).

[4] Gasson, "Human ICT Implants: From Restorative Application to Human Enhancement" (2012), p. 25.

[5] Gasson (2012), p. 14.

[6] See Prestes et al., "Towards a Core Ontology for Robotics and Automation" (2013).

[7] See "FIPA Device Ontology Specification" (2002).

acteristics of neuroprosthetic devices that are relevant for enterprise architecture. In part, the lack of such formal tools is an effect of the relatively low level of interaction and collaboration between the fields of enterprise architecture and biomedical engineering – a phenomenon that is not surprising, given the historical paucity of areas of convergence (such as biometric security and ergonomics) in which organizational pressures drive the disciplines to collaborate.[8]

Developing a Neuroprosthetic Device Ontology for Enterprise Architecture

In this text we present a template for a neuroprosthetic device ontology that is both broad and detailed enough to serve as a resource for the design of enterprise architectures for organizations deploying neuroprosthetic technologies.[9] The ontology draws on a review of current literature and practice and expected future developments in the fields of neuroprosthetics, implantable computing, and enterprise architecture and on existing typologies of neuroprosthetic devices and ontologies developed for other technologies (such as mobile devices more generally).

It should be noted that because many neuroprosthetic devices are also implantable medical devices or mobile devices, any generalized ontology for use in enterprise architecture that is applicable to IMDs or mobile devices will also apply to the relevant kinds of neuroprosthetic devices; however, such ontologies do not explicitly account for the unique device features and capacities that make the deployment of neuroprostheses attractive for certain organizations.[10] The specialized ontology developed here can be understood as a more focused extension of such generalized ontologies that analyzes the form and functioning of neuroprostheses as computing devices that must be accounted for by an organization's enterprise architecture.[11]

---

[8] This resembles the low degree of collaboration historically seen, e.g., between the fields of biomedical engineering and information security. See Clark & Fu, "Recent Results in Computer Security for Medical Devices" (2012).

[9] An earlier variant of this ontology developed for use in maintaining the information security of neuroprosthetic devices and systems can be found in Gladden, *The Handbook of Information Security for Advanced Neuroprosthetics* (2015).

[10] Among the primary early adopters of such technologies within an organizational context are expected to be military agencies and departments. Regarding potential military applications of neurotechnologies for human enhancement, see, e.g., Schermer, "The Mind and the Machine. On the Conceptual and Moral Implications of Brain-Machine Interaction" (2009); Brunner & Schalk, "Brain-Computer Interaction" (2009); Coker, "Biotechnology and War: The New Challenge" (2004); Graham, "Imagining Urban Warfare: Urbanization and U.S. Military Technoscience" (2008), p. 36; Krishnan, "Enhanced Warfighters as Private Military Contractors" (2015); Falconer, "Defense Research Agency Seeks to Create Supersoldiers" (2003); Moreno, "DARPA On Your Mind" (2004); Clancy, "At Military's Behest, Darpa Uses Neuroscience to Harness Brain Power" (2006); and Kourany, "Human Enhancement: Making the Debate More Productive" (2013), pp. 992-93.

[11] For an overview of the ways in which different organizational computing technologies are treated by enterprise architecture frameworks, see, e.g., Rohloff, "Framework and Reference for Architecture Design" (2008).

Fig. 1: An Ontology of the Neuroprosthesis as Computing Device

I. Context and relations

A. Human agents
1. Designer
2. Manufacturer
3. Regulators
4. Owner
5. Host
6. User or operator

B. Factors impacting availability
1. Licensing and legality
2. Cost
3. Required expertise
4. Required maintenance
5. Required user customization
6. Reusability

C. Physiological context
1. Operational lifespan
2. Physical situation in the body
   a. Physical visibility and discoverability
   b. Physical access to manipulate the device
   c. Remote discoverability
3. Level of neurocognitive interface
4. Health sensitivity and criticality

II. Physical components

A. Device morphology
1. Identity and unitarity
2. Size
3. Materials
4. System participation

B. Input mechanisms: physical receivers of matter, energy, and information
1. Power supply
2. Physical ports and controls
3. Environmental sensors
   a. Specialized data reception mechanisms
   b. Unpurposeful receptors
4. Neuronal input mechanisms

C. Output mechanisms: physical emitters of matter, energy, and information
1. Physical effectors
2. Environmental emitters
   a. Specialized data transmission mechanisms
   b. Unpurposeful emitters
3. Neuronal output mechanisms

D. Computational substrate
1. CPU-based computers
   a. Processors
   b. Memory
2. Physical neural networks

III. Processes

A. Computational processes and computer control
  1. Level of autonomy
     a. Capacity for control by its human host
     b. Capacity for control via teleoperation
  2. Computer programs
     a. Operating system
     b. Applications
     c. Software update methods
  3. Neural computing
     a. Patterns of connection
     b. Learning mechanism
     c. Activation function

B. Input modalities and signals interpretable by a device
  1. Machine communication protocols
  2. Environmental signals
  3. Natural human communication
     a. Textual verbal input
     b. Oral verbal input
     c. Nonverbal input
  4. The host's biological processes
     a. Host's cognitive activity and patterns
     b. Host's motor activity and actions

C. Output modalities and signals expressible by a device
  1. Machine communication protocols
  2. Environmental signals
  3. Natural human communication
     a. Textual verbal output
     b. Oral verbal output
     c. Nonverbal output
  4. The host's biological processes
     a. Host's cognitive activity and patterns
     b. Host's motor activity
     c. Control of other organs of the host

IV. Data, information, and knowledge received, generated, stored, and transmitted

A. Data regarding a device's status and diagnostics
  1. Received by a device
  2. Generated by the device
  3. Stored on the device
  4. Transmitted by the device

B. Environmental data
  1. Received by a device
  2. Generated by the device
  3. Stored on the device
  4. Transmitted by the device

C. Data regarding a host's physical status and biological processes
  1. Received by a device
  2. Generated by the device
  3. Stored on the device
  4. Transmitted by the device

D. Data regarding or information contained in a host's cognitive activity
  1. Received by a device
  2. Generated by the device
  3. Stored on the device
  4. Transmitted by the device

E. Procedural knowledge
  1. Received by a device
  2. Generated by the device
  3. Stored on the device
  4. Transmitted by the device

F. Declarative knowledge
  1. Received by a device
  2. Generated by the device
  3. Stored on the device
  4. Transmitted by the device

As illustrated in Figure 1, our ontology is organized into four areas which allow neuroprostheses to be categorized according to their context and relations; their physical components; their processes; and the data, information, and knowledge that is received, generated, stored, and transmitted by the devices. Each area comprises a number of individual characteristics which are addressed in detail in the following sections. Together, a device's values for these variables constitute the profile of that particular model of neuroprosthesis. The ontology presumes that a neuroprosthetic device is an implantable unit, but it can also be applied to non-implantable external systems, in which case some of the characteristics may not be relevant.

I. First Area: Context and Relations

This first part of the ontology describes the physical, socioeconomic, and political context within which a posthuman neuroprosthesis is being deployed. This context helps determine which individuals receive neuroprostheses and for what purposes they may be used. Specific characteristics in this area are:

A. Human Agents

The human agents and organizations comprising human agents who are in different ways responsible for a neuroprosthetic device's functioning[12] include its:

1. Designer

The designer[13] of a neuroprosthesis develops the plans for its physical structure and its core processes and behaviors; together, these elements determine the basic capabilities of the device. Such designers may work for a military research agency, a university research laboratory, a medical facility, a commercial firm, or some other organization.

2. Manufacturer

A neuroprosthetic device's manufacturer physically instantiates the device's design by producing one or more copies of the neuroprosthesis.[14] In the

---

[12] For an overview of such human agents and their roles in the development and deployment of neuroprostheses, see Gladden, "Information Security Concerns as a Catalyst for the Development of Implantable Cognitive Neuroprostheses" (2016).

[13] Regarding the importance of a neuroprosthetic device's designer, see, e.g., Clark & Fu (2012); McCullagh et al., "Ethical Challenges Associated with the Development and Deployment of Brain Computer Interface Technology" (2013); and *Content of Premarket Submissions* (2014), p. 1.

[14] For the significance of a medical device's manufacturer, see, e.g., *Content of Premarket Submissions* (2014), p. 1. For a device ontology that incorporates manufacturer information, see "FIPA Device Ontology Specification" (2002). Each manufacturer possesses its own unique history of

case of highly sensitive posthuman neuroprostheses produced by a commercial contractor for use by military, police, or other governmental organizations, the manufacturer may not be authorized to offer such devices for sale to other potential customers; in other cases, a device's manufacturer may be free to develop and implement a strategy of providing such devices to other organizations (or even to the general public) as part of its overall business model.

3. Regulators

Regulators exercise legal and political power at the local, national, or international level.[15] They may require or forbid a neuroprosthesis to possess certain physical components, processes, or functions and may mandate or restrict its distribution and use for certain classes of users. Multiple jurisdictions may be involved, since a neuroprosthetic device's human host can travel between and within geographically dispersed locations and devices can potentially transmit data to or be controlled from remote locations that are physically removed from the site at which the device and its host are currently located.

4. Owner

The owner of a neuroprosthesis may or may not be the human host in whom the device is implanted, may or may not oversee the device's operation and maintenance once it is deployed, and may or may not own the rights to intellectual property that is produced by or stored in it.[16] In the case of neuroprostheses deployed by governmental organizations, a single organization may serve as a device's owner and operator and potentially also its designer.

---

business practices which may include a track record of periodically producing flawed products and responding effectively or ineffectively to address such flaws; see *NIST Special Publication 800-100: Information Security Handbook: A Guide for Managers* (2006).

[15] On the role of regulators and regulation, see McCullagh et al. (2013); Patil & Turner, "The Development of Brain-Machine Interface Neuroprosthetic Devices" (2008); and Kosta & Bowman, "Implanting Implications: Data Protection Challenges Arising from the Use of Human ICT Implants" (2012).

[16] The word 'owner' refers here to the entity that enjoys the right to possess, use, and transfer a device and its economic benefits. This differs from the specialized use of the term 'owner' in the field of information security to refer to a responsible party within an organization. In that context, an 'information system owner' is the individual within an organization who is "responsible for the overall procurement, development, integration, modification, and operation and maintenance of the information system," while an 'information owner' is an individual who has been given "authority for specified information and is responsible for establishing the controls for information generation, collection, processing, dissemination, and disposal." See *NIST SP 800-100* (2006), p. 69.

If a device is leased commercially to an end user rather than sold, the device's manufacturer (or an intermediary firm) may remain the device's owner.

5. Host

A critical actor is the human host into whose neural circuitry a neuroprosthesis has been temporarily or permanently integrated – typically through a process of surgical implantation. Depending on the nature of the device, the host may or may not realize that it has been implanted[17] and may or may not have an ability to consciously control or exploit the device's functioning; thus the device's host may be a different individual than its user or operator. From the perspective of enterprise architecture, the active participation of a device's human host may or may not be required in order to maximize the device's organizational impact. Human hosts who are unaware of a device's presence or who have been poorly trained in its use may inadvertently undermine its operation; a host who is aware of a device's existence may also intentionally compromise the device's secure and effective functioning for a number of reasons.[18]

6. User or Operator

The user (or operator) of a neuroprosthesis is an agent who possesses the ability to monitor and purposefully control or exploit the device's functioning.[19] Such an operator might collectively be a hospital's medical staff who remotely monitor and control a device's operation, rather than the human host in whose body the neuroprosthesis has been implanted.[20] A military or-

---

[17] Regarding the possibility of unwitting hosts, see Gasson (2012); Kraemer, "Me, Myself and My Brain Implant: Deep Brain Stimulation Raises Questions of Personal Authenticity and Alienation" (2011); and Van den Berg, "Pieces of Me: On Identity and Information and Communications Technology Implants" (2012).

[18] For psychological, social, and cultural factors that might cause the host of an implanted device to intentionally ignore, disable, or otherwise subvert its security features and mechanisms – despite the host's awareness that this may put him or her at greater risk of harm – see Denning et al., "Patients, pacemakers, and implantable defibrillators: Human values and security for wireless implantable medical devices" (2010).

[19] Regarding the roles of the operators of neuroprostheses and similar medical devices, see Gasson (2012). Regarding the unique types of medical and engineering expertise that device operators must possess, see Clark & Fu (2012) and Fairclough, "Physiological Computing: Interfacing with the Human Nervous System" (2010).

[20] See Gasson (2012). While the use of remotely controlled implantable devices creates the possibility of illicit surveillance of a device's human host by medical personnel, such remote monitoring may also have positive effects – e.g., by making it easier to gather needed medical data from individuals who consider implanted devices less bothersome than external systems; see Lorence et al., "Transaction-Neutral Implanted Data Collection Interface as EMR Driver: A Model

ganization may be the operator of devices implanted in its personnel to expand their capacities and enhance their performance. A government intelligence agency, corporate competitive intelligence unit, or cybercriminal organization may be the operators of devices that have been implanted in the members of its own or a rival organization for use in surveillance, intelligence-gathering, or espionage or that have been hijacked to serve such a purpose. It is theoretically possible for a neuroprosthesis to generate some passive enhancement of its host's sensory, cognitive, or motor capacities without a need for monitoring, maintenance, or direct management of the device's functionality[21] – in which case the device might possess a 'beneficiary' (in the form of its human host) but lack a 'user' or 'operator.'

B. Factors Impacting Availability

Factors determining the extent to which a neuroprosthesis (and its related external system components, replacement parts, software, and technical manuals) may be available for acquisition by intended or unintended users include its:

1. Licensing and Legality

Within particular jurisdictions, laws, regulations, and licensing requirements[22] may ban certain kinds of neuroprosthetic devices altogether and require certification as a skilled medical professional for those implanting other kinds of neuroprostheses. Particular neuroprostheses may be authorized for implantation only upon issuance of a prescription by a qualified medical professional confirming that use of a neuroprosthesis is medically indicated for a specific human host. The early-adopter deployment of posthuman neuroprostheses for use in human enhancement by specialized military, police, and other governmental organizations may be facilitated by the ability of such organizations to satisfy, modify, or obtain exemptions from such legal requirements.

---

for Emerging Distributed Medical Technologies" (2009). Implantable devices can allow remote care for those in underserved remote areas and potentially increase efficiency while decreasing cost and aggravations associated with in-hospital care; see Reynolds et al., "Device Therapy for Remote Patient Management" (2008).

[21] For approaches to classifying information and communications technology (ICT) implants as 'active' or 'passive' with regard to their functionality, see Roosendaal, "Implants and Human Rights, in Particular Bodily Integrity" (2012), and Gladden, *The Handbook of Information Security for Advanced Neuroprosthetics* (2015), pp. 30-31.

[22] See McGee, "Bioelectronics and Implanted Devices" (2008), and Kosta & Bowman (2012).

2. Cost

The costs associated with a posthuman neuroprosthetic device, its implantation surgery, and its ongoing maintenance may place significant practical limits on the ability of individuals or organizations to acquire neuroprostheses or their components or maintenance equipment.[23] The cost of neuroprostheses for human enhancement will be especially high for early adopters (such as specialized military organizations) that may be required to bear the full costs of research and development for a new device in addition to the direct cost of manufacturing the devices required for implantation – without benefitting from the economies of scale that would result from the mass production and distribution of neuroprostheses within the general public as medical devices or consumer products.

3. Required Expertise

The required type and level of expertise needed to acquire, implant, and operate neuroprosthetic devices may significantly limit their potential use.[24] In particular, the small number of doctors and medical facilities capable of performing neuroprosthetic implantations serves as a constraining factor on the rate of adoption of invasive neuroprosthetic technologies.[25] Such limitations may form less of an obstacle for early adopters of posthumanizing neuroprostheses such as specialized military organizations that have access to highly qualified medical personnel and which only seek to implant a relatively small number of neuroprostheses in a select group of personnel. The future development of non-invasive neuroprostheses that can match the range and degree of functionality of surgically implanted devices would also reduce the skills needed for device installation.

---

[23] See McGee (2008) and Park et al., "The Future of Neural Interface Technology" (2009). For example, conventional cochlear implant surgery may currently cost between $40,000-$100,000 per person (see "Cochlear Implant Quick Facts" (2016)), while DBS implantation surgery costs between $35,000-$100,000 per recipient (Okun, "Parkinson's Disease: Guide to Deep Brain Stimulation Therapy" (2014)).

[24] Regarding the expertise that must be possessed by a device's institutional operators, see Clark & Fu (2012) and Fairclough (2010). For issues relating to the training of individual hosts and users of BCI systems, see Neuper & Pfurtscheller, "Neurofeedback Training for BCI Control" (2009).

[25] For example, even in those countries where DBS implantation surgery is funded by government health agencies, the lack of trained surgeons qualified to perform such complex and risky operations may result in waiting lists of 2-3 years for eligible recipients. See, e.g., Fayerman, "Funding, doctors needed if brain stimulation surgery to expand in B.C." (2013).

4. Required Maintenance

The ongoing and complex maintenance processes[26] that are required to keep a neuroprosthesis in safe working order after its implantation may effectively restrict the organizational deployment of such devices to institutions that possess appropriate biomedical device maintenance facilities and supply chains.[27]

5. Required User Customization

The ability to deploy a particular model of neuroprosthesis may be diminished if such a device must be extensively customized for each host or user[28] – whether through physical fittings of prostheses, adaptation of device software to a host's unique neurological or behavioral characteristics, or assurance of biological and genetic compatibility for devices with biological components.[29] Similarly, the ability to use devices may be limited if the anatomical structures, biological processes, or psychological activity of potential hosts or users must be 'customized' prior to or after implantation in order to allow operation of a device.

6. Reusability

Some neuroprostheses may be impossible to extract from their location within a host's body without destroying a device or otherwise rendering it inoperable; other neuroprostheses (e.g., non-invasive types) might easily be removed from their current host and integrated into the neural circuitry of a new host. The ability of a particular model of device to be reused by an additional human host after it has been disposed of by its original human host can increase the availability of such neuroprostheses. The potential for such devices to be recycled by a subsequent host (which may or may not occur with the original host's knowledge) requires effective procedures to be put in

---

[26] Regarding the role of maintenance in maintaining information system availability, see *NIST SP 800-100* (2006) and *NIST Special Publication 800-53, Revision 4: Security and Privacy Controls for Federal Information Systems and Organizations* (2013), p. F-112. Third-party organizations providing maintenance or upgrade services for implanted neuroprostheses may enjoy only a limited ability to access relevant device functions and data, due to legal restrictions regarding the privacy of personal health information that bind the devices' installers and operators. See Gladden, "Information Security Concerns as a Catalyst" (2016), and *ISO 27799:2016, Health informatics – Information security management in health using ISO/IEC 27002* (2016).

[27] Regarding supply chain considerations for information system components, see *NIST SP 800-53* (2013).

[28] For issues relating to device customization, see Merkel et al., "Central Neural Prostheses" (2007); Fairclough (2010); and Patil & Turner (2008).

[29] Regarding the possibility of neuroprosthetic devices involving biological components, see Merkel et al. (2007).

place for the disposal of devices to ensure the security of hosts' personal health information and confidential organizational data.[30]

C. Physiological Context

The relevant aspects of a neuroprosthetic device's context and situation within its host's body include its:

1. Operational Lifespan

The period of service or operational lifespan[31] that a device is expected to demonstrate before it requires replacement[32] or invasive maintenance is a critical characteristic of implantable neuroprostheses for which replacement or maintenance may require a complex, risky, and expensive surgical procedure. Drawing on practices utilized for computerized devices more generally, a neuroprosthetic device's operational lifespan can be quantified as its 'reliability' or mean time to failure (MTTF), the average length of time that a system will remain continuously in operation before experiencing its next failure. Similarly, the mean time to repair (MTTR) is the average length of time needed to detect and repair the failure and return the system to operation.[33]

2. Physical Situation in the Body

A neuroprosthetic device's physical situation includes its specific location within or upon its host's body and its physical connections to the host's bodily organs. This bodily context also affects the device's:

    a. **Physical visibility and discoverability**[34] that allow a device's presence and activity to be observed by its human host and by other human beings in the vicinity of the host.

    b. **Physical access to manipulate the device,**[35] as the mere fact that a neuroprosthesis is visibly present does not necessarily entail the possibility of being able to physically access the device's key components without harming the device's host, rendering the device inoperable, or destroying data contained within it.

---

[30] For procedures relating to the sanitization of information systems and their components prior to their disposal, see *NIST SP 800-100* (2006), p. 24, and *NIST SP 800-53* (2013), pp. F-122-F-123.

[31] See Merkel et al. (2007).

[32] See Gasson (2012).

[33] Grottke et al., "Ten fallacies of availability and reliability analysis" (2008).

[34] Regarding the significance of a device's physical discoverability, see Rao & Nayak, *The InfoSec Handbook* (2014), and Merkel et al. (2007).

[35] Regarding physical access controls, see *NIST SP 800-53* (2013), p. F-129; Rao & Nayak (2014); and Merkel et al. (2007).

c. **Remote discoverability,**[36] or the ability of external devices such as X-ray scanners, millimeter wave scanners, Wi-Fi routers, or Bluetooth devices to detect the existence of a neuroprosthetic device and localize its position from outside of its host's body, even when the device is not visible to the unaided eyes of external human observers.

3. Level of Neurocognitive Interface

The range of unique organizational roles that a neuroprosthetic device's human host can fill depends to a significant degree on the level of the neurocognitive interface between the device and the conscious awareness and unconscious cognitive activities of its host.[37] This includes the extent to which the device's host (who may be different from its primary user) is consciously aware[38] of the device's presence, status, and activity and the extent to which the device's functioning is determined by the host's neural activity.

4. Health Sensitivity and Criticality

A neuroprosthetic device that can directly affect, for example, the functioning of its host's heart or brain requires more safeguards than one that has no such capacity.[39]

II. Second Area: Physical Components

The second part of the ontology describes the key physical components of a neuroprosthetic device that have implications for enterprise architecture. Specific characteristics of this area are:

A. Device Morphology

A device's basic physical morphology includes its:

1. Identity and Unitarity

A comprehensive ontology must describe a neuroprosthetic device's degree of physical identity and unitarity – i.e., whether the device is a single, clearly identifiable physical unit or whether it comprises a large number of

---

[36] For the significance of a device's wireless or remote discoverability, see *NIST SP 800-53* (2013), pp. F-191, F-211, F-194, and Gasson (2012).

[37] See Gasson (2012); "FIPA Device Ontology Specification" (2002); and Merkel et al. (2007).

[38] See Fairclough (2010).

[39] For the health impacts that can be generated, e.g., by implantable medical devices such as neuroprostheses, see Ankarali et al., "A Comparative Review on the Wireless Implantable Medical Devices Privacy and Security" (2014).

small physical components (such as a nanorobot swarm[40]) whose nature and location cannot easily be determined by observers.

2. Size

Neuroprosthetic devices may display a wide range of sizes,[41] from those of microscopic nanorobotic units that can circulate within a host's bloodstream and manipulate individual neurons to those of large external units resembling contemporary MRI machines that are immobile and occupy an entire room. The size and degree of mobility of a neuroprosthesis significantly affect the kinds of roles that it can play within an enterprise architecture.

3. Materials

The materials from which a neuroprosthetic device is constructed[42] are relevant for enterprise architecture. While some neuroprosthetic devices may be wholly conventional electromechanical devices, others may include organic (and perhaps even living[43]) components. The deployment of devices containing biological components may require the implementation of new processes and support systems for organizations that have previously only utilized conventional electronic devices such as desktop computers.

4. System Participation

While some invasive neuroprostheses may function as standalone devices, others may simply be the implanted portion of a larger system[44] that also includes monitoring and control equipment outside of the host's body[45] or external effectors (as in the case of a robotic prosthetic arm) or entire external systems (such as an exoskeleton, vehicle, or 3D printer) that are remotely controlled by a neuroprosthetic device.[46] Multiple implanted neuroprosthetic

---

[40] See McGee (2008).

[41] Regarding device dimensions, see "FIPA Device Ontology Specification" (2002).

[42] See McGee (2008).

[43] For the possibility of neuroprostheses involving biological components, see Merkel et al. (2007). For an example of a hybrid biological-electronic interface device (or 'cultured probe') that includes a network of cultured neurons on a planar substrate, see Rutten et al., "Neural Networks on Chemically Patterned Electrode Arrays: Towards a Cultured Probe" (2007). Hybrid biological-electronic interface devices are also discussed by Stieglitz in "Restoration of Neurological Functions by Neuroprosthetic Technologies: Future Prospects and Trends towards Micro-, Nano-, and Biohybrid Systems" (2007).

[44] See Prestes et al. (2013) and NIST SP 800-100 (2006).

[45] See Gasson (2012) and Tarín et al., "Wireless Communication Systems from the Perspective of Implantable Sensor Networks for Neural Signal Monitoring" (2009).

[46] See Widge et al., "Direct Neural Control of Anatomically Correct Robotic Hands" (2010); Gladden, "Neural Implants as Gateways to Digital-Physical Ecosystems and Posthuman Socioeco-

devices may also interact to form an implanted body area network (BAN).[47] Blocking or jamming an implanted neuroprosthetic device's communications with such external systems – for example, by enclosing the device's host in RF shielding that severs the device's wireless link – may disrupt the device's functioning, especially if maintained for an extended period of time.

B. Input Mechanisms: Physical Receivers of Matter, Energy, and Information

A neuroprosthetic device's physical mechanisms for receiving input in the form of matter, energy, or information[48] may include its:

1. Power Supply

Important are both the nature of a device's primary power source (whether an internal battery, external wireless or wired power supply, or power supplied by the host's organism), the ability to utilize any backup or alternative power sources, and the behavior that the device will demonstrate if its primary (or only) supply of power is lost.[49]

2. Physical Ports and Controls

Physical ports and controls that allow input to reach a device from its external environment may include on/off switches, microSD card slots, or micro-USB or proprietary communication ports.[50]

3. Environmental Sensors

Environmental sensors through which a neuroprosthesis receives input from the environment may include:

---

nomic Interaction" (2016); and Gladden, "Enterprise Architecture for Neurocybernetically Augmented Organizational Systems" (2016).

[47] Regarding BANs, see Sayrafian-Pour et al., "Channel Models for Medical Implant Communication" (2010).

[48] See "FIPA Device Ontology Specification" (2002).

[49] Regarding power supplies for neuroprostheses and IMDs, see Gasson (2012); Merkel et al. (2007); and Li et al., "Advances and Challenges in Body Area Network" (2011). Potential 'energy harvesting' technologies that might allow implantable devices to draw power from biological phenomena such as their host's body heat, kinetic energy, or blood glucose are discussed in Mitcheson, "Energy harvesting for human wearable and implantable bio-sensors" (2010); Zebda et al., "Single glucose biofuel cells implanted in rats power electronic devices" (2013); and MacVitte et al., "From 'cyborg' lobsters to a pacemaker powered by implantable biofuel cells" (2013).

[50] Regarding the significance of such ports and controls, see Rao & Nayak (2014); Merkel et al. (2007); and *NIST SP 800-53*, p. F-212.

a. **Specialized data reception mechanisms**[51] such as photoreceptors, ultra-sonic sensors, or radio receivers that are intentionally designed to detect particular signals for processing by the device.

b. **Unpurposeful receptors** in the form of components that were not designed to function as sensors but which can nevertheless be affected by environmental phenomena such as electromagnetic radiation, heat, or acceleration. Many, if not all, of a device's components have the potential to be unpurposeful receptors (in addition to whatever other intended roles they fill), insofar as their structure or performance can be affected by external forces or phenomena.[52]

4. Neuronal Input Mechanisms

Neuronal input mechanisms[53] allow a neuroprosthesis to receive electro-chemical signals or other input from afferent neurons, efferent neurons, or interneurons within the host's body, potentially through a brain-computer interface (BCI).[54] Some neuroprostheses detect the collective activity of large groups of neurons, while others may be capable of detecting the behaviors of a single neuron.[55] Some neuroprostheses may undergo direct electrical stimulation by neurons in their host's body, while others may infer the occurrence of certain types of neural activity on the basis of other biological indicators.[56]

C. Output Mechanisms: Physical Emitters of Matter, Energy, and Information

A neuroprosthetic device's physical mechanisms for generating output in the form of matter, energy, or information may include:[57]

---

[51] Such mechanisms are discussed in "FIPA Device Ontology Specification" (2002); Gasson (2012); Park et al. (2009); Merkel et al. (2007); Lebedev (2014); Warwick & Gasson, "Implantable Computing" (2008); and *Implantable Sensor Systems for Medical Applications*, edited by Inmann & Hodgins (2013).

[52] Regarding the potential for electronic computing devices and their components to generate 'soft errors' as a result of being affected by cosmic rays or other environmental phenomena, see, e.g., Borkar, "Designing reliable systems from unreliable components: the challenges of transistor variability and degradation" (2005); Wilkinson & Hareland, "A cautionary tale of soft errors induced by SRAM packaging materials" (2005); Srinivasan, "Modeling the cosmic-ray-induced soft-error rate in integrated circuits: an overview" (1996); and KleinOsowski et al., "Circuit design and modeling for soft errors" (2008).

[53] See Fairclough (2010) and Park et al. (2009).

[54] See Gasson (2012) and Merkel et al. (2007).

[55] See Widge et al. (2010) and Lebedev (2014).

[56] For example, fMRI equipment is able to map the level of neural activity in different regions of the brain indirectly by detecting areas of increased cerebral blood flow.

[57] See "FIPA Device Ontology Specification" (2002), pp. 6-8, for an example of how such mechanisms may be described and categorized.

1. Physical Effectors

Physical effectors are often found in robotic prostheses (such as artificial arms or legs) that are controlled through a neuroprosthetic interface.[58] Audio speakers used to broadcast synthesized speech[59] are also, in a sense, highly specialized mechanical effectors.

2. Environmental Emitters

Environmental emitters through which a neuroprosthetic device affects the environment outside of its host's body[60] may include:

a. **Specialized data transmission mechanisms**[61] such as radio, optical, and ultrasonic transmitters and electromagnetic induction mechanisms that have been intentionally designed.

b. **Unpurposeful emitters** in the form of components that were not designed to function as transmitters but which can nevertheless affect the external environment (e.g., by producing audible sounds or electromagnetic radiation).[62]

3. Neuronal Output Mechanisms

Neuronal output mechanisms[63] enable a neuroprosthesis to transmit electrochemical signals to afferent neurons, efferent neurons, or interneurons within its host's body (potentially through a BCI) or to otherwise stimulate, impede, or manipulate neural activity.

D. Computational Substrate

In their role as computing devices, some neuroprostheses may display a physical computational architecture resembling that of contemporary desktop computers or mobile devices, utilizing a conventional Von Neumann ar-

---

[58] See Gasson (2012); McGee (2008); Merkel et al. (2007); Lebedev (2014); and Widge et al. (2010).

[59] See Patil & Turner (2008).

[60] See Fairclough (2010).

[61] For discussion of such mechanisms, see "FIPA Device Ontology Specification" (2002); Lebedev (2014); and Gasson (2012). Regarding the importance of reliable data-transport systems for transmitting data from an implanted device to other implanted devices or external systems, see Fernandez-Lopez et al., "The Need for Standardized Tests to Evaluate the Reliability of Data Transport in Wireless Medical Systems" (2012).

[62] Such mechanisms might conceivably be harnessed and used to exfiltrate information from a neuroprosthetic device in a manner that will not be noticed by its operator or host; the practice of covert channel analysis attempts to identify unintended ways in which computing systems may transmit information into their environment. See *NIST SP 800-53* (2013), pp. F-206-F-207.

[63] See Gasson (2012); Fairclough (2010); Park et al. (2009); and Merkel et al. (2007).

chitecture that comprises memory, I/O devices, and one or more central processing units that are connected by a communication bus;[64] such a neuroprosthetic device would execute software programs by means of one or more serial processors. Other neuroprostheses might take the form of a physical neural network that resembles the brain of a living organism and does not execute conventional 'programs.'

In order to reflect such details regarding the nature of a neuroprosthetic device's physical computational substrate, an ontology should describe the specifications of a device's:

1. CPU-based Computers

This portion of the ontology is applicable to a neuroprosthetic device that utilizes a conventional CPU-based architecture as a substrate for data processing. A neuroprosthesis possessing such a computational substrate can be used to execute traditional types of software programs. Such a device's processing abilities are constrained by the characteristics of the system's:

a. **Processor(s),**[65] which might take the form of a single central processing unit (CPU), a multicore computer, a CPU-based cluster or grid, or other architectures yet to be developed.

b. **Memory,**[66] which might include volatile RAM used as primary storage, non-volatile RAM used as longer-term secondary storage, non-volatile ROM used for storing firmware or data that should not be altered, or other possible future storage mechanisms.

2. Physical Neural Networks

This portion of the ontology is applicable to a posthuman neuroprosthetic device that utilizes a physical neural network as a substrate for data processing. A neuroprosthesis whose computational substrate is a physical neural network includes or comprises a large number of artificial neurons that transmit signals to one another through artificial synapses and that store data in the form of activation patterns within the network.[67] While capable of pro-

---

[64] See Dumas, *Computer Architecture: Fundamentals and Principles of Computer Design* (2006), and Friedenberg, *Artificial Psychology: The Quest for What It Means to Be Human* (2008), pp. 27-29.

[65] See "FIPA Device Ontology Specification" (2002) for ways of describing a device's processor.

[66] See "FIPA Device Ontology Specification" (2002) for ways of describing a device's various memory components. See also Gasson (2012) for memory-related issues in implantable devices.

[67] Regarding artificially intelligent systems that utilize physical neural networks, see, e.g., Snider, "Cortical Computing with Memristive Nanodevices" (2008); Versace & Chandler, "The Brain of a New Machine" (2010); *Advances in Neuromorphic Memristor Science and Applications*, edited by

cessing input, making decisions, generating output, and learning, such networks do not execute traditional computer programs. A physical neural network's computational capacities are determined by the type, quantity, network topology, and patterns of interaction of neurons that constitute the network.

III. Third Area: Processes

The third part of the ontology describes the key computational or cognitive processes carried out by a neuroprosthetic device that have implications for enterprise architecture. Specific characteristics of this area are:

A. Computational Processes and Computer Control

Neuroprostheses frequently incorporate some means by which human users can exercise at least partial control over a device after its implantation – whether for the purpose of controlling a device's regular operation or for periodically conducting special diagnostic, calibration, and maintenance procedures. However, even a device that is ultimately controlled by a human being will typically still carry out internal data-processing activities in order to select among possible actions and govern at least some of its own behavior.[68] Moreover, it is possible that future neuroprostheses may operate with full autonomy or may (in the case of passive neuroprosthetic devices) be subject to control by the unconscious biological processes of their human host at a biochemical level. The kinds of computational processes and control by internal or external computers available to a neuroprosthetic device will depend on the kind of physical substrate upon which the processes are performed. The ontology includes specifications for a device's:

1. Level of Autonomy

As is true for robotic systems, a neuroprosthesis might function autonomously, semiautonomously, or as a telepresence device under the full and direct control of its operator.[69] The level of autonomy is related to a neuroprosthetic device's:

Kozma et al. (2012); Pino & Kott, "Neuromorphic Computing for Cognitive Augmentation in Cyber Defense" (2014); and Lohn et al., "Memristors as Synapses in Artificial Neural Networks: Biomimicry Beyond Weight Change" (2014).

[68] See Lebedev (2014).

[69] Bekey, *Autonomous Robots: From Biological Inspiration to Implementation and Control* (2005), p. 1, explains that "For computerized devices such as robots, autonomy can be understood as the state of being "capable of operating in the real-world environment without any form of external control for extended periods of time." Regarding various degrees of semiautonomy and teleoperation, see Murphy, *Introduction to AI Robotics* (2000). For discussion of autonomous actions by

a. **Capacity for control by its human host**, as a device might be designed to respond to instructions delivered by its host via means such as oral verbal commands, instructions typed into a keypad or into the device's housing, or electrochemical signals from individual neurons or groups of neurons.[70]

b. **Capacity for control via teleoperation**, which may allow a neuroprosthetic device to be remotely controlled by external devices or by a user other than its human host, such as members of a medical support team.[71]

2. Computer Programs

This portion of the ontology is applicable to a neuroprosthetic device that utilizes a conventional CPU-based architecture as a substrate for data processing. Such a neuroprosthesis can run traditional software programs. Depending on a device's design, the programs that it is capable of executing might include customized software written particularly for the device (or that model of device) that will not run on other computers, general-purpose software that can run on a wide range of computers, and computer worms or viruses that have been prepared by an adversary and illicitly introduced into the device's computer. The ontology's specifications for a neuroprosthesis should describe its:

a. **Operating system**, which is typically designed or installed by the device's manufacturer and may include built-in device control mechanisms, diagnostic programs, and security features.[72] Some neuroprostheses may run a generic operating system such as Android, Windows, or Linux; others may run proprietary firmware whose capabilities are closely tailored to a device's unique hardware and operational requirements.

b. **Applications** that can be stored and executed by a device to expand its functionality and which may be produced by its manufacturer, user, host, or third-party software developers (which may include

implantable computers, see Gasson (2012).

[70] See "FIPA Device Ontology Specification" (2002) for descriptions of user controls that may be built into devices. For discussion of robotic neuroprosthetic devices designed to be controlled by their human hosts, see Widge et al. (2010).

[71] For possible use of implantable computers as teleoperated medical devices, see Gasson (2012) and Lorence et al. (2009). As previously noted, implantable devices may allow the provision of telemedicine services to patients in underserved remote areas and potentially increase efficiency while decreasing the cost and inconvenience associated with in-hospital care; see Reynolds et al. (2008).

[72] Regarding the significance of device operating systems, see "FIPA Device Ontology Specification" (2002) and Clark & Fu (2012).

not only legitimate software developers but also developers of malware targeted at neuroprosthetic devices).[73]

c. **Software update methods** that allow the device's operating system and applications to be modified after the device has been implanted in its human host.[74]

3. Neural Computing

This portion of the ontology is applicable to neuroprostheses that utilize neural networks for data processing. Here we use 'neural computing' to describe the manner in which a neural network assimilates input and utilizes the information contained in its activation patterns to generate decisions and actions; this represents a means of computation different from the execution of traditional computer programs. A device that utilizes a physical neural network will thereby instantiate a neural computing process.[75] Some CPU-based systems may execute software that simulates a neural network in order to process data;[76] however, at their most fundamental level, such systems' computational processes are conventional computer programs and not neural computing processes as defined here. At a minimum, the specifications of a neural computing process will describe the:

a. **Patterns of connection** between neurons.

b. **Learning mechanism** or means by which connection weights are updated.

c. **Activation function** by which an individual neuron converts input to output.

---

[73] For the possible installation of applications on IMDs, see Clark & Fu (2012). Regarding the potential vulnerability of neuroprostheses and IMDs to malware, see Denning et al., "Neurosecurity: Security and Privacy for Neural Devices" (2009); Ankarali et al. (2014), pp. 247-48; "Cybersecurity for Medical Devices and Hospital Networks: FDA Safety Communication," 2013, p. 1; and Gladden, *The Handbook of Information Security for Advanced Neuroprosthetics* (2015).

[74] Regarding the ability to update an implantable computer's software after its implantation, see Gasson (2012) and McGee (2008). For the role of software updates in maintaining a device's information security, see *NIST SP 800-53* (2013), p. F-216, and *Postmarket Management of Cybersecurity in Medical Devices: Draft Guidance for Industry and Food and Drug Administration Staff* (2016).

[75] For a valuable discussion of the distinction between a physical neural network and the neural computational processes that may be occurring within it, see Mizraji et al., "Dynamic Searching in the Brain" (2009).

[76] See Merkel et al. (2007) and Lebedev (2014).

B. Input Modalities and Signals Interpretable by a Device

A neuroprosthesis is typically capable of processing and interpreting certain kinds of signals received through its input mechanisms to extract or synthesize information that can potentially be stored, transmitted, compressed, transformed, and used to inform or control the device's operation.[77]

Note that while a neuroprosthetic device might be able to receive and store many other kinds of input, only those forms of input whose modalities are interpretable by the device will typically have the potential to trigger purposeful and targeted responses by the device's computational processes. For example, a neuroprosthesis might include a designated unit of non-volatile flash memory (similar to that used in smartphones or digital cameras) that its human user can wirelessly access and use to store any type of binary digital files, such as images, video, audio, text, or executable programs; such generic storage capacity may exist independently of – and have no way of influencing – the device's core functionality.

A neuroprosthetic device may be able to recognize and extract meaning from input that arrives in the form of:

1. Machine Communication Protocols

Machine communication protocols that can be interpreted by a neuroprosthesis might include TCP/IP, Bluetooth, 3G, and proprietary NFC formats.[78]

2. Environmental Signals

If a neuroprosthetic device possesses sufficiently advanced AI, it might be able to extract meaning directly from sensory input that it receives from the environment outside of its host's body. For example, it may convert raw sense data into modalities analogous to those that are available to human beings, identifying sights, sounds, touches, tastes, and smells that fall within the human range of perception.[79]

---

[77] See Merkel et al. (2007) and Gladden, *The Handbook of Information Security for Advanced Neuroprosthetics* (2015).

[78] See "FIPA Device Ontology Specification" (2002) for ways of representing such capacities within a device ontology. For the use of machine communication protocols by IMDs, see Li et al. (2011) and Tarín et al. (2009).

[79] See Merkel et al. (2007) and Lebedev (2014).

3. Natural Human Communication

Natural forms of human communication interpretable by a neuroprosthesis[80] might include input provided by the device's human host or user in the form of:

a. **Textual verbal input**[81] such as typed or emailed instructions that can be parsed and interpreted by a device using some form of artificial intelligence.

b. **Oral verbal input**[82] such as spoken natural-language instructions that are identified through a device's speech recognition software.

c. **Nonverbal input**[83] such as gestures, eye gaze, or vocal intonation on the part of a device's human host or external operator.

4. The Host's Biological Processes

The biological processes of a neuroprosthetic device's human host contain data such as those reflected in brain activity, cardiac rhythms, and blood chemistry[84] that can potentially be interpreted by the neuroprosthesis in a way that recognizes particular patterns within the data and triggers some predetermined response. The patterns of biological behavior found in such data may or may not be under the conscious control of the device's host and may, for example, comprise the:

a. **Host's cognitive activity and patterns,**[85] which may include emotional states, memories, volitions, and sensory percepts.

b. **Host's motor activity and actions,** which may manifest the presence of particular volitions or other cognitive activity in the form of muscle contractions that a device can detect through the use of electromyography.[86]

---

[80] The development of artificial devices possessing such communication capacities is a focus of the field of social robotics; see, e.g., Breazeal, "Toward sociable robots" (2003); Gockley et al., "Designing Robots for Long-Term Social Interaction" (2005); Kanda & Ishiguro, *Human-Robot Interaction in Social Robotics* (2013); *Social Robots and the Future of Social Relations*, edited by Seibt et al. (2014); *Social Robots from a Human Perspective*, edited by Vincent et al. (2015); and *Social Robots: Boundaries, Potential, Challenges*, edited by Nørskov (2016).

[81] See Fairclough (2010).

[82] See Merkel et al. (2007).

[83] See Fairclough (2010). The field of biometrics employs the detection and analysis of such behaviors by computerized systems in order to maintain information security; see Rao & Nayak (2014).

[84] See Fairclough (2010); Lebedev (2014); and Gladden, *The Handbook of Information Security for Advanced Neuroprosthetics* (2015).

[85] See McGee (2008) and Fairclough (2010).

[86] See Gasson (2012) and Fairclough (2010).

C. Output Modalities and Signals Expressible by a Device

In its role as a computer, a neuroprosthetic device is typically capable of generating output in forms that can be interpreted by a human being or external device and which may carry specific meanings and have the potential to influence or control the behavior of external entities in purposeful ways.[87]

Information that a neuroprosthetic device transmits using such output modalities may not have been formulated by the device itself; for example, it may have been generated by the device's human host or user or stored within the device by its designer. Alternatively, it might comprise a stream of 'real' sensory data from the external world that has been altered or augmented in some way by the device, or it may be wholly fabricated within the neuroprosthetic device in order to create a virtual reality environment to be experienced by its host. Such capacities could potentially be used in manipulative or harmful ways[88] if a device's functioning has been compromised by malware or hacked by an unauthorized user.

A neuroprosthesis may be able to generate such output that is transmitted by the device via:

1. Machine Communication Protocols

As was true for a device's input modalities, machine communication protocols by means of which a neuroprosthesis might transmit data may include TCP/IP, Bluetooth, 3G, or proprietary NFC formats.

2. Environmental Signals

If a neuroprosthetic device possesses sufficiently sophisticated computational processes, it might be able to directly encode messages using a channel of physical output that is released into the environment beyond its host's body. Such messages may be directed at external observers or at the device's host; they might be transmitted, for example, through physical displays[89] such as a small OLED display visible on the exterior surface of a robotic artificial arm. Alternatively, a neuroprosthesis might utilize augmented or virtual reality technologies to present output to its host through sensory pathways that convey or create sights, sounds, touches, tastes, or smells interpretable by human beings.

---

[87] See Merkel et al. (2007).

[88] Regarding possibilities of neuroprostheses being used to provide false data or information to their hosts or users, see McGee (2008), p. 221, and Gladden, *The Handbook of Information Security for Advanced Neuroprosthetics* (2015).

[89] See "FIPA Device Ontology Specification" (2002) for an example of display specifications.

3. Natural Human Communication

A neuroprosthesis can potentially transmit information using natural forms of human communication which might include:

a. **Textual verbal output**[90] provided to the device's human host (e.g., either as messages appearing on a physical display screen on the device's external housing or text displayed within an augmented or virtual reality environment) or to an external user (e.g., as messages displayed on an external accessory device or sent as ordinary emails).

b. **Oral verbal output,**[91] which might take the form of synthesized speech that is broadcast aloud through an external physical speaker, that is heard by the device's host through the direct transmission of signals to the host's cochlear nerve or to interneurons in the brain, that is heard by participants within some virtual reality environment, or that might be generated by the host's own speech organs, if the device is able to directly stimulate them.

c. **Nonverbal output** such as gestures, eye gaze, or tone of voice, which could be displayed either by means of the device's control over its host's natural biological bodily organs, through the device's control of prosthetic components in the host's body, or through the device's manifestation as a digital avatar with human-like features within a virtual reality environment.[92]

4. The Host's Biological Processes

A neuroprosthetic device can potentially influence or control the biological processes of its human host to produce phenomena such as specific brain activity, cardiac rhythms, or changes in blood chemistry that contain and transmit information[93] and which can be noticed and interpreted by the device's host or user or by external systems. Such patterns might be generated within the:

a. **Host's cognitive activity and patterns,** which may include the generation or alteration in the human host of emotional states, memories, volitions, and sensory percepts.[94]

---

[90] See Fairclough (2010).

[91] See Patil & Turner (2008) and Fairclough (2010).

[92] As noted previously, the development of artificial systems that can appropriately generate nonverbal communications understandable by human beings is a focus of the field of social robotics.

[93] See Gasson (2012) and McGee (2008).

[94] See Gasson (2012); McGee (2008); Patil & Turner (2008); and Gladden, "Neural Implants as

   b. **Host's motor activity,**[95] through action of a device upon the host's efferent neurons or directly upon muscles.

   c. **Control of other organs of the host**, whose functioning a device may influence or control either through direct electrochemical stimulation or through the manipulation of other systems within the host's body to trigger particular responses by organs.[96]

IV. Fourth Area: Data, Information, and Knowledge Received, Generated, Stored, and Transmitted

The enterprise architecture profile of a neuroprosthesis is shaped not only by the processes through which the device manipulates information or the physical locations in which the information is stored but also by the kinds of information involved.[97] As noted earlier, a neuroprosthetic device may be able to receive, generate, store, and transmit[98] types of information that the device itself cannot directly interpret or utilize (e.g., encrypted files that have been saved to the device by its user). This fourth part of the ontology specifies a device's capacity for reception, generation, storage, and transmission of:

A. Data Regarding a Device's Status and Diagnostics

A neuroprosthetic device typically handles data regarding its own internal status and functioning that is useful for performance analysis, troubleshooting, and other routine tasks. Such data may be:

   1. **Received by a device**, either from components internal to the device or from external monitors.

   2. **Generated by the device**, such as during malware or vulnerability scans.[99]

   3. **Stored on the device**, as in the form of logfiles.

Gateways" (2016).

[95] See Gasson (2012); McGee (2008); and Fairclough (2010).

[96] Regarding the neuroprosthetic stimulation of one type of bodily organ (e.g., a sense organ) in order to stimulate some desired behavior within another organ (e.g., the adrenal glands), see Gladden, "Neuromarketing Applications of Neuroprosthetic Devices: An Assessment of Neural Implants' Capacities for Gathering Data and Influencing Behavior" (2016).

[97] Regarding distinctions between information types, see Rao & Nayak (2014).

[98] Transmission of information by such devices is discussed, e.g., in Li et al. (2011).

[99] Regarding such practices, see *NIST SP 800-53* (2013), pp. F-218, F-153, and Rao & Nayak (2014).

4. **Transmitted by the device**, perhaps to remote external components of the neuroprosthetic system or to medical personnel who are monitoring the device's status to ensure proper functioning.[100]

B. Environmental Data

An implantable neuroprosthesis may handle data regarding the environment existing outside of its human host's body that is comparable to data gathered by the human body's natural biological sense organs.[101] Such data may be:

1. **Received by a device** through its own sensors or from afferent neurons in the host's body.

2. **Generated by the device.**

3. **Stored on the device** before or after undergoing processing.

4. **Transmitted by the device** – for example, to its human host as though it were sense data obtained directly by the host's own sensory organs, or to external systems or individuals who might use the data in teleoperation of the device.

C. Data Regarding a Host's Physical Status and Biological Processes

A neuroprosthesis may handle highly sensitive data regarding the identity, location, health, and biological processes of its human host.[102] Such data may be:

1. **Received by a device**, such as through sensors directly monitoring the host's biological processes.

2. **Generated by the device.**

3. **Stored on the device** after being gathered and processed by the device.

4. **Transmitted by the device** – for example, through wireless communication to medical personnel who are remotely monitoring the host's medical condition and manipulating the device to administer telemedicine services.

---

[100] For a discussion of the remote administration of such devices, see Reynolds et al. (2008).

[101] See McGee (2008). A retinal prosthesis is an example of such a device; for a discussion of such technologies, see Linsenmeier, "Retinal Bioengineering" (2005); Weiland et al., "Retinal Prosthesis" (2005); and Viola & Patrinos, "A Neuroprosthesis for Restoring Sight" (2007).

[102] Regarding the collection, use, and transmission of such information by neuroprostheses and IMDs, see Fairclough (2010); Kosta & Bowman (2012); and Gasson (2012). The use of such information in the provisioning of eHealth services is discussed in Lorence et al. (2009) and Reynolds et al. (2008).

D. Data Regarding or Information Contained in a Host's Cognitive Activity

In addition to data regarding more general biological processes, a neuro-prosthesis with the appropriate form of neural connectivity could potentially receive, generate, store, and transmit in at least limited fashion data regarding – or information explicitly manifesting the contents of – its host's cognitive processes,[103] including emotional states, memories,[104] and volitions.[105] Such information might be:

1. **Received by a device,** such as through a BCI that can remotely detect activity in the host's brain or which interacts directly with interneurons in the brain through artificial synapses.

2. **Generated by the device,** perhaps through its analysis of other data.[106]

3. **Stored on the device** for later processing and analysis, even if the device itself is incapable of performing such analysis.

4. **Transmitted by the device**[107] – for example, to external medical systems for purposes of analyzing the user's neurological condition or to remote systems where the neural activity can be used to control prosthetic limbs or other robotic devices.

E. Procedural Knowledge

Procedural knowledge is the body of information needed to perform particular tasks; in the context of a neuroprosthesis, such knowledge might be:

1. **Received by a device,** such as in the form of downloadable packages that provide a device with instructions (e.g., including linguistic, social, and cultural information) that allow a device to interact effectively with human hosts or operators in particular countries.[108]

---

[103] For a general discussion of the use of neuroprostheses in gathering data regarding cognitive activity, see Gasson (2012); Fairclough (2010); Tarín et al. (2009); and Gladden, "Neuromarketing Applications of Neuroprosthetic Devices" (2016).

[104] For the possibility of developing mnemoprosthetics, see Han et al., "Selective Erasure of a Fear Memory" (2009), and Ramirez et al., "Creating a False Memory in the Hippocampus" (2013).

[105] The ability to detect a host's volitions is discussed in McGee (2008).

[106] As noted previously, an fMRI device generates information about cognitive activity through its analysis of data regarding cerebral blood flow.

[107] For a discussion of such activities, see Tarín et al. (2009).

[108] Attempts to develop artificially intelligent systems possessing such cultural knowledge are discussed in Mascarenhas et al., "Social Importance Dynamics: A Model for Culturally-Adaptive Agents" (2013); Obaid et al., "Cultural Behaviors of Virtual Agents in an Augmented Reality Environment" (2012); Rehm & Nakano, "Some Pitfalls for Developing Enculturated Conversational Agents" (2009); and Rehm et al., "From observation to simulation: generating culture-specific behavior for interactive systems" (2009).

2. **Generated by the device**, perhaps as a result of its own internal learning processes.

3. **Stored on the device**, such as in the form of software[109] that guides the device's routine functioning and specifies how it should carry out particular actions in response to unique events or circumstances.

4. **Transmitted by the device** – for example, in the form of device drivers or other files sent to external devices to allow them to connect with the device and utilize its resources.

F. Declarative Knowledge

A neuroprosthesis may handle descriptive or declarative knowledge in a format that is readily comprehensible to (and may have been prepared by) human beings.[110] This might include information contained in everyday correspondence like emails or text messages or the contents of documents downloaded from the Internet. Such information may be:

1. **Received by a device**, such as through electronic messages composed by the device's user, downloaded from websites through wireless Internet connections, or recorded in the form oral conversations that undergo speech-to-text processing.

2. **Generated by the device**.

3. **Stored on the device**, perhaps as ordinary text files.

4. **Transmitted by the device** – for example, in the form of messages or alerts sent to the device's human host or content downloaded from websites and available on demand for transmission to the host's sensory system for the host to experience.

Note that the presence of declarative knowledge may not be apparent to a device itself. For example, consider an artificial eye that records high-fidelity video of its host's environment and streams a live feed of that video to an external display where it may be viewed by other human beings: a human observer watching the video might instantly recognize the text printed on a roadside billboard or handwritten note or the screen of a mobile device belonging to some passerby and immediately comprehend the declarative knowledge conveyed by that text. However, the artificial eye itself would not be able to recognize the existence of such text or decode its meaning, if the eye had not been designed to possess appropriate OCR capabilities.

---

[109] See "FIPA Device Ontology Specification" (2002). Automated incident handling processes are discussed in *NIST SP 800-53* (2013), p. F-105.

[110] See McGee (2008).

Conclusion

It is possible to analyze the structure and behavior of a neuroprosthetic device from various perspectives: for example, one might describe from a biocybernetic perspective the device's impact on the sensory, cognitive, or motor functioning of its human host or analyze the way in which a neuroprosthesis affects the capacities and relationships of its human host at a social and economic level.

In this text, we have formulated an ontology that can be employed to describe the fundamental characteristics of a neuroprosthesis in its role as a *computing device*. The ontology describes four key aspects that shape the functioning of a neuroprosthesis as computing device, which are the device's external context, its physical components, the processes utilized by the device, and the types of information that it generates or handles. Through the use and further refinement of such an ontology, the functionality of a neuroprosthesis as a computing device may be more easily planned or analyzed, and interoperability between neuroprosthetic devices, their human hosts and users, and external computer systems can become easier to achieve.

Chapter Two

Integrating Neuroprostheses into Human Sensory, Cognitive, and Motor Processes:
A Biocybernetic Ontology of the Host-Device System

Abstract. In this text, we develop an ontology that envisions, captures, and describes the full range of ways in which a neuroprosthesis may participate in the sensory, cognitive, and motor processes of its human host. By considering anticipated future developments in neuroprosthetics and adopting a generic biocybernetic approach, the ontology is able to account for therapeutic neuroprostheses already in use as well as future types of neuroprostheses expected to be deployed for purposes of human enhancement.

The ontology encompasses three areas. First, a neuroprosthesis may participate in its host's processes of *sensation* by (a) detecting stimuli such as photons, sound waves, or chemicals; (b) fabricating sense data, as in the case of virtual reality systems; (c) storing sense data; (d) transmitting sense data within a neural pathway; (e) enabling its host to experience sense data through a sensory modality such as vision, hearing, taste, smell, touch, balance, heat, or pain; or (f) creating mappings of sensory routes – e.g., in order to allow sensory substitution. Second, a neuroprosthesis may participate in processes of *cognition* by (a) creating a basic interface between the device and the host's conscious awareness or affecting the host's (b) perception, (c) creativity, (d) memory and identity, or (e) reasoning and decision-making. Third, a neuroprosthesis may participate in processes of *motion* by (a) detecting motor instructions generated by the host's brain; (b) fabricating motor instructions, as in the case of a medical device controlled by software algorithms rather than its host's volitions; (c) storing motor instructions; (d) transmitting motor instructions, as within a neural pathway; (e) effectuating physical action within effectors such as natural biological muscles and glands, synthetic muscles, robotic actuators, video screens, audio speakers, or wireless transmitters; (f) allowing the expression of volitions through motor modalities such as language, paralanguage, and locomotion; or (g) creating mappings of motor routes. The use of such an ontology allows easier, more systematic, and more robust analysis of the biocybernetic role of a neuroprosthesis within its host-device system.

Introduction

Having formulated in the previous chapter an ontology of the neuroprosthesis as a computing device, in this chapter we develop an ontology of the neuroprosthesis as a biocybernetic instrument that becomes integrated into the neural circuitry of a human organism in order to participate in processes of sensation, cognition, and motor action. This investigation of the nature of neurocybernetic technologies will then be carried further in the next chapter by developing an ontology of the neuroprosthesis as a means for the 'cyborgization' of human beings that shapes how such individuals experience posthumanized digital-physical ecosystems.

Fundamental Characteristics of Neuroprostheses

A neuroprosthesis can be defined as *an artificial device that is integrated into the neural circuitry of a human being* to create a neurocybernetic host-device system with both human and computerized elements.[1] In principle, neuroprostheses may be either 'invasive' (i.e., surgically implanted in the brain of a human host) or 'non-invasive' (e.g., consisting of an external device worn by a human host); however, at present it is difficult to develop non-invasive technologies that can become truly integrated into the neural circuitry of a human being.[2] According to the definition used in this text, contemporary neuroprostheses are thus typically identified with 'neural implants'; devices utilizing non-invasive technologies such as EEG or fMRI are likely to be classified more generally as brain-computer interfaces (BCIs) or brain-machine interfaces (BMIs) rather than neuroprostheses.

Contemporary neuroprosthetic devices are typically classified as either sensory, motor, bidirectional sensorimotor, or cognitive neuroprostheses.[3] As shall be explored within this text, however, the distinctions between these different categories of neuroprostheses are not clear-cut; for example, an artificial eye that is primarily considered to be a 'sensory neuroprosthesis' because it restores or enhances the sight of its human host might also possess strong cognitive or motor aspects.

Neuroprostheses are currently used primarily for therapeutic purposes, as a means of restoring some capacity that is absent as a result of injury or ill-

---

[1] See Lebedev, "Brain-Machine Interfaces: An Overview," *Translational Neuroscience* 5, no. 1 (2014), and Gladden, "Enterprise Architecture for Neurocybernetically Augmented Organizational Systems" (2016).

[2] See Gasson, "Human ICT Implants: From Restorative Application to Human Enhancement" (2012), p. 14, and Panoulas et al., "Brain-Computer Interface (BCI): Types, Processing Perspectives and Applications" (2010).

[3] See Lebedev (2014).

ness: thus cochlear implants and retinal prostheses are used to restore sensory functionality to those who are deaf or blind, robotic prosthetic limbs are used to replace natural biological limbs that have been amputated, and deep brain stimulation (DBS) devices are used to treat tremors in those suffering from Parkinson's disease.[4] However, efforts are underway to develop and deploy neuroprostheses whose purpose is not to restore some capacity that is found in typical human beings but to grant their human hosts sensory, cognitive, and motor capacities that far exceed those that are possible for natural biological human beings.[5]

Developing an Ontology of the Neuroprosthesis as Participant in Human Sensory, Cognitive, and Motor Processes

In the following sections, we develop an ontology of the neuroprosthesis as a participant in human processes of sensation, cognition, and motor action. The ontology is biocybernetic in nature, insofar as it focuses primarily on patterns of communication and control within the host-device system.[6] Unlike some contemporary schemas for classifying neuroprostheses, this ontology is intended not only to allow the description, categorization, and analysis of those therapeutic neuroprostheses that are already in use; by taking into account anticipated future developments in the field of neuroprosthetics and envisioning as broadly as possible the potential biocybernetic capacities of neuroprosthetic devices, this ontology is designed to be generic enough to account for the characteristics of the full spectrum of future neuroprostheses that may be developed for purposes of human enhancement.

As illustrated in Figure 1, our ontology is organized into three areas, which allow neuroprostheses to be understood according to their participation in human processes of sensation, cognition, and motion. These aspects of the ontology are developed in detail in the following sections.

---

[4] See, e.g., Gasson et al., "Human ICT Implants: From Invasive to Pervasive" (2012); Ochsner et al., "Human, non-human, and beyond: cochlear implants in socio-technological environments" (2015); Weiland et al., "Retinal Prosthesis" (2005); Viola & Patrinos, "A Neuroprosthesis for Restoring Sight" (2007); and *Deep Brain Stimulation for Parkinson's Disease*, edited by Baltuch & Stern (2007).

[5] See, e.g., McGee, "Bioelectronics and Implanted Devices" (2008); Warwick & Gasson, "Implantable Computing" (2008); Gasson (2012); Gladden, "Neural Implants as Gateways to Digital-Physical Ecosystems and Posthuman Socioeconomic Interaction" (2016); and Gladden, "Enterprise Architecture for Neurocybernetically Augmented Organizational Systems" (2016).

[6] The field of cybernetics offers a transdisciplinary theoretical framework and vocabulary that allows insights to be shared and translated between the diverse range of disciplines that study patterns of communication and control in machines, living organisms, or social systems. For this classic definition of cybernetics, see Wiener, *Cybernetics: Or Control and Communication in the Animal and the Machine* (1961).

Fig. 1: An Ontology of the Neuroprosthesis as Participant in Sensory, Cognitive, and Motor Processes

I. Sensation

A. Detection of stimuli and transduction into cellular action potentials
   1. Electromagnetic reception
      a. Photons
      b. Electric current or charge
      c. Magnetic fields
   2. Mechanoreception
      a. Sound waves
      b. Particle kinetic energy
      c. Gravity
      d. External bodily pressure
      e. Internal bodily pressure
      f. Excessive pressure
   3. Chemoreception
      a. Solutes and other substances in liquid mixtures
      b. Atmospheric particulates
      c. Internal body chemistry
   4. Activity cycles
   5. Xenostimuli
B. Fabrication of electrochemical signals in a sensory neural pathway
C. Storage of sense data
   1. Memory type
      a. Volatile memory
      b. Non-volatile memory
   2. Information format
      a. Digital data files
      b. Neural network activation patterns
D. Transmission of action potentials in a sensory neural pathway

E. Modalization of sensory experience
   1. Vision
   2. Hearing
   3. Taste
   4. Smell
   5. Touch
   6. Posture and bodily extension (proprioception)
   7. Balance
   8. Heat and cold
   9. Energy level (interoception)
   10. Pain
   11. Passage of time
   12. Acoustic contextualization
   13. Magnetoreception
   14. Electroreception
F. Mapping of sensory routes
   1. Routing biologically receivable sense data to its biologically normal modality
   2. Routing fabricated sense data to an arbitrary modality
   3. Remapping biologically receivable sense data to an atypical but natural modality
   4. Employing a non-human xenosensory mapping
      a. Mapping sense data from xenostimuli to an arbitrary modality
      b. Mapping arbitrary sense data to a xenesthetic modality

II. Cognition

A. Basic interface with the mind
   1. With the host's conscious awareness
   2. Without the host's conscious awareness
B. Perception
   1. Percepts
   2. Consciousness and self-awareness
   3. Sleep
   4. Emotion
   5. Autonomous mental reactions
C. Creativity
   1. Ideation and imagination
   2. Dreaming

D. Memory and identity
   1. Memory
      a. Sensory memory (SM)
      b. Short-term memory (STM)
      c. Long-term memory (LTM)
   2. Personality
   3. Habit
   4. Belief
E. Reasoning and decision-making
   1. Working memory (WM)
   2. Attention control
   3. Volition
   4. Metavolition and conscience
   5. Internal monologue
   6. Savant skills

III. Motion

A. Detection of action potentials or other motor instructions in a motor neural pathway

B. Fabrication of cellular action potentials in a motor neural pathway

C. Storage of instructions for (repeated) motor execution
  1. Memory type
     a. Volatile memory
     b. Non-volatile memory
  2. Information format
     a. Digital data files
     b. Neural network activation patterns

D. Transmission of cellular action potentials in a motor neural pathway

E. Effectuation of physical action
  1. Stimulation of natural biological effectors
     a. Contraction and relaxation of muscles
        i. Voluntary functions
        ii. Autonomic functions
     b. Regulation of gland cells' production and release of chemicals
  2. Control of artificial effectors
     a. Electromagnetic emissions
        i. Photons
        ii. Electric current or charge
        iii. Magnetic fields
     b. Mechanical effectors
        i. Sound waves
        ii. Particle kinetic energy
        iii. Motion of an actuator
     c. Chemical production and release
        i. Release of airborne chemicals into atmosphere
        ii. Release of chemicals into a liquid mixture
     d. Xenoactive phenomena

F. Modalization of motor action
  1. Symbolic action
     a. Language classified by its mechanism of production
        i. Oral language
        ii. Written language
        iii. Sign language
     b. Language classified by its historical origin
        i. Natural human language
        ii. Synthetic language
        iii. Language of thought
  2. Nonsymbolic action
     a. Paralanguage
     b. Oculesics
     c. Kinesics
     d. Haptic communication
     e. Proxemics
     f. Internal organ activity

G. Mapping of motor routes
  1. Routing natural motor instructions to their biologically normal effector
  2. Routing fabricated motor instructions to an arbitrary effector
  3. Remapping natural motor instructions to a biologically atypical but natural effector
  4. Employing a non-human xenoexpressive mapping
     a. Mapping motor instructions to a xenergetic effector
     b. Mapping motor instructions to manifest a xenopractic motor modality

I. Sensation

Many neuroprosthetic devices perform or participate in behaviors designed to sense details about their host's body[7] or the surrounding external environment. For some of these devices, the act of sensing may be incidental to their primary function; for example, a cognitive or motor neuroprosthesis might employ sensors to aid with device calibration or troubleshooting, or it might receive software updates through the use of radio receivers.

Other devices are **sensory neuroprostheses** whose primary function is to fill some role in the sensory processes of their human host.[8] Many sensory neuroprostheses present sense data to the conscious awareness of their host's mind, where it is perceived and consciously experienced. Other neuroprostheses may provide sense data to a host's spinal cord or brain in a way that triggers some autonomous reflex response in a manner that is not perceived by the host's conscious awareness.

A sensory neuroprosthesis performs tasks like acquiring sense data from distal stimuli such as physical objects and sources of energy and transducing it into cellular action potentials; fabricating cellular action potentials within a sensory neural pathway; storing sense data; transmitting proximal stimuli toward the brain of the device's human host in the form of cellular action potentials within a sensory neural pathway; modalizing sensory experience by presenting it to the mind in a form that it can perceive;[9] or mapping the sensory routes that connect sense data, sense organs, and sensory modalities. Below we consider in more detail these ways in which a neuroprosthesis can participate in its host's processes of sensation.[10]

A. Detection of Stimuli and Transduction into Cellular Action Potentials

Some neuroprosthetic devices serve as receptor organs that acquire sense data from distal stimuli such as physical objects and sources of energy and utilize a process of transduction to translate that input into the form of cellular action potentials that can be transmitted to postsynaptic neurons in the

---

[7] Theoretical and practical issues relating to such devices are discussed in *Implantable Sensor Systems for Medical Applications*, edited by Inmann & Hodgins (2013).

[8] For an overview of sensory neuroprostheses, see, e.g., Troyk & Cogan, "Sensory Neural Prostheses" (2005); Merkel et al., "Central Neural Prostheses" (2007); and Lebedev (2014).

[9] Neuroprostheses that generate or alter perceptions primarily by manipulating interneurons within the brain (rather than sensory organs or nerves) and which relate to mental phenomena such as memory and imagination are classified as cognitive rather than sensory neuroprostheses and are discussed in a later section.

[10] A comprehensive investigation of the natural biological systems that perform sensory roles within the human organism can be found, e.g., in Smith, *Biology of Sensory Systems* (2008), and Møller, *Sensory Systems: Anatomy and Physiology* (2014).

body of a device's human host.[11] Below we consider a range of stimuli found within a host's body or the external environment which neuroprostheses may be capable of detecting through various receptor mechanisms.

1. Electromagnetic Reception

A neuroprosthetic device may be capable of registering the presence of electromagnetic radiation, a magnetic field, or electric charge.

a. Photons

A photoreceptive neuroprosthesis registers electromagnetic radiation – most commonly from the environment external to the host's body – by detecting photons. In a natural biological human being, such sensors include at least three types of photoreceptor cells embedded in the retina. Rods do not distinguish colors and provide poor spatial resolution, but they are highly sensitive and can be activated by a single photon, offering a degree of monochromatic vision in very low-light conditions. Cones detect specific wavelengths of light (i.e., colors) and are only activated upon absorbing a large number of photons, which means that colors are not naturally distinguishable in extremely low-light conditions. Finally, photosensitive ganglion cells are also activated by absorbing photons; however, the information that they generate is not consciously experienced by the mind through the modality of vision. Instead they are believed to play a role in the generation of reflex responses and the regulation of circadian rhythms.[12]

A retinal prosthesis may attempt to mimic the natural functioning of the human eye by registering light that falls within the spectrum normally visible to human beings – or it may register electromagnetic radiation of a frequency higher than that of visible light (e.g., ultraviolet light, X-rays, or gamma rays) or lower than that of visible light (e.g., infrared light or radio waves). Such devices could potentially be employed, for example, to allow a human host to 'see' in infrared[13] or to visually detect the presence of Wi-Fi hubs or mobile devices that emit radio waves. Photoreceptive neuroprostheses may also take forms other than that of an artificial eye. For example, an implantable smartphone that has been surgically implanted within a host's abdomen, is integrated into the host's nervous system, and contains a radio receiver would also technically be a photoreceptive neuroprosthesis. Types of photoreceptive

---

[11] Natural biological processes of transduction are discussed in Smith (2008), pp. 1-30, and Møller (2014), pp. 29-62.

[12] The structure and functioning of biological photoreceptors is discussed in Smith (2008), pp. 315-49, and Møller (2014), pp. 42-43, 269-71.

[13] Regarding neuroprosthetically enabled infrared vision, see Merkel et al. (2007).

neuroprosthetic devices already in existence include experimental retinal prostheses.[14]

b. Electric current or charge

An electroreceptive neuroprosthesis registers the presence of an electric charge in the form of static electricity or an electric current. In a natural biological human being, sensory and motor neurons can be indiscriminately stimulated by electricity from the external environment (as in the case of an electric shock), and the ability of neurons to be stimulated electrically (e.g., through the use of artificial electrode) makes possible the functioning of many types of neuroprosthetic devices. However, human beings are not known to possess specialized electroreceptors of the sort that are found, for example, in some aquatic or amphibious animals.[15]

c. Magnetic fields

A magnetoreceptive neuroprosthesis registers the presence of a magnetic field. It is not clear whether natural biological human beings possess any such specialized cells; the supposed existence of magnetosensitive bones or brain cells is a matter of ongoing debate.[16]

2. Mechanoreception

A neuroprosthetic device may incorporate a mechanoreceptive component that allows it to register physical motion or pressure. Such neuroprostheses can detect a range of phenomena.

a. Sound waves

Some neuroprosthetic devices contain a specialized form of mechanoreceptor that registers sound waves produced in the external environment or within a host's body. In a natural biological human being, such a sensory system includes hair cells in the ear.[17] An audioreceptive neuroprosthesis may

---

[14] See Linsenmeier, "Retinal Bioengineering" (2005); Weiland et al. (2005); and Viola & Patrinos (2007).

[15] Electroreceptors in animals are discussed in Smith (2008), pp. 428-33. Note that not all electroreceptors possessed by neuroprostheses are designed to detect naturally occurring electric current or electric fields present in an environment: a neuroprosthetic device could also be said to possess 'electroreceptors' of a specialized sort if it includes a mini-stereo audio input jack or a power supply input port into which an AC or DC power cable can be inserted to charge the device's internal battery.

[16] For magnetic sensitivity in animals, see Smith (2008), pp. 434-35. In the case of neuroprosthetic devices, the detection of magnetic fields might be used, e.g., in a specialized way to detect the presence of an external hardware token that provides emergency access to a device. Such token-based access schemes for neuroprostheses are discussed in Gladden, *The Handbook of Information Security for Advanced Neuroprosthetics* (2015).

[17] Such biological mechanisms are discussed in Smith (2008), pp. 123-46, and Møller (2014), pp.

attempt to mimic the natural hearing process by registering only those sound frequencies and volumes that are detectable by the human ear, or it might provide augmented sensory capacities by registering sound waves possessing higher or lower frequencies or a lower volume than the ear can naturally detect. Existing types of such neuroprosthetic devices include cochlear implants and auditory brainstem implants.[18]

b. Particle kinetic energy

A thermoreceptive neuroprosthesis contains a specialized form of mechanoreceptor that registers the particle kinetic energy of matter adjoining it, which is transferred in the form of heat. (Note that this would not include a device utilizing the remote thermometric technique of pyrometry.) In a natural biological human being, there are different thermoreceptors specialized to detect heat and cold.[19]

c. Gravity

A neuroprosthesis may contain a specialized form of mechanoreceptor that registers the orientation of gravitational force exerted on the device (and thus – in a terrestrial environment – the direction toward the ground). In a natural biological human being, this system includes hair cells in the ears and vestibular nuclei in the brainstem that assist in maintaining balance and an upright posture.[20]

d. External bodily pressure

A tactioceptive neuroprosthesis contains a specialized form of mechanoreceptor that registers pressure from objects or phenomena external to the host's body that are touching the receptor. Such a neuroprosthesis may attempt to mimic the natural human sense of touch by registering only levels of force that fall within the range of what is typically perceptible by the fingers, skin, or other body organs; alternatively, it may possess a finer degree of sensitivity which allows it to register force too small to be detected by the unaided human body, or it may be capable of safely and accurately measuring very great physical forces which would result in incapacitating pain or the destruction of tissue if applied to an unmodified human body.

In a natural biological human being, this system includes mechanoreceptors in the somatosensory system, such as those in the skin, muscles, and

---

191-260.

[18] See Cervera-Paz et al., "Auditory Brainstem Implants: Past, Present and Future Prospects" (2007); Merkel et al. (2007); and Bostrom & Sandberg, "Cognitive Enhancement: Methods, Ethics, Regulatory Challenges" (2009), p. 321.

[19] Thermosensitivity in biological organisms is discussed in Smith (2008), pp. 415-22.

[20] The role of hair cells in detecting bodily orientation is discussed in Smith (2008), pp. 123-31.

body organs.[21] Existing neuroprosthetic devices possessing such functionality include experimental bidirectional prosthetic hands that utilize such mechanisms to detect pressure applied to or by a hand's robotic fingers.[22]

e. Internal bodily pressure

A neuroprosthetic device may contain a similar sort of specialized mechanoreceptor that registers the degree of pressure that muscles and other internal body components exert on one another depending on their relative position. In a natural biological human being, this system includes mechanoreceptors in the skeletal striated muscles and joints that allow the brain to detect the body's posture and spatial orientation.

f. Excessive pressure

A neuroprosthetic device may possess specialized mechanoreceptors that detect levels of pressure which are so great that they are likely to damage or destroy the device or other natural or artificial components of the host's body. In this case, the primary purpose of a mechanoreceptor is not to detect excess pressure as a quantity that must be precisely measured for abstract data-gathering reasons but as an immediate physical danger that must be eliminated. While such a mechanoreceptor may be capable of measuring the current pressure with extreme sensitivity, it might instead only be activated once a certain threshold pressure has been detected and may thus only be capable of registering the binary states of 'non-excessive pressure' and 'excessive pressure.'

Rather than utilizing mechanoreceptors to directly detect excess pressure applied to it, a neuroprosthetic device could potentially detect excessive pressure (or other harmful forces) indirectly – for example, by utilizing chemoreceptors that detect chemical substances within the bloodstream whose presence indicates that some of the body's biological cells are being damaged or destroyed. In a natural biological human being, such excessive pressure is perceived through the sensory modality of pain, employing a mechanism involves nociceptors in the somatosensory system, including those in the skin, muscles, and body organs.[23]

3. Chemoreception

A neuroprosthetic device may possess chemoreceptors that detect the presence of particular kinds of chemical substances, typically on the basis of

---

[21] Such biological systems are described in Smith (2008), pp. 99-122, and Møller (2014), pp. 159-90.

[22] For example, see Hoffmann & Micera, "Introduction to Neuroprosthetics" (2011), pp. 792-93.

[23] Biological mechanisms relating to pain are discussed in Smith (2008), pp. 437-62, and Møller (2014), pp. 337-65.

their molecular structure. Chemical substances might reach such receptors in a variety of forms.

a. Solutes and other substances in liquid mixtures

A neuroprosthesis might incorporate a specialized form of chemoreceptor that registers the presence of particular chemical substances dissolved, suspended, or otherwise present in a liquid medium adjoining the receptor.[24] In a natural biological human being, such sensors include chemoreceptors found on the tongue's taste buds.[25]

b. Atmospheric particulates

Alternatively, a neuroprosthesis might include a specialized form of chemoreceptor that detects the presence of particular types of solid particulates, liquid droplets, or other substances present in an aerosol, suspended in the atmosphere, or otherwise contained in a gaseous mixture adjoining the receptor.[26] In a natural biological human being, such sensors include olfactory receptor neurons found in the nose.[27]

c. Internal body chemistry

A neuroprosthetic device may contain a specialized form of chemoreceptor that detects within the body in solid, liquid, or gaseous form particular chemical substances whose presence or absence provides information regarding the health or functionality of specific biological systems or the status of biological processes.

In a natural biological human being, such chemoreceptors (and mechanoreceptors) located in various body organs detect stimuli that are conveyed to the mind through the sensory modality of interoception and are experienced as feelings such as hunger, thirst, or drowsiness.[28]

---

[24] For challenges with the development of chemoreceptive neuroprostheses to restore the sense of taste, see Merkel et al. (2007).

[25] Such chemoreceptors are described in Smith (2008), pp. 187-218, and Møller (2014), pp. 303-22.

[26] The difficulty of designing chemoreceptive neuroprostheses for restoring the sense of smell are noted in Merkel et al. (2007).

[27] Chemoreceptors of this sort are discussed in Smith (2008), pp. 219-44, and Møller (2014), pp. 303-22.

[28] Regarding the nature of interoception in the human organism, see Cameron, *Visceral Sensory Neuroscience: Interoception* (2001). For the possibility of neuroprosthetic devices that relate, e.g., to their host's need for or experience of sleep, see Claussen & Hofmann, "Sleep, Neuroengineering and Dynamics" (2012), and Kourany, "Human enhancement: Making the debate more Productive" (2013), pp. 992-93.

4. Activity Cycles

A neuroprosthetic device may measure the passage of time by counting instances of some regular phenomenon such as electron transitions (in the case of atomic clocks), the oscillation of an electrically charged silicon dioxide tuning fork (in the case of quartz clocks), or diurnal cycles of light and darkness (in the case of some circadian oscillators found within natural biological organisms). In a natural biological human being, time is measured at a number of different scales by a complex array of systems including the cerebral cortex, cerebellum, and basal ganglia.[29]

5. Xenostimuli

A neuroprosthetic device might incorporate sensors that register the presence of some physical object, source of energy, activity, or other phenomenon that is not normally detectable in a meaningful way by the sense organs of a natural biological human being.[30] Examples might include a photoreceptive device that can receive radio transmissions from a Wi-Fi router or a chemoreceptive device that can detect the presence of radon gas. In comparison to the limited sensory capacities of a natural biological human being, such a neuroprosthetic device can be said to detect 'xenostimuli' through the use of 'xenoreceptors.'[31]

---

[29] For a discussion of biological timekeeping mechanisms that are employed by living organisms at the intracellular and intercellular level to maintain both relatively short ultradian rhythms (which primarily regulate an organism's internal intracellular processes) and longer circadian rhythms (which primarily regulate an organism's interaction with its external environment), see Lloyd, "Biological Time Is Fractal: Early Events Reverberate over a Life Time" (2008).

[30] From one perspective, the body of a natural biological human being might be said to possess sensors that are capable of 'detecting' the presence of, e.g., X-rays or odorless airborne carcinogenic chemicals – insofar as cells within the body that are exposed to such phenomena may eventually manifest in response some recognizable behavior (such as cell death or the production of tumors) that indicate to the human being the presence of the phenomena. However, the structures contained within the body that react to the presence of such stimuli are not normally described as 'sensory receptors' (and the processes by which they register the presence of such phenomena are not described as the reception of 'sense data'), insofar as the structures do not constitute mechanisms that are specialized to serve that purpose and do not transmit information about the phenomena's presence to the spinal cord or brain through typical sensory pathways.
A liminal case is that of ultraviolet radiation: in itself, such radiation is invisible and cannot be directly detected by the human eye; however, its presence in sufficient quantities is 'detected' by melanocytes in the epidermis, which visibly signal the presence of UV radiation when the melanin contained within the cells undergoes a process of oxidation that darkens the pigment. An individual's (retrospective) realization that he or she has been exposed to an excessive quantity of UV radiation by means of the appearance of a sunburn is a not uncommon occurrence.

[31] The use of neuroprostheses to grant human hosts the ability to detect new kinds of stimuli not detectable to natural human beings is discussed in Warwick, "The Cyborg Revolution" (2014); Gasson et al. (2012); and Merkel et al. (2007).

B. Fabrication of Electrochemical Signals within a Sensory Neural Pathway

One possible behavior for a sensory neuroprosthesis is for the device to accurately detect the presence of environmental or bodily stimuli using specialized receptors, transduce those stimuli into a pattern of electrochemical signals whose nature faithfully reflects the contents of the stimuli, and then transmit those signals to adjoining postsynaptic neurons so that they can be conveyed to the spinal cord or brain. That is the approach normally employed with contemporary sensory neuroprostheses, whose purpose is generally to restore as authentically as possible forms of biologically typical sense perception that are absent in a device's human host due to injury, illness, or some other medical condition.

However, rather than accurately detecting environmental or bodily stimuli that exist objectively as part of the primary physical world[32] and transducing them into electrochemical signals, another possible behavior is for a sensory neuroprosthesis to fabricate electrochemical signals that correspond to a particular set of distal stimuli (and which will be interpreted by the host's brain as such), despite the fact that those distal stimuli do not actually exist at that moment within the primary physical world. For example, a cochlear implant containing a sufficiently sophisticated computer could provide its human host with a real-time stream of auditory sense data that accurately reflects all of the sounds that the device is detecting in the surrounding environment: nearby conversations, the sound of traffic in the street outside, the whir of the fan in the host's desktop computer, the sound of the host's own breathing. However, by drawing on data stored within itself or accessed through the Internet, the implant could alternatively supply its host with the sounds of a Mozart symphony (to enhance the host's focus and productivity while immersed in some critical work-related task), soothing white noise (to help the host fall asleep in a distracting auditory environment), a 'replay' of a conversation that the host had engaged in an hour earlier and which the device had recorded (to refresh the host's memory regarding forgotten details of the conversation), or turn-by-turn navigation directions to a destination

---

[32] Here, the term 'primary physical world' is used to refer to what is commonly described as the 'real' physical world, whose contents possess an objective, independent existence; that world can be contrasted with 'secondary physical worlds,' or virtual worlds whose contents are determined by the computational processes of a computerized virtual reality system. The contents of such virtual worlds are arbitrary, insofar as they are not constrained by the organization of the primary physical world and can be dramatically altered at will by a virtual world's human designer or world-management algorithms. Even secondary physical worlds are still 'real' and 'physical,' though, insofar as the structure of their contents is maintained within real physical objects (e.g., the hard drives or ROM chips of a VR computer system) and are experienced by their inhabitants through the mediation of real physical stimuli (such as electrons or chemical neurotransmitters used to stimulate neurons in a host's sensory system or brain).

while the host is driving an automobile (to avoid disturbing passengers with such continual interruptions or to avoid revealing to the passengers that the host does not already know the way to the destination).

In such cases, the neuroprosthetic device is not simply a passive receiver that attempts to detect environmental stimuli in the form of sound waves and convey them accurately and instantaneously to the mind of its host: instead, the device employs various specialized algorithms or processes to determine the auditory percepts that the host should experience; accesses audio files or other data stored within the device or acquired from a remote location (e.g., through the Internet) that contain the information needed to correctly assemble the stream of auditory sense data; computes the correct pattern of electrochemical signals that the cochlear nerve should receive in order for the host to ultimately experience the intended auditory percepts; and physically generates the correct electrochemical stimuli and transmits them to the cochlear nerve. From a biocybernetic perspective, a neuroprosthetic device that possesses such an ability to fabricate sense data is of a qualitatively different sort than one that is only capable of transducing and transmitting stimuli that currently exist within the real physical environment.[33]

C. Storage of Sense Data

A sensory neuroprosthesis may store sense data within itself either temporarily or indefinitely; the data stored may be either 'real' sense data that corresponds to actual sets of stimuli that were detected in the host's primary physical environment, sense data that was detected in some remote environment, or 'fabricated' sense data that was procedurally generated by the device. Such storage can employ different memory types and information formats.

1. Memory Type

a. Volatile memory

A sensory neuroprosthesis such as a cochlear implant will typically need to store newly received sense data within itself at least briefly in order to process the data and compute the pattern of electrical stimulation that will be applied to its host's neurons in order to transmit the data. Such data may be

---

[33] The potential for sensory neuroprostheses to provide their hosts with sensory experiences that do not correspond to physical stimuli existing in the external primary physical world is discussed in Koops & Leenes, "Cheating with Implants: Implications of the Hidden Information Advantage of Bionic Ears and Eyes" (2012), and McGee (2008), p. 221.

stored using random-access memory (e.g., a DRAM or SRAM chip) that requires a continuous supply of electrical power in order to preserve the data. A loss of power to the device would result in the loss of such data.[34]

In principle, RAM can be used by a neuroprosthetic device to store sense data for a lengthy and indefinite period of time, as long as the device does not power down. However, in practice, RAM is used in mobile devices to store data that is being actively utilized for a brief period of time. This means that an implantable sensory neuroprosthetic device's use of electronic volatile memory technologies to store sense data would be functionally comparable to the mechanisms of iconic memory, echoic memory, haptic memory, and other forms of sensory memory (SM) that are employed by the human brain to store sense data for a brief period (e.g., a few hundred milliseconds) before it is processed by short-term memory (STM) and working memory (WM).[35]

b. Non-volatile memory

A sensory neuroprosthesis may also store sense data using non-volatile mechanisms that allow the data to be preserved even if the device is powered down. For example, a cochlear implant that is designed to allow its user to replay noteworthy conversations days or weeks after they occurred may store a copy of raw auditory sense data using flash memory that will maintain the data even if the device is restarted or its battery fails. From a functional perspective, such artificial memory technologies are comparable to the biological mechanisms for long-term memory (LTM) found in the human brain, which are able to store sense data in a relatively stable and permanent (if compressed and incomplete) form.[36]

2. Information Format

a. Digital data files

A neuroprosthetic device may store sense data in the form of binary digital files of the sort that can easily be exchanged with desktop computers or other external conventional computing systems. Data may be stored either in a raw format that preserves all available information but requires large file sizes or a processed and compressed format that preserves only selected information.

[34] Regarding different memory mechanisms for computerized devices, see Dumas, *Computer Architecture: Fundamentals and Principles of Computer Design* (2006).

[35] Such memory mechanisms utilized by the human brain are discussed, e.g., in Baars, *In the Theater of Consciousness* (1997), and Baddeley, "The episodic buffer: a new component of working memory?" (2000).

[36] The human brain's mechanisms for long-term memory are discussed in Dudai, "The Neurobiology of Consolidations, Or, How Stable Is the Engram?" (2004). Efforts to develop artificially intelligent systems that can mimic such memory processes are described in Friedenberg, *Artificial Psychology: The Quest for What It Means to Be Human* (2008).

A neuroprosthetic device's use of artificial technologies for the storage of digital data can be seen as roughly equivalent to the natural biological mechanisms by which the human body stores genetic information in the form of DNA.[37]

b. Neural network activation patterns

A neuroprosthetic device may store sense data within the components and processes of a physical artificial neural network that mimics to a greater or lesser degree the functioning of the human brain's biological neural network and which may directly interface with, complement, or partially replace neurons and neural structures within the brain.[38] Such a neural network stores information within its topology and activation patterns in a form that may be difficult or impossible for external computing systems to detect and decode and which may only be interpretable to the neural network in which the information is stored[39] – in effect, potentially creating a form of encryption that possesses the trait of information-theoretic security.

D. Transmission of Cellular Action Potentials within a Sensory Neural Pathway

It is possible for a neuroprosthetic device to serve as a 'bridge' that receives electrochemical signals that were generated by a natural biological sense organ and retransmits them along their path toward the brain of the device's host. For example, in a case where a secondary sensory neuron within the spinal cord has been damaged or destroyed (e.g., as a result of illness or injury), a neuroprosthesis might detect electrochemical signals produced by the related primary sensory neuron that correspond to sense data from mechanoreceptors in the skin and then retransmit that sense data to the tertiary neuron that carries it to the primary sensory cortex.[40] In healthy individuals,

---

[37] And, indeed, the use of DNA by artificial devices as a mechanism for data storage and computation is a technology whose development is being actively pursued. See Church et al., "Next-generation digital information storage in DNA" (2012).

[38] The development of memristors represents one approach to fashioning such a neural network. See e.g., Snider, "Cortical Computing with Memristive Nanodevices" (2008); Versace & Chandler, "The Brain of a New Machine" (2010); *Advances in Neuromorphic Memristor Science and Applications,* edited by Kozma et al. (2012); and Lohn et al., "Memristors as Synapses in Artificial Neural Networks: Biomimicry Beyond Weight Change" (2014).

[39] Regarding the difficulty of analyzing and interpreting the current state of information contained within an artificially intelligent system, especially that of a distributed artificial intelligence (DAI) displaying emergent behavior, see Friedenberg (2008), pp. 31-32.

[40] For a discussion of such sensory pathways, see Smith (2008), pp. 19-30, and Møller (2014), pp. 63-158. While they relate to processes of memory rather than sensation, the 'memory prostheses' described in Soussou & Berger, "Cognitive and Emotional Neuroprostheses" (2008), operate on a similar principle by serving as a bridge between neurons that spans a damaged area within the

such neuroprosthetic bridges could potentially be employed to allow sense data to reach the spinal cord or brain faster than would naturally be possible – but without altering the contents of the sense data.

E. Modalization of Sensory Experience

Some neuroprosthetic devices may sense details of their environment for internal purposes (e.g., to enable device calibration or the wireless download-ing of routine software updates) without the results of those sensing pro-cesses being shared directly with the conscious awareness of their human hosts. Those sensory neuroprostheses that detect stimuli in order to present them to their hosts' conscious awareness must do so using a specific **sensory modality**, a means by which the human mind perceives, consciously experi-ences, and interprets sense data.[41]

For natural biological human beings, each of the body's sensory organs normally presents sense data to the mind using a particular fixed sensory mo-dality: for example, the arrival of photons stimulating the retina is revealed to the mind through the modality of vision, while pressure applied to mech-anoreceptors in the skin is presented through the modality of touch. How-ever, atypical connections between a particular form of sense data and a par-ticular sense modality are sometimes found in nature: for example, an indi-vidual suffering from projective chromesthesia[42] might see a particular color whenever his or her ears detect the presence of a particular frequency of sound waves; in such a case, auditory sense data detected by the ears is pre-sented to the mind through the modality of vision.

A neuroprosthetic device may present authentic or fabricated sense data to the mind of its human host using any of a number of sensory modalities. For example, a neuroprosthesis that is capable of continually monitoring its host's blood glucose level could alert the host to the fact that his or her blood glucose has fallen to a dangerously low level by stimulating the cochlear nerve to present the host's conscious awareness with an audible tone (i.e., using the modality of hearing), by stimulating the optic nerve to present a flashing light in the host's field of vision (i.e., using the modality of vision), or by stimulat-ing nociceptors in a designated part of the host's body to create a piercing

---

hippocampus.

[41] Sensory modalities and related phenomena are discussed in Smith (2008), pp. 31-40.

[42] For an overview of such phenomena, see Cytowic, *Synesthesia: A Union of the Senses* (1989), and *Synaesthesia: Theoretical, artistic and scientific foundations*, edited by De Córdoba Serrano et al. (2014).

sensation (i.e., using the modality of pain).[43] Key sensory modalities are discussed below.

1. Vision

A neuroprosthetic device may present sense data to the conscious awareness of its human host using the naturally existing sensory modality of vision. The sense data thus presented might correspond to visible light detected in the host's primary physical environment, visual patterns that represent the appearance of some fabricated object within a virtual world in which the host is immersed, or another form of sense data that has been mapped to the modality of vision.[44] In a natural biological human being, the perception of sense data through the modality of vision involves the optic nerve and regions of the brain such as the visual cortex.[45]

It should be noted that while the rods and cones present in a natural biological retina might be loosely understood as forming a field of vision with a 'resolution' comprising a certain number of 'pixels' with regard to the retina's ability to detect the arrival photons, the manner in which visual data is transmitted to the brain and perceived by the mind is vastly more complex. While a human eye contains far more than one million sensors (i.e., individual rods and cones), there are only about one million output neurons (in the form of ganglion cells) that transmit data from the eye to the brain, with roughly half of the visual data coming from the tiny fovea, or focal point at the center of the field of vision.[46] Moreover, different neurons within areas V1 and V2 of the visual cortex are responsible for detecting edges, corners, spatial orientation, colors, and motion, while other neurons may detect patterns corresponding to the face of a particular person or a certain type of object. This allows the mind to experience the visual environment not as an array of pixels that

---

[43] It may also be possible for a neuroprosthetic device to present sense data to the conscious awareness of its human host in a manner that bypasses the body's sensory systems altogether: e.g., a neuroprosthesis that is capable of manipulating the processes of short-term or long-term memory may be able to provide its host with the (likely hazy and imperfect) *recollection* of having seen or heard something, despite the fact that the host had never originally experienced the sight or sound as incoming sense data. Such possibilities are considered in this text's later sections regarding memory as a form of cognition.

[44] Regarding the possibility of such sensory substitution, see Wiener (1961), loc. 2784ff; Lebedev (2014), p. 106; and Warwick (2014). For existing types of visual neuroprostheses, see See Linsenmeier (2005); Weiland et al. (2005); and Viola & Patrinos (2007). For the possibility of visual cortical implants, see Thanos et al., "Implantable Visual Prostheses" (2007).

[45] Vision in human beings is discussed in Smith (2008), pp. 281-378, and Møller (2014), pp. 261-302.

[46] See Linsenmeier (2005) for a discussion of some such issues.

changes from moment to moment but as a collection of recognizable objects and phenomena displaying various behaviors.

Simulations suggest that a retinal prosthesis would need to possess a visual resolution of 600-1,000 pixels in order to provide a level of vision that would allow its human host to perform functions such as reading ordinary text or recognizing faces.[47] It has been estimated that in order to fully replicate the visual acuity of a natural biological human eye, the photoreceptors of a retinal prosthesis would need to receive roughly 200 Gbps of raw sense data from the external environment and would transmit roughly 200 Mbps of processed sense data to the brain of the device's human host.[48]

2. Hearing

A neuroprosthesis may present sense data to the conscious awareness of its human host that is experienced through the naturally occurring sensory modality of hearing. The sense data experienced in this way might correspond to audible sound waves detected in the host's primary physical environment, sounds emanating from some fabricated object within a virtual world, or another type of sense data that has been mapped to the modality of hearing.[49] It has been estimated that in order to replicate the auditory sensitivity of a natural biological human ear, the mechanoreceptors of an auditory prosthesis would need to receive roughly 4 Mbps of raw sense data from the external environment and transmit 2 Mbps of processed sense data to the brain of the device's host.[50] In a natural biological human being, sensory experiences employing the modality of hearing are created by systems including the cochlear nerve and auditory cortex.[51]

3. Taste

A neuroprosthetic device may present sense data to the mind of its human host using the natural sensory modality of taste. Sense data presented in this way might correspond to the chemical properties of some object existing in the host's primary physical environment, to the 'taste' of some fabricated object existing within a virtual environment within which the host is immersed, or other sense data that has been mapped to the modality of taste. It has been

---

[47] Weiland et al. (2005).

[48] See Berner, *Management in 20XX: What Will Be Important in the Future – A Holistic View* (2004), pp. 45-46.

[49] Existing types of auditory neuroprostheses include cochlear implants and auditory brainstem implants. See Bostrom & Sandberg (2009), p. 321, and Cervera-Paz et al. (2007).

[50] See Berner (2004), pp. 45-46.

[51] Hearing in human beings is discussed in Smith (2008), pp. 123-186, and Møller (2014), pp. 191-261.

estimated that fully replicating the sensitivity of the human gustatory system would require the chemoreceptors of a gustatory neuroprosthesis to receive roughly 150 Mbps of raw sense data from the substance being tasted and would require the device to transmit 11 Mbps of processed sense data to the brain of its host.[52] In a natural biological human being, experiences of the modality of taste involve certain cranial nerves and the gustatory cortex.[53]

4. Smell

A neuroprosthesis may provide sense data to the conscious awareness of its human host in such a way that it is experienced through the naturally occurring sensory modality of smell. The sense data thus experienced might correspond to the chemical properties of airborne substances found in the host's primary physical environment, the 'smell' of substances within a virtual environment occupied by the host, or another type of sense data that has been mapped to the modality of smell. It has been estimated that in order to replicate the sensitivity of the human gustatory system, the chemoreceptors of an olfactory neuroprosthesis would need to receive roughly 20 Gbps of raw sense data from the gaseous mixture being smelled and transmit 30 Mbps of processed sense data to the brain of the device's host.[54] In a natural biological human being, components involved in the experience of smell include the olfactory bulb and olfactory cortex.[55]

5. Touch

A neuroprosthetic device may present sense data to the mind of its human host through the natural sensory modality of touch. Sense data presented in this manner might correspond to the shape, texture, and pressure of some object existing the host's primary physical environment, to the 'touch' of some fabricated object existing in a virtual environment within which the host is immersed, or to other sense data that has been mapped to the modality of touch. It has been estimated that in order to fully replicate the acuity of the human sense of touch, the mechanoreceptors of a neuroprosthetic 'skin' would need to receive roughly 1.5 Gbps of raw sense data and transmit 10 Mbps of processed sense data to the brain of the device's host.[56] In a natural

---

[52] See Berner (2004), pp. 45-46.

[53] Sensations of taste in human beings are discussed in Smith (2008), pp. 208-18, and Møller (2014), pp. 303-22.

[54] See Berner (2004), pp. 45-46.

[55] The sense of smell in human beings is described in Smith (2008), pp. 225-24, and Møller (2014), pp. 303-22.

[56] See Berner (2004), pp. 45-46.

biological human being, the experience of touch involves systems such as the primary somatosensory area in the parietal lobe.[57]

6. Posture and Bodily Extension (Proprioception)

A neuroprosthetic device may present sense data to the mind of its human host using the natural sensory modality of proprioception, which provides an individual with an awareness of the position of his or her limbs in relation to his or her head and torso.[58] In a natural biological human being, components contributing to the experience of proprioception include sensory neurons in the inner ear and ligaments as well as the cerebrum and cerebellum. The capacity for proprioception allows one to sense whether one's arms are extended to the sides, raised overhead, or pointed forward even when one is standing in a darkened room and unable to see one's arms. Proprioceptive sense data presented by means of a neuroprosthesis might correspond to the current posture and extension of the host's primary physical body (including biological and artificial components), the posture and extension of the virtual (and potentially radically nonhuman[59]) body possessed by the host in some virtual environment, or another type of sense data that has been mapped to the modality of proprioception.

The use of a neuroprosthetic device to create the proprioceptive sensation of a virtual body part that does not exist as a real physical object is in some ways comparable to the phantom limb sensation that causes amputees to feel the presence of a limb that no longer physically exists.[60]

7. Balance

A neuroprosthesis may provide sense data to the conscious awareness of its human host that is experienced through the sensory modality of balance, which reveals the orientation and motion of the host's body relative to the ground. This modality allows an individual to know whether he or she is standing upright and motionless, spinning, or falling even when located in a darkened room in which it is impossible to see one's body or the surrounding environment. In a natural biological human being, the experience of balance is produced through joint action of components such as the vestibular system

---

[57] Human beings' sense of touch is discussed in Smith (2008), pp. 112-22, and Møller (2014), pp. 159-90.

[58] See Siegel & Sapru, *Essential Neuroscience* (2006), pp. 259-60.

[59] For such possibilities, see Gladden, "Cybershells, Shapeshifting, and Neuroprosthetics: Video Games as Tools for Posthuman 'Body Schema (Re)Engineering'" (2015).

[60] See *Amputation, Prosthetic Use, and Phantom Limb Pain: An Interdisciplinary Perspective*, edited by Murray (2010).

(including the inner ear, cerebellum, and thalamus), visual system, and proprioceptive system.[61]

If a neuroprosthetic device is employed, the sense data experienced through this modality might correspond to the orientation of the host's real physical body relative to the primary physical environment, the orientation of the host's virtual body relative to the infrastructure of a virtual world in which the host's conscious awareness is immersed, or another type of sense data that has been mapped to the modality of balance. The employment of a neuroprosthesis to create a sense of balance, motion, or motionlessness that does not correspond to the actual state of a host's real physical body is in some ways comparable to the phenomenon of subjective vertigo that causes an individual to feel as though his or her body is spinning when in fact that is not the case.[62]

8. Heat and Cold

A neuroprosthetic device may present sense data to the mind of its human host through the natural sensory modalities of heat and cold. Sense data presented in this fashion might reflect the temperature of the host's real physical body or the primary external physical environment, of the host's virtual body or the fabricated environment of a virtual world in which the host is immersed, or other sense data that has been mapped to the modalities of heat and cold. In a natural biological human being, the experience of heat and cold is regulated by systems including portions of the thalamus.[63]

9. Energy Level (Interoception)

A neuroprosthetic device may provide sense data to the conscious awareness of its human host that is experienced through the sensory modalities of interoception, which (among other things) allows the host to feel his or her internal energy level and the degree to which he or she needs to replenish his or her operating ability by inhaling or ingesting physical resources (such as oxygen, water, and food) or obtaining a sufficient amount of sleep. Such modalities may take the form of feelings such as hunger, thirst, drowsiness, or suffocation. In a natural biological human being, the modalities of interoception are regulated by various internal organs and regions of the brain.[64] If an interoceptive neuroprosthesis is employed, the sense data experienced through the modalities of interoception might correspond to the energy or

---

[61] The sense of balance is discussed in Smith (2008), pp. 129-31.

[62] See Brandt, *Vertigo: Its Multisensory Syndromes* (2003).

[63] Thermosensitivity is discussed in Smith (2008), pp. 415-22.

[64] Regarding the nature of interoception in the human organism, see Cameron (2001). For the possibility of neuroprosthetic devices that relate, e.g., to their host's need for or experience of sleep, see Claussen & Hofmann (2012) and Kourany (2013), pp. 992-93.

resource level of the host's real physical body, the energy or resource level of the host's virtual body within an immersive virtual environment, or another type of sense data that has been mapped to one of the modalities of interoception.

10. Pain

A neuroprosthesis may present sense data to its host's mind using the naturally existing sensory modality of pain. In a natural biological human being, the experience of pain often results from the presence of an excess of skin pressure, kinetic energy, light, sound waves, or chemical substances that are causing damage or destruction to biological tissue and which at lower quantities would be experienced through a different modality (such as that of touch, heat, vision, hearing, or hunger). In a natural biological human being, the experience of pain may involve portions of the brainstem, among other systems.[65]

If a neuroprosthetic device participates in these processes, the sense data presented through the modality of pain might reflect excessive stimulation of or damage to biological or artificial components of its host's real physical body, the host's virtual body within an immersive virtual environment, or some other type of sense data that has been mapped to the modality of pain.[66]

11. Passage of Time

A neuroprosthetic device may present sense data to the mind of its human host in such a way that creates the experience of the passage of time. In a natural biological human being, such perception of the passage of time is regulated by a complex array of systems including the cerebral cortex, cerebellum, and basal ganglia.[67] If a neuroprosthetic device is involved, the sense data presented in this fashion might reflect the passage of time as measured within the primary physical environment in which the host's real physical body is located, the passage of time within some virtual environment in which the host is mentally immersed, or potentially some other form of sense data (e.g., the presence of high temperatures or a particular airborne chemical) which causes the neuroprosthetic device to present stimuli to the host's mind at an accelerated or decelerated rate.

If a neuroprosthetic device creates a fully immersive virtual reality environment in which visual, auditory, and other sense data are provided to its

---

[65] The experience of pain by human beings is discussed in Smith (2008), pp. 437-62, and Møller (2014), pp. 337-65.

[66] For different kinds of neuroprosthetic devices that have an analgesic effect of reducing the experience of pain or producing a feeling of euphoria for their human host, see Denning et al., "Neurosecurity: Security and Privacy for Neural Devices" (2009).

[67] See The Nature of Time: Geometry, Physics and Perception, edited by Buccheri et al. (2003).

host's conscious awareness by the device, it may be possible for the device to cause time to pass more slowly or quickly in the virtual world than in the real physical world in which the host's body is located. (For example, the host might be presented with the experience of sitting in an auditorium and listening to a lecture that lasts 50 minutes – while in real life the lecture had lasted 55 minutes but was recorded and played back to the host's conscious awareness at an accelerated rate.) Depending on the senses involved and the difference in speed between the real and experienced worlds, a host might feel as though: 1) the world's activities were continuing to occur at their regular speed but the host's cognitive and physical activities had been 'slowed down' or 'speeded up'; 2) the host's cognitive and physical activities were continuing to occur at their normal rate but the activity of the external world had been 'slowed down' or 'speeded up'; or 3) there was no detectable difference in the host's experience of the passage of time.

12. Acoustic Contextualization

A neuroprosthetic device may provide sense data to the conscious awareness of its human host that is experienced through the auditory sensory modality that allows one to 'feel' the size, shape, and contents of the environment in which one's body is located, even when the environment is completely dark and it is not possible to judge its extent and contents through vision or touch. When an individual intentionally produces a sound in order to judge the size and features of the environment based on the sound's echo, this modality takes the form of echolocation. When it relies on the characteristics of ambient sounds already present in the environment, it takes the more general form of acoustic wayfaring or contextualization. To some extent, all human beings possess a limited ability to exploit the information conveyed through this sensory modality: for example, a person might instinctively navigate his or her unlit home at night without bumping into walls by relying on subtle auditory clues provided by the hum of a refrigerator or water heater or the echoes produced by the sound of his or her footsteps on a hardwood floor. The skill of recognizing the information conveyed through this modality might be more explicitly trained and developed by, for example, one who is blind or who regularly works in a lightless environment.[68]

Some bat species possess specialized inner ears organs that are structured to be especially sensitive and accurate in detecting the particular frequency

---

[68] For a discussion of echolocation and acoustic wayfaring, see Goldstein, *Sensation and Perception* (2014), pp. 312-13; Teng & Whitney, "The acuity of echolocation: spatial resolution in the sighted compared to expert performance" (2011); and Fiehler et al., "Neural correlates of human echolocation of path direction during walking" (2015).

of sound wave generated by the bat's echolocation process.[69] In human beings, there do not appear to be specialized sensory structures dedicated for use in echolocation and acoustic contextualization. Instead, hair cells within the cochlea are continuously receiving a wide range of generalized auditory sense data and transmitting it to the brain – where most of it is experienced through the modality of hearing but where some of it may (consciously or subconsciously) be interpreted to provide information about the size, shape, and contents of the surrounding environment.

If an auditory neuroprosthetic device is employed to participate in such processes, the sense data experienced through the modalities of echolocation or acoustic contextualization might reflect the features of the primary physical environment in which its host's real physical body is located, the fabricated features of a virtual reality environment in which the host is immersed,[70] or perhaps another type of sense data that has been mapped to these modalities.

13. Magnetoreception

A number of invertebrates, mammals, and birds (e.g., the homing pigeon) show an ability to detect the presence and orientation of magnetic fields, which they may use as an aid in navigation or for other purposes.[71] The hypothesis that human beings may possess some natural mechanism for magnetoreception that allows the brain to perceive (even if subconsciously) the existence of magnetic fields is highly controversial,[72] although research has found, for example, that transcranial magnetic stimulation of the brain can have significant effects on emotions and other cognitive behavior.[73]

---

[69] See Neuweiler, "Evolutionary aspects of bat echolocation" (2003).

[70] Robust and correct handling of the phenomenon of acoustic contextualization is necessary in order to create an immersive VR environment of the highest possible sensory fidelity. When fabricating the contents of a VR environment, it is not sufficient to accurately calculate and present the frequency, volume, and point-source direction of obvious environmental sounds (such as the voice of a conversation partner or the engine of a passing automobile); it is also necessary to determine, for example, the particular way in which these sounds are partially absorbed and reflected by a desk near which the listener is standing, depending on whether the desk is made of metal, wood, or some other substance. The fact that echolocation data can be successfully interpreted by human subjects in recorded (rather than live) audio sources is demonstrated, e.g., in Fiehler et al. (2015).

Efforts to develop artificial technologies to enhance or expand human beings' capacities for echolocation and acoustic wayfaring are described, e.g., in Sohl-Dickstein et al., "A device for human ultrasonic echolocation" (2015), and Davies, *Audification of Ultrasound for Human Echolocation* (2008).

[71] See Smith (2008), pp. 434-35, and Rozhok, *Orientation and Navigation in Vertebrates* (2008).

[72] See Hand, "Maverick scientist thinks he has discovered magnetic sixth sense in humans" (2016).

[73] See *A Clinical Guide to Transcranial Magnetic Stimulation*, edited by Holtzheimer & McDonald

Given these scientific uncertainties, it may or may not be possible for a neuroprosthesis to present sense data to its host's mind using a sense modality that involves the perception of magnetic fields. If it is possible, sense data presented through such a modality might reflect the characteristics of the magnetic fields (typically dominated by the earth's natural magnetic field) of the primary physical environment in which the host's physical body is located or an artificial magnetic field that has been created specifically to override the naturally occurring ambient magnetic fields and create the perception that the host's body is oriented in a different direction than it actually is; the latter arrangement might be used, for example, to make the magnetic fields experienced by a host who is immersed in a virtual reality environment consistent with the visual or auditory clues that the host is receiving regarding the direction of his virtual body's orientation. Note that if, for example, an individual possessed an artificial eye that displayed a computer-generated compass within the host's field of vision in order to indicate the direction of magnetic north, such a presentation would constitute a use of the natural modality of vision – not the use of a hypothesized perception of magnetic fields.

14. Electroreception

Some animals (including sharks, electric eels, and the platypus) possess a dedicated sensory system for electroreception that allows them to detect the presence of electric fields – often in order to locate prey.[74] In a natural biological human being, the perception of electrical charge as such is not known to exist as a separate sensory modality. While specialized electroreceptors of the sort found in fish are absent in human beings, a sufficiently large electric charge present in the environment can indiscriminately stimulate sensory and motor neurons in a way that produces the sensation of an electric shock; however, this is typically experienced through existing sensory modalities such as those of touch, vision, hearing, heat, and pain. Nevertheless, the ability to detect an electrical charge is a critical component of neurons' capacities for communication and control, and the basic functioning of many types of sensory and cognitive neuroprostheses is premised on the fact that that the application of an electric charge can generate desired responses in individual neurons or the brain as a whole. While the potential for a neuroprosthetic device to present sense data using a hypothesized sensory modality involving

---

(2014); *Transcranial Magnetic Stimulation in Clinical Psychiatry*, edited by George & Belmaker (2007); and *Magnetic Stimulation in Clinical Neurophysiology*, edited by Hallett & Chokroverty (2005).

[74] See Smith (2008), pp. 428-33; *Electroreception*, edited by Bullock et al. (2005); and *Sensory Evolution on the Threshold: Adaptations in Secondarily Aquatic Vertebrates*, edited by Thewissen & Nummela (2008), pp. 425-432.

the perception or experience of electric charge as such is thus unclear, it may merit further future research.

F. Mapping of Sensory Routes

A sensory neuroprosthetic device creates a 'mapping' by which a particular type of sensory stimulus is detected by a particular sense organ and experienced by the mind of the device's host through a particular sensory modality. Within the organism of a natural biological human being, many such sensory routes or mappings already exist: for example, photons present in the environment excite photoreceptors in the retina, which produce electrochemical signals that are transmitted to the visual cortex and experienced by the mind as particular sights. In that way, the stimulus of visible light is detected by the sense organ of the eye and experienced through the sensory modality of vision. It is possible for a neuroprosthetic device to participate in such a straightforward mapping that attempts to replicate or complete a mapping found naturally in the human body – such as when a cochlear implant detects sound waves that fall within the normal audible range of roughly 20 to 20,000 Hz and converts that stimulus into a pattern of electrical impulses that are transmitted to the cochlear nerve and the cochlear nucleus in the brainstem and perceived through the modality of hearing.

On the other hand, it is possible for a neuroprosthetic device to participate in a 'non-standard' sensory mapping. As noted earlier, the possibility of such mappings is suggested by the existence of the rare but naturally occurring neurological condition of synesthesia, in which sense data typically experienced through one sensory modality are instead (or also) experienced through a different modality – as is the case for individuals who see a color or taste a flavor upon hearing a particular sound.[75] Discussion of the potential use of neurocybernetic technologies to artificially 'remap' the processes of human sensation dates back at least as far as the 1940s, when Wiener considered the possibilities, for example, of rerouting sense data received by the eyes so that it would be experienced by the mind through the modality of hearing – or rerouting sense data received by the ears so that it would be perceived through the modality of vision.[76] Such neuroprosthetically aided sensory remapping could potentially have therapeutic uses: for example, a

[75] Such phenomena occur, e.g., in cases of chromesthesia and lexical-gustatory synesthesia. Some research (e.g., into the phenomenon of ideasthesia) suggests that while in some cases of synesthesia it superficially appears as though one kind of sensory experience has directly triggered the other, it may in fact be the mind's internal semantic representation of elements contained in the first sensory experience (and not the sense data itself) which has triggered the second. See Cytowic (1989); *Synaesthesia: Theoretical, artistic and scientific foundations* (2014); and Jürgens & Nikolić, "Ideaesthesia: conceptual processes assign similar colours to similar shapes" (2012).

[76] See Wiener (1961), loc. 2784ff. For more recent discussions, see Lebedev (2014), p. 106, and Warwick (2014).

neuroprosthetic device that includes a microphone and which stimulates the optic nerve or visual cortex could provide someone who is deaf with visual cues when certain kinds of environmental stimuli (e.g., the sound of a ringing doorbell or approaching siren) are detected. Such non-standard mappings will always be the case with neuroprosthetic devices that detect xenostimuli which the human body's natural biological sensory organs are incapable of detecting, as – by definition – such stimuli have no default modality through which the mind normally experiences them.

Below we consider different ways in which a sensory neuroprosthetic device can tie together sense data with a sense organ and sensory modality to create a unified process of sensing for its human host.

1. Routing Biologically Receivable Sense Data to its Biologically Normal Sensory Modality

A neuroprosthetic device may detect stimuli that are typically detected by a natural biological sense organ and present them to its host's mind using the modality that is naturally associated with that form of sense data. An example would be a robotic prosthetic hand that detects the pressure of external objects against its fingers and conveys that sense data to its host's mind through the modality of touch.[77]

2. Routing Fabricated Sense Data to an Arbitrary Sensory Modality

Instead of detecting stimuli that exist within its host's body or the external environment and conveying them through the electrochemical stimulation of neurons, a neuroprosthetic device might instead transmit electrochemical signals that do not correspond to any physical stimuli existing in the environment and detected by the device. A neuroprosthesis might fabricate such sense data, for example, in order to provide its host with an immersive sensory experience of a virtual world (or 'secondary physical world') that does not objectively exist as such in the primary physical environment.[78]

Such a mapping differs from human beings' natural biological sensory mappings, because the origin of the sense data lies not in some natural physical phenomena that occur independently of a neuroprosthetic device and are then detected by it but in digital constructs or objects that are purposefully fabricated by the device by means of its own internal computational processes and for some particular ends.

---

[77] See Hoffmann & Micera (2011), pp. 792-93.

[78] Regarding the distinction between the primary physical world (or 'real world') and secondary physical worlds (or 'virtual worlds'), see the footnote in the earlier section on neuroprostheses that fabricate electrochemical signals within a sensory neural pathway.

3. Remapping Biologically Receivable Sense Data to a Biologically Atypical but Natural Sensory Modality

Another possibility is for a neuroprosthetic device to detect stimuli that are normally detectable by the human body's natural biological sensory organs but to present them by means of a sensory modality which – while naturally occurring in human beings – is not the modality through which such sense data is normally experienced. Examples might include an artificial eye that can detect the current air temperature and uses augmented reality to display this as a set of numbers in its host's field of vision.[79]

One variation of such a routing is for a neuroprosthesis is to present sense data that relate to the external environment using a sensory modality that is normally a means of experiencing phenomena originating within the host's body. An example would be a neuroprosthesis whose external photoreceptor measures the amount of visible light present in the external environment and – when the level drops below a certain threshold (e.g., after the sun has set) – stimulates the relevant neurons to cause the device's host to experience that change through the modality of interoception as a growing sense of drowsiness. A complementary variation is for a neuroprosthetic device to present sense data relating to its host's internal bodily processes by means of a sensory modality that normally conveys sense data relating to the external environment. An example would be a neuroprosthesis that detects pressure within the body's muscles and joints in order to construct a representation of the body's current posture and displays an image of the body in the host's field of vision by stimulating the optic nerve.[80]

4. Routing Sense Data Using a Non-human Xenosensory Mapping

It is possible for a neuroprosthetic device to incorporate into its sensory mapping some non-natural 'xenosensory' element that is not found in the sensory mappings of natural biological human beings. The use of such technologies provides a human host with superhuman sensory capacities that are

---

[79] The hypothesized technologies for sensory substitution discussed by Wiener would constitute such neuroprosthetic devices; see Wiener (1961), loc. 2784ff.

[80] In some ways, such a mapping is functionally comparable to forms of exteroception seen in animals such as the octopus, whose brain does not possess a mechanism for determining the precise current location and shape of each of its tentacles through its sense of proprioception (which would be an intensive and complex information-processing task, given the large number of limbs and infinite degrees of freedom possessed by each of them); instead, the brain gathers such information by visually observing its limbs. See Wells, M.J., *Octopus: Physiology and Behaviour of an Advanced Invertebrate* (1978); Zullo et al., "Nonsomatotopic organization of the higher motor centers in octopus" (2009); Niven "Invertebrate neurobiology: Visual direction of arm movements in an octopus" (2011); and Gutnick et al., "Octopus vulgaris uses visual information to determine the location of its arm" (2011).

'extrasensory' in the sense of exceeding the natural limits of the human organism (although not in the sense of being paranormal).

a. Mapping sense data from non-human xenostimuli to an arbitrary sensory modality

One type of xenosensory mapping involves a neuroprosthetic device that is capable of detecting 'xenostimuli' (either in the external environment or within the host's body) that are not normally detectable by the human body's sense organs.[81] By definition, the augmented or wholly artificial sense organs used to register such sense data differ from those of natural biological human beings, and the mind does not have a natural default sense modality through which it consciously perceives such data; the designer of the device must decide which sensory modality will be used for that purpose.

b. Mapping arbitrary sense data to a non-human xenesthetic sensory modality

Another type of xenosensory mapping would involve a neuroprosthesis that presents sense data to the mind of its host using a sensory modality that is not normally experienced by a natural biological human being. It is not clear whether such a mapping is possible: the scientific knowledge and technological capacities needed to map xenostimuli to an existing natural sensory modality are relatively straightforward (and, indeed, such neuroprosthetic mappings have already been implemented on an experimental basis[82]); however, it is much more difficult to imagine how one might effect the creation of a new sensory modality within the mind – or the 'unlocking' of a natural biological modality that is typically latent. While practical avenues for the development or exploitation of such experiences of 'xenesthesia' are not immediately obvious, such a possibility should be acknowledged at a conceptual level as a subject meriting further research. Possible directions for research might involve studying atypical but natural phenomena such as synesthesia, hallucinations, or *déjà vu*, to explore whether they might provide a basis for the creation of new sensory modalities.

II. Cognition

A cognitive neuroprosthesis restores, augments, or otherwise participates in processes that are internal to the mind or brain of its host and which do not directly involve sensory organs or motor organs. In comparison to that of

---

[81] Examples would include neuroprostheses that provide their host with infrared vision or the ability to hear ultrasonic phenomena. See, e.g., Warwick (2014); Gasson et al. (2012); and Merkel et al. (2007).

[82] For example, Warwick (2014) describes a successful experiment in which a surgically implanted microelectrode array was used to grant its human host 'extrasensory' perception in the form of the ability to detect ultrasonic phenomena.

sensory and motor neuroprostheses, the development of cognitive neuro-prostheses that can affect phenomena such as memory, emotion, and person-ality is still in its early experimental stages – although rapid advances are be-ing made.[83]

A cognitive neuroprosthesis can be analyzed according to the extent to which it establishes a basic interface with the conscious awareness of its host, its impact on the contents of autonomous cognitive processes, and its impact on the execution of processes that involve conscious decisions by its host. Below we discuss these aspects of cognitive neuroprostheses in more detail.

A. Basic Interface with the Mind

Cognitive neuroprostheses can be classified according to whether a device interfaces with its host's cognitive processes at a conscious or subconscious level. A single device may act on both levels simultaneously.

1. With the Host's Conscious Awareness

It is possible for the action of a cognitive neuroprosthesis to directly in-form, affect, be experienced by, or be controlled by its host's conscious aware-ness. An example would be a mnemoprosthetic device that stores long-term memories as engrams and which allows its host to recall and consciously ex-perience a particular memory through an act of will.[84]

2. Without the Host's Conscious Awareness

Alternatively, a cognitive neuroprosthesis may be integrated with the neu-ral circuitry of its human host in such a way that the host cannot consciously control the device and does not have a direct conscious experience of the de-vice's operation. An example would be a sensor implanted in the brain that wirelessly transmits real-time data regarding the host's neural activity to an external computer, which is able to decode some aspects of what the host is consciously experiencing in that moment (e.g., whether he or she is imagin-ing an oceanfront vista, is attempting to mentally solve a math problem, or is mentally rehearsing the melody of a favorite song), without the host knowing whether or not the device is in operation at any given moment.

Note that in the case of some neuroprosthetic technologies – such as deep brain stimulation (DBS) implants – a host may not possess direct and certain

---

[83] For an overview of such technologies, see Soussou & Berger, "Cognitive and Emotional Neuro-prostheses" (2008), and Gladden, "Enterprise Architecture for Neurocybernetically Augmented Organizational Systems" (2016).

[84] For discussion of future neuroprosthetic devices that could allow 'playback' of recorded or previously experienced information, see Merkel et al. (2007); Robinett, "The consequences of fully understanding the brain" (2002); and McGee (2008), p. 217.

knowledge of the fact that that the device has been activated but may be able to infer as much from the conscious experience of the device's impacts, such as certain muscle behaviors or emotional effects.[85]

B. Perception

A neuroprosthetic device may access or affect those cognitive processes that enable or shape an individual's perception and experience of his or her own mind and body and the external environment that he or she inhabits. Such a device might relate to its host's mental percepts, degree of consciousness and self-awareness, periodic transition into and out of sleep and other states of unconsciousness, experience of emotions, and autonomous processes that induce experiences such as that of surprise.

1. Percepts

A percept is not sense data as received by the brain but as experienced and interpreted by the mind. Two people viewing the same inkblot or optical illusion may receive identical visual sense data from their eyes but perceive the data to present very different objects. Similarly, two people who listen to identical recordings of a work of classical music may experience the same auditory sense data in very different ways, if one person has a rich knowledge of music theory and the work's history and the other person does not. A percept frequently constitutes an integrated mental experience that combines several different types of sense data and cognitive processes; for example, the experience of conversing with another human being and one's interpretation of that experience may draw on auditory sense data (e.g., the sound of identifiable words being spoken as well as paralinguistic features such as tone of voice), visual sense data (e.g., the sight of the other person's gestures and facial expressions), emotion (e.g., one's attitude of fondness or dislike for the person who is speaking), and long-term memory (e.g., one's recollection of the meaning of the words being heard and the historical context and significance of the information being conveyed by one's conversation partner).[86]

A neuroprosthetic device that can affect the brain at the level of percepts – for example, causing its host to interpret an optical illusion in one way rather than another – would participate in the final stages of perception rather

---

[85] For a discussion of the effects of DBS, see Kraemer, "Me, Myself and My Brain Implant: Deep Brain Stimulation Raises Questions of Personal Authenticity and Alienation" (2011), and Van den Berg, "Pieces of Me: On Identity and Information and Communications Technology Implants" (2012).

[86] For an overview of human perception, see Goldstein (2014); Rookes & Willson, *Perception: Theory, Development and Organisation* (2000); and *The Oxford Handbook of Philosophy of Perception*, edited by Matthen (2015).

than the early stages of receiving raw sense data, and it may thus be understood primarily as a cognitive rather than sensory neuroprosthesis.[87]

2. Consciousness and Self-awareness

A neuroprosthesis may directly affect its host's experience of consciousness, sapience, and self-awareness. For example, it is conceivable that a neuroprosthetic device could impair or even destroy its host's ability to experience conscious self-awareness – such as by temporarily suppressing the activity of or destroying particular regions of the cerebral cortex or the brainstem's reticular activating system (RAS) to induce a coma-like state.[88] It can be speculated that a sufficiently sophisticated (and highly futuristic) neuroprosthetic system possessing adequate sensory, cognitive, and motor interfaces with its host might even suppress its host's conscious awareness without this being immediately obvious to outside observers, if the neuroprosthesis were able to cause its host's body to produce appropriate motor actions (such as speech or gestures) in response to sensory stimuli from the external environment.

3. Sleep

A neuroprosthesis may affect its host's processes of sleep – either by creating a sense of drowsiness and desire for sleep, directly inducing a state of unconsciousness (as is done by a general anesthetic), or artificially reducing an existing state of drowsiness and keeping an individual awake (as is done by some stimulants).[89] Efforts to develop neurotechnologies that can extend an individual's period of wakefulness are being undertaken by military research agencies; such technologies could be employed, for example, to allow soldiers to operate for extended periods of time without sleep when in hostile territory.[90]

---

[87] Many kinds of neuroprosthetic devices relating to sensory perception already exist; see Lebedev (2014); Merkel et al. (2007); and Gladden, "Enterprise Architecture for Neurocybernetically Augmented Organizational Systems" (2016).

[88] Regarding the nature of comas in human beings, see *Coma Science: Clinical and Ethical Implications*, edited by Laureys et al. (2009), and *Comas and Disorders of Consciousness*, edited by Schnakers & Laureys (2012). Techniques such as transcranial magnetic stimulation have been used by researchers to temporarily take particular regions of the brain 'offline,' making it possible to study how the brain's performance of certain tasks is impacted by the impairment or disabling of those regions; see Ariely and Berns, "Neuromarketing: The Hope and Hype of Neuroimaging in Business" (2010).

[89] For the possibility of neuroprosthetic devices relating to sleep, see Claussen & Hofmann (2012) and Kourany (2013), pp. 992-93.

[90] Regarding efforts by the DARPA military research agency and other researchers to develop neurotechnologies for increasing soldiers' alertness and reducing their need for sleep, see, e.g., Falconer, "Defense Research Agency Seeks to Create Supersoldiers" (2003); Moreno, "DARPA On

4. Emotion

A neuroprosthesis may alter its host's immediate emotional state or influence the host's long-term patterns of emotional behavior.[91] Such a device could be used to treat emotional disorders or to optimize emotional behaviors for individuals (such as actors, politicians, teachers, counselors, police officers, or spies) whose work requires them to engage in highly effective social interactions.

5. Autonomous Mental Reactions

A neuroprosthesis may participate in cognitive processes that are autonomous, insofar as they are involuntary and take place beyond the conscious control of the mind of the device's host. Such processes might include the generation of feelings of surprise or dread in response to particular stimuli or a spontaneous irrational presumption or drawing of conclusions that results from a naturally occurring human cognitive bias.[92] Such autonomous cognitive processes are distinct, for example, from those processes of the autonomic nervous system (ANS) whose actions are often regulated by the brainstem and result in motor activity rather than internal cognitive effects as well as from those flexor withdrawal reflexes regulated by neurons in the spinal cord.[93]

C. Creativity

A neuroprosthetic device may access or affect cognitive processes that manifest mental creativity, including those relating to ideation and imagination and those that generate dreams while a device's host is asleep.

Your Mind" (2004); Clancy, "At Military's Behest, Darpa Uses Neuroscience to Harness Brain Power" (2006); and Wolf-Meyer, "Fantasies of extremes: Sports, war and the science of sleep" (2009).

[91] For the possibility of developing emotional neuroprosthetics, see Soussou & Berger (2008); Hatfield et al., "Brain Processes and Neurofeedback for Performance Enhancement of Precision Motor Behavior" (2009); Kraemer (2011); and McGee (2008), p. 217.

[92] For an overview of human cognitive biases – especially with regard to their impact in organizational settings – see Kinicki & Williams, *Management: A Practical Introduction* (2010), pp. 217-19. Note that if a neuroprosthetic device is involved in regulating biological processes such as breathing, digestion, or circulation, it may be more likely to be a motor neuroprosthesis rather than a cognitive neuroprosthesis. The device's classification depends on the nature of the output or impact produced by the device and the system upon which it acts.

[93] Regarding such phenomena, see *Primer on the Autonomic Nervous System*, edited by Robertson et al. (2012), and Muscolino, *Kinesiology: The Skeletal System and Muscle Function* (2017), pp. 609-38.

1. Ideation and Imagination

A neuroprosthetic device may participate in the process of imagination and the mind's generation of new ideas. Some aspects of that process lie beyond an individual's conscious control, while other aspects can be consciously steered or nurtured in an effort to generate certain kinds of new ideas. Anecdotal accounts of increased creativity have been reported, for example, among patients who have received treatment for medical conditions with DBS devices.[94]

2. Dreaming

A neuroprosthetic device could potentially affect the dreams of its user – for example, by suppressing the individual's ability to dream, intensifying the experience of dreams (e.g., by facilitating lucid dreaming in which an individual is consciously aware of being in a dream state), or modifying or controlling the contents of dreams. A sufficiently sophisticated neuroprosthesis may also allow at least some features of the contents of a sleeping individual's dreams to be detected and interpreted by external systems.[95]

D. Memory and Identity

A neuroprosthetic device may access or affect those cognitive processes by which its host encodes, stores, and retrieves memories and creates a consistent identity that persists over time. Such processes include those relating to sensory, short-term, and long-term memory; personality; habit; and belief.

1. Memory

A neuroprosthetic device may access, modify, control, or otherwise relate to its host's processes of encoding, storing, and retrieving memories. Such 'mnemoprostheses' may involve one or more type of memory.

a. Sensory memory (SM)

A neuroprosthetic device may affect or access the contents of iconic memory, echoic memory, haptic memory, and other forms of sensory memory that the human brain employs to briefly (e.g., for a few hundred milliseconds) store incoming sense data before it is processed by short-term and working memory.[96]

---

[94] See Cosgrove, "Session 6: Neuroscience, brain, and behavior V: Deep brain stimulation" (2004); Gasson (2012); and Gladden, "Neural Implants as Gateways" (2016).

[95] For an overview of the neuroscience of dreaming, see Kalat, *Biological Psychology* (2007), pp. 286-91, and "Section X: Dreaming" in *The Neuroscience of Sleep*, edited by Stickgold & Walker (2009).

[96] For an overview of the biological basis of sensory memory, see Chapter 4, "Sensory and Short-Term Memory," in Radvansky, *Human Memory* (2016).

b. Short-term memory (STM)

A neuroprosthesis may affect or access the contents of its host's short-term memory, which is capable of storing a small number of items (e.g., a handful of numerical digits) in memory for a matter of a few seconds. A device's ability to enhance short-term memory could compensate for short-term memory problems resulting from neurodegeneration caused by aging, injury, or illness.[97]

c. Long-term memory (LTM)

The notion of developing neuroprosthetic devices that can access and manipulate the contents of a host's long-term memory is one whose theoretical possibility and practical feasibility are much contested. Recent achievements in artificially modifying the memories of mice[98] suggest that a limited ability to alter some of a human subject's memories in a fairly non-targeted way may someday be possible. However, if certain hypotheses regarding the holographic nature of long-term memory storage in the brain[99] were to be demonstrated to be correct, it might be difficult or impossible to access or manipulate particular complex long-term memories without decoding or disrupting the complete contents of an individual's long-term memory; it may also be impossible to edit the information stored in the brain's long-term memory structures and processes through the relatively gross electrical stimulation of neurons that is employed by technologies such as those for DBS or to access and fully decode such information using conventional monitoring and imaging technologies such as EEG, MRI, CT, and PET, each of which possesses limitations that make it inadequate for such purposes.[100]

Regardless of whether the targeted creation, editing, or deletion of long-term memories through the use of neuroprosthetic devices might someday prove feasible, mnemoprostheses might be able to affect the structures and processes of long term-memories in a more generalized way – for example,

---

[97] See Chapter 4, "Sensory and Short-Term Memory," in Radvansky (2016).

[98] See Han et al., "Selective Erasure of a Fear Memory" (2009), and Ramirez et al., "Creating a False Memory in the Hippocampus" (2013).

[99] Such models have been described, e.g., in Longuet-Higgins, "Holographic Model of Temporal Recall" (1968); Westlake, "The possibilities of neural holographic processes within the brain" (1970); and Pribram, "Prolegomenon for a Holonomic Brain Theory" (1990). An overview of conventional contemporary models of long-term memory is found in Rutherford et al., "Long-Term Memory: Encoding to Retrieval" (2012).

[100] Regarding various approaches to the neuroimaging of memory mechanisms, see, e.g., *Neuroimaging and Memory*, edited by Foster (1999); Greve & Henson, "What We Have Learned about Memory from Neuroimaging" (2015); Rugg et al., "Encoding and Retrieval in Episodic Memory: Insights from fMRI" (2015); and Peigneux, "Neuroimaging Studies of Sleep and Memory in Humans" (2015).

by providing increased storage capacity, encoding and retention of memories in a manner that is less compressed and does not degrade over time, or enhanced recall speed and ability.[101]

2. Personality

A neuroprosthesis may alter its host's long-term formation and storage and short-term manifestation of distinct personality traits.[102]

3. Habit

A neuroprosthetic device may participate in the processes by which the mind of its host forms, stores, and is influenced by subconscious habits or preferences of which the mind may or may not be consciously aware.

4. Belief

In many cases, beliefs are tied to specific long-term memories: for example, an individual may believe that he or she is an attorney because he or she possesses a distinct memory of passing the bar exam and working in a law firm as an attorney for the past several years. In other cases, belief may also involve the contents of sensory memory, short-term memory, and working memory: for example, a person may believe that a bluebird is sitting outside his or her windowsill because he or she is experiencing certain visual sense data that correspond to his or her recollection of the appearance of a bluebird. (Implicit in such a belief is the host's additional supposition that he or she is not hallucinating and is not being supplied fabricated sense data by an artificial eye that has been hacked by some adversary.)

In other cases, beliefs may not be tied directly to specific memories (and may even be contradicted by such memories); such beliefs might include an individual's conviction that he or she is a talented singer, the belief in a particular theological or philosophical doctrine whose truth has not been directly demonstrated through a sensory experience, an intuitive feeling that someone is glancing over one's shoulder, or the belief that the mission of

---

[101] A device might potentially allow enhanced storage in a manner that is longer-lasting and more accurate than documented cases of eidetic memory and less constrained in subject matter than hyperthymesia (or 'superior autobiographical memory') – perhaps resembling the phenomenon of 'photographic memory' whose existence in natural biological human beings has yet to be demonstrated. Regarding hyperthymesia, see Taylor, "Hyperthymesia" (2013); regarding cases of supposed eidetic memory, see Moxon, *Memory* (2000), p. 15, and Schwartz, *Memory: Foundations and Applications* (2014), p. 172.

The neuroprosthetic hippocampal bridge described in Soussou & Berger (2008) that is designed to span a damaged area of the hippocampus and restore normal memory functionality represents an example of a mnemoprosthetic device that supports memory processes generically without directly altering the contents of specific long-term memories.

[102] For the possibility of developing personality-related neuroprosthetics, see Soussou & Berger (2008).

one's employer is just. It may be possible for a neuroprosthesis to directly affect such immediate or long-term beliefs in a way that is distinct from the typical manipulation of memories or sense data.

E. Reasoning and Decision-making

A neuroprosthetic device may access, modify, control, or otherwise relate to those executive functions by which the mind of its host reasons, makes decisions, and embarks on a course of action. Such processes include those relating to working memory, attention control, volition, metavolition and conscience, and a host's internal monologue.

1. Working Memory (WM)

While short-term memory stores a small number of items that are the subject of one's mental focus, working memory is the executive function that allows those items to be manipulated by one's mind. From a cybernetic perspective, working memory is thus better understood as a part of the mind's reasoning and decision-making apparatus than as a conventional form of memory through which information is encoded, stored, and retrieved.[103]

While the existence and importance of working memory is generally accepted, different scientific theories attribute somewhat different roles and limits to working memory and describe its relationship to short-term and long-term memory in various ways. A neuroprosthetic device that affects processes of working memory would impact its host's most fundamental ability to think and interact with the external environment.

2. Attention Control

A neuroprosthetic device may participate in its host's experience of attentiveness and ability to focus attention on one sensation or cognitive phenomenon rather than another at a particular instant in time.[104] Such a device may affect the phenomenon of 'metacognition' by which the mind recognizes and corrects its periodically wandering and wayward attention.

3. Volition

In the philosophical sense employed here, a particular 'volition' combines a desire and a belief.[105] For example, an individual may wish to reach down to

---

[103] For different perspectives on working memory and reasoning processes, see, e.g., Kyllonen & Christal, "Reasoning ability is (little more than) working-memory capacity?!" (1990); Baddeley (2000); Baddeley, "Working memory: theories, models, and controversies" (2012); and Ma et al., "Changing concepts of working memory" (2014).

[104] For an overview of processes of attention control, see Mangun, *The Neuroscience of Attention: Attentional Control and Selection* (2012).

[105] For a discussion of the nature of volition (and its relationship to conscience and moral responsibility), see Calverley, "Imagining a non-biological machine as a legal person" (2008).

grasp a drinking glass and raise it to his or her lips and believe that this can be accomplished by willing the muscles in his or her arm to contract in a particular sequence. A neuroprosthetic device may participate in its host's processes of forming and storing volitions as well as acting on those volitions whose effects are internal to the mind and do not entail the execution of motor activity.

4. Metavolition and Conscience

A neuroprosthetic device may participate in its host's process of forming volitions *about* the kinds of volitions that he or she wishes to possess. This phenomenon of 'metavolition' often manifests itself as a desire to change one's personal habits or preferences and is closely tied to the notion of conscience.[106] Insofar as a neuroprosthesis affects an individual's ability to form desires and beliefs, it will affect the nature of his or her metavolitions – and may either strengthen or weaken the individual's ability to form and follow the guidance of his or her own conscience.

5. Internal Monologue

A neuroprosthetic device may participate in or affect its host's processes of carrying out an 'internal monologue,' which is used both for conceptualizing and formulating speech that is to be verbally articulated for sharing with others, for giving verbal internal form to one's imaginings, and for reasoning internally about some question or problem and attempting to arrive at a conclusion.[107]

6. Savant Skills

A small percentage of natural biological human beings possess savant skills that allow them to instantaneously perform extraordinary mental feats of calculation, memory, or creativity that are not possible for typical human beings. Such skills are discussed here as a form of reasoning, insofar as the most common savant skills are those involving the ability to calculate the day of the week of calendar dates in the distant past or to solve other mathematical problems; however, other savant skills (such as the ability to recall and perform a just-heard musical work note for note) are closely related to perception and memory.[108]

---

[106] Regarding metavolition and conscience, see Negoescu, "Conscience and Consciousness in Biomedical Engineering Science and Practice" (2009), and Gladden, *Sapient Circuits and Digitalized Flesh* (2016), p. 120.

[107] For an overview of such cognitive processes, see "The Internal Monologue" in Butler, *Rethinking Introspection* (2013).

[108] The nature of savant skills is discussed in Treffert, "The savant syndrome: an extraordinary condition. A synopsis: past, present, future" (2009), and Rodriguez, *Autism Spectrum Disorders*

Such savant skills naturally occur primarily in individuals suffering from autism or another medical condition (such as illness or injury) affecting the left temporal lobe. The neurological phenomena that grant individuals savant skills appear to do so by providing the human mind with conscious access to and the ability to control cognitive processes at a profound level of detail that is normally hidden from us.[109] Savants' ability to directly access the 'inner workings' of perception and memory might be understood as roughly analogous to the ability to directly manipulate the operations of a desktop computer through the use of low-level binary machine code: most computer programmers lack such a skill and must instead use a higher-level programming language (or at least, assembly language) in which the ultimate details of processor instructions are hidden from the programmer's view. While writing programs directly in machine code offers some advantages (such as the ability to maximally optimize some performance characteristics), the fact of being forced to operate at the finest level of detail and unable to create complex persistent objects or to more holistically manipulate abstract information structures severely limits the feasibility of such an approach.[110]

Individuals undergoing damage to the left temporal lobe as adults can experience acquired savant syndrome, in which they suddenly acquire such savant skills,[111] and recent research suggests that it is possible to temporarily grant savant skills to individuals by temporarily disrupting the functioning of the left anterior temporal lobe through the use of neurotechnologies such as transcranial magnetic stimulation.[112] An implantable neuroprosthetic device that is capable of briefly disrupting the behavior of the left temporal lobe in such a way might be able to temporarily provide its host with savant skills when desired, while at other times avoiding the deficits in shared attention, social cognition, and empathy that often accompany damage to the left temporal lobe.

III. Motion

Some neuroprostheses execute or participate in behaviors designed to produce a physical action or effect within the world outside of their host's

---

(2011), pp. 36-39.

[109] See Snyder, Allan, "Explaining and inducing savant skills: privileged access to lower level, less-processed information" (2009).

[110] See Dandamudi, *Introduction to assembly language programming: from 8086 to Pentium processors* (1998); Dumas (2006); and Streib, *Guide to Assembly Language: A Concise Introduction* (2011).

[111] Regarding acquired savant syndrome, see Treffert, "Accidental Genius" (2014).

[112] See Snyder et al., "Savant-like skills exposed in normal people by suppressing the left fronto-temporal lobe" (2003), and Snyder (2009).

brain: either in some other part of the host's body or in the external environment. **Motor neuroprostheses** are those neuroprosthetic devices whose primary purpose is to fill such a role; they currently include systems such as robotic prosthetic limbs and wheelchairs whose movements can be controlled by their user's thoughts by means of a neuroprosthetic interface.[113]

Many motor neuroprostheses function by detecting a host's thoughts or volitions (such as the desire to reach out with his or her arm and grasp for an object) and translating them into electrical or electrochemical signals that induce a corresponding action in some biological motor organ (such as a muscle or gland) or in an artificial effector (such as an electric motor, electroactive polymer, piezoelectric bimorph, audio speaker, or display screen). Other motor neuroprostheses – such as those designed to treat sleep apnea or epileptic seizures[114] – themselves autonomously determine when to execute a physical act and what sort of act to execute. In such cases it is the neuroprosthesis and not its host's brain that serves as the direct initiator of motor action.

Below we consider in more detail these different ways in which a neuroprosthesis can participate in its host's processes of motor action.

A. Detection of Action Potentials or Other Motor Instructions within a Motor Neural Pathway

Many motor neuroprostheses include electrodes or other sensors that allow them to detect motor instructions that have arisen within a motor neural pathway of their host's natural biological organism. Such devices may, for example, detect motor instructions as they are being generated within the brain's motor cortex, as they are being carried toward a motor organ by the upper and lower motor neurons of the corticospinal tract, or at the neuromuscular junction synapse where the signals are transmitted electrochemically from a motor neuron to a muscle.

The detection of motor instructions may be accomplished with varying degrees of speed and accuracy using technologies such as EMG, EEG, MEG, fMRI, and PET.[115] The use of targeted muscle reinnervation (TMR) surgery

---

[113] Such motor neuroprostheses are discussed in Edlinger et al., "Brain Computer Interface" (2011); Lebedev (2014); Merkel et al. (2007); and Widge et al., "Direct Neural Control of Anatomically Correct Robotic Hands" (2010).

[114] For existing or anticipated neuroprostheses of this sort, see Taylor, "Functional Electrical Stimulation and Rehabilitation Applications of BCIs" (2008), and Van Drongelen et al., "Seizure Prediction in Epilepsy" (2005).

[115] Regarding various technologies that can be used for this purpose, see Patil & Turner, "The Development of Brain-Machine Interface Neuroprosthetic Devices" (2008); Principe & McFarland, "BMI/BCI Modeling and Signal Processing" (2008); Ayaz et al., "Assessment of Cognitive

allows a motor nerve that had previously controlled a limb or organ that is now missing to be rerouted so that it innervates a remaining muscle in a different part of the body, thereby allowing motor instructions sent by the brain to the missing component to be detected as contractions in the muscle to which the nerve has been rerouted.[116]

B. Fabrication of Cellular Action Potentials within a Motor Neural Pathway

While some motor neuroprostheses detect motor instructions that have been generated by the brain of their human host and transmit those instructions to a biological motor organ or artificial effector, other motor neuroprostheses autonomously fabricate motor instructions – perhaps because the brain of their host is unable to do so (e.g., as a result of injury or illness) or because the neuroprosthesis can do so more effectively (e.g., when creating motor instructions for some non-human motor organ that the human brain is not well-suited to control). For example, while some invasive neuroprostheses (e.g., DBS devices) treat Parkinson's disease and essential tremor by stimulating regions of the brain, other less invasive motor neuroprostheses treat the same condition by gyroscopically detecting the occurrence of a tremor and then electrically stimulating muscles to stabilize a limb through muscle co-contraction.[117] In the latter case, a device's actions are not determined by instructions consciously or unconsciously generated by its host's brain but by instructions generated by the device's autonomous internal computer.

C. Storage of Instructions for (Repeated) Motor Execution

A motor neuroprosthesis may store motor instructions within itself temporarily or permanently; the instructions stored may correspond either to naturally generated instructions produced in the host's brain or spinal cord or to fabricated instructions that were procedurally generated by the device itself. Such storage can employ different memory types and information formats.

Neural Correlates for a Functional Near Infrared-Based Brain Computer Interface System" (2009); Miller & Ojemann, "A Simple, Spectral-Change Based, Electrocorticographic Brain–Computer Interface" (2009); and Lebedev (2014).

[116] See *Targeted Muscle Reinnervation: A Neural Interface for Artificial Limbs*, edited by Kuiken et al. (2014).

[117] See Gallego et al., "A neuroprosthesis for tremor management through the control of muscle co-contraction" (2013).

1. Memory Type

a. Volatile memory

A motor neuroprosthesis such as a robotic prosthetic arm will typically need to store newly detected motor instructions within itself (or within its external controller, if the device's data processing is performed by an external computer) at least briefly in order to process the data and compute the pattern of motor actions that will be initiated in order to carry out the instructions. Such data may be stored using random-access memory (such as a DRAM or SRAM chip) that requires a continuous electrical power supply in order to preserve the data. A loss of power to the device would result in loss of the data.[118] In principle, RAM can be used by a neuroprosthesis to store motor instructions for an indefinite period of time, as long as the device does not power down.

b. Non-volatile memory

A motor neuroprosthesis may also store motor instructions using non-volatile systems that allow the data to be preserved even after the device is powered down. There is generally less need for a motor neuroprosthesis to possess such functionality than there is for a sensory neuroprosthesis to be able to store a permanent non-volatile record of incoming sense data: while ordinary persons might periodically find it useful to replay the audio recording of a conversation in one's mind hours after the conversation first took place or to relive the visual experience of an important event, motor actions are often performed in response to some immediate environmental stimulus – in order to manipulate the external environment as it exists at a particular moment in time.

NEUROPROSTHETICALLY AIDED EXECUTION AND REPETITION OF PHYSICAL ACTIONS

The ability to 'record' a wave of one's arm or the creation of a facial expression and then to repeat that exact movement at a later moment may thus, for most people, have limited utility. However, for some types of individuals, such capacities might prove highly useful: for example, a professional pianist might record the muscle movements that were executed during the flawless performance of a work, so that the performance might be repeated in the future. Similarly, a professional athlete might record the motor instructions involved with the perfect execution of a kick or throw, so that the action might be executed again with perfect consistency whenever needed in the future. A soldier or aircraft pilot might possess a motor neuroprosthesis that has been programmed with a full suite of recorded movements that can be executed flawlessly by the device even under the most difficult circumstances – for example, allowing the individual to call for help, activate an emergency medical

---

[118] Regarding different memory mechanisms for computerized devices, see Dumas (2006).

device, or even engage in hand-to-hand combat, fire a weapon, or land a plane when he or she is in shock or unconscious as a result of injuries sustained.

However, the importance of real-time feedback to the successful execution of motor movements must not be overlooked. While it might be possible for a sensory neuroprosthesis to store received sense data in a lossless digital form that does not degrade over time and is experienced in an identical fashion every time it is replayed, the recreation of neuroprosthetically stored motor instructions by a natural biological muscle or organ is subject to greater variation. Even an identical set of motor instructions will be executed by biological components in a slightly different fashion every time it is reenacted: the performance of 'the same' physical gesture may vary from one instance to the next, depending on factors such as a muscle's degree of fatigue, the weight and construction of the clothing being worn by a device's host at the moment, or the current orientation of the host's body (and corresponding direction of gravitational forces). It would be possible to achieve much greater precision and invariability when a stored set of motor instructions is repeatedly executed by an artificial limb or organ utilizing electromechanical components.

2. Information Format

a. Digital data files

A neuroprosthesis may store motor instructions in the form of binary digital files that can be readily exchanged with desktop computers or other external conventional computing systems. Data may be stored in a raw format that preserves all available information (e.g., the activity detected in the motor cortex at a particular moment) but requires large file sizes or in a processed and compressed format that preserves only selected information (e.g., the sequence of stimuli to be transmitted to a lower motor neuron) with correspondingly smaller file sizes.

b. Neural network activation patterns

A neuroprosthesis may store motor instructions within the components and processes of a physical artificial neural network that mimics the functioning of the brain's biological neural network and which may directly interface with, complement, or partially replace neurons and neural structures within its host's brain.[119] As noted earlier in the discussion of the storage of sense data by sensory neuroprostheses, a neural network of this sort stores information within its topology and activation patterns in a manner that external computing systems may be unable to detect and decode and that may be interpretable only by the neural network in which it is stored – potentially

---

[119] The use of memristors is one approach to developing such a neural network. See e.g., Snider (2008); Versace & Chandler (2010); *Advances in Neuromorphic Memristor Science and Applications* (2012); and Lohn et al. (2014).

implementing a form of information-theoretic security that surpasses the security of common encryption methods. Such a system would resemble those mechanisms of procedural memory (or the somewhat misleadingly named 'muscle memory') by which the brain learns to perform complex actions such as playing a musical instrument, manipulating a video game controller, or typing letters on a keyboard without conscious attention.[120]

D. Transmission of Cellular Action Potentials within a Motor Neural Pathway

A neuroprosthesis can serve as a 'bridge' that receives motor instructions in the form of electrochemical signals that were generated by the brain or spinal cord of its human host and retransmits them along their path to a motor organ such as a muscle or gland. For example, an upper motor neuron lesion can disrupt the transmission of motor instructions through the nerve, resulting in muscle weakness and a loss of motor control in the corresponding muscles in a limb (i.e., upper motor neuron syndrome);[121] in such a case, a neuroprosthetic device might detect electrochemical signals produced by the motor cortex that correspond to motor instructions and then bypass the damaged upper motor neuron, retransmitting the instructions directly to the lower motor neuron that innervates the relevant muscle in the limb.[122] Even in healthy individuals that do not require these types of devices in order to enjoy typical motor functionality, such neuroprosthetic bridges could potentially prove useful for purposes of human enhancement, allowing motor instructions to travel from the brain or spinal cord to the relevant muscle or organ faster than is possible with natural biological mechanisms of neurotransmission (e.g., through the use of a wireless or hardwired electronic signal transmission) – thereby increasing the speed of an individual's reflexes and voluntary actions.

E. Effectuation of Physical Action

Motor neuroprostheses are generally designed to effect some physical action in the world – either within the body of their host or in the external environment. Within a natural biological human organism, the effectuation of

---

[120] Such phenomena are discussed in, e.g., Ericsson & Charness, "Expert performance: Its structure and acquisition" (1994), and Chafe & O'Modhrain, "Musical muscle memory and the haptic display of performance nuance" (1996).

[121] See *Upper Motor Neurone Syndrome and Spasticity: Clinical Management and Neurophysiology*, edited by Barnes & Johnson (2008).

[122] For a discussion of current efforts to develop neuroprostheses of this sort, see, e.g., Sakas et al., "An introduction to operative neuromodulation and functional neuroprosthetics, the new frontiers of clinical neuroscience and biotechnology" (2007), and Logan, "Rehabilitation techniques to maximize spasticity management" (2011).

motor action typically takes one of two forms: the stimulation of muscles causing their contraction or the stimulation of glands causing them to secrete a hormone or other substances.[123] When motor neuroprostheses are employed, the range of available physical actions becomes much wider; for example, a device might be designed to produce specialized patterns of visible light, sound waves, radio signals, heat, electric current, or other effects.

Below we discuss in more detail ways in which a motor neuroprosthesis might produce some action in the world, either through the stimulation of the natural biological components of a host's body or the control of artificial effectors.

1. Stimulation of Natural Biological Effectors

A motor neuroprosthesis may produce two main types of physical effects in the natural biological components of its host's body: the contraction or relaxation of muscles (through the excitation of muscle fibers or inhibition of motor neurons) or the regulation of glands' secretion of hormones and other substances.

a. Contraction and relaxation of muscles

A motor neuroprosthetic device may control the physical activity of natural biological muscle tissue by exciting muscle fibers (which results in muscle contraction) or inhibiting the motor neurons innervating muscle fibers (which results in muscle relaxation).

I. VOLUNTARY FUNCTIONS

The somatic nervous system regulates the voluntary movement of skeletal muscles, allowing the purposeful movement of body parts such as limbs, the neck, eyes, and the mouth and tongue. The motor instructions governing such actions may be generated within the brain or (in the case of some reflex actions) the spinal cord.[124]

II. AUTONOMIC FUNCTIONS

The sympathetic and parasympathetic branches of the autonomic nervous system (ANS) regulate physical actions involving muscles such as an increase or decrease in heart rate, dilation and constriction of pupils, dilation or constriction of bronchi in the lungs, coughing, sneezing, swallowing, vomiting, an increase or decrease in muscular activity of the digestive organs, relaxation or constriction of the urinary bladder, and the muscular aspects of sexual function.[125]

---

[123] See Rosenbaum, *Human Motor Control* (2010), and "The Endocrine System" in Starr & McMillan, *Human Biology* (2016).

[124] For an overview of such structures and mechanisms, see Rosenbaum (2010).

[125] See *Primer on the Autonomic Nervous System* (2012) and Starr & McMillan (2016).

b. Regulation of gland cells' production and release of chemicals

Within a natural biological human organism, many endocrine and exocrine glands are not directly regulated by signals received from neurons; their production and secretion of hormones and other substances is instead regulated by other phenomena, such as the presence or lack of particular chemical substances in the bloodstream. However, the functioning of other glands is indeed regulated by electrochemical signals from neurons that innervate a gland; for example, the adrenal gland's behavior is regulated by stimuli from preganglionic neurons of the sympathetic nervous system. In such ways, the autonomic nervous system at least partially controls the production and secretion of substances such as epinephrine (or adrenaline), saliva, and gastric acid.[126]

A motor neuroprosthesis might be able, for example, to artificially stimulate the release of epinephrine in order to treat urgent medical conditions such as anaphylaxis, cardiac arrest, and asthma attacks; dilate the airway in order to prepare the body for a period of intense physical activity; induce fear and a fear response; or enhance the consolidation of the long-term memory of a significant event.[127]

2. Control of Artificial Effectors

A motor neuroprosthesis may produce various types of physical effects through its control and utilization of artificial effectors; the effects generated by such effectors may mimic those typically produced by the natural biological components of a human body, or they may display characteristics that are impossible for a body's natural biological effectors. Below we discuss a range of physical products or phenomena that may be generated by a neuroprosthetic device through the use of artificial effectors.

a. Electromagnetic emissions

A neuroprosthetic device may generate various types of electromagnetic phenomena.

I. PHOTONS

A neuroprosthesis may emit photons which – depending on their frequency – may take forms such as those of X-rays, ultraviolet radiation, visible

---

[126] See Sherwood, *Fundamentals of Human Physiology* (2012), pp. 70-105, 494-543, and Starr & McMillan (2016), pp. 197-266, 287-306.

[127] The wide-ranging effects of epinephrine are discussed in *Epinephrine in the Central Nervous System*, edited by Stolk et al. (1988), and Goodman, *Basic Medical Endocrinology* (2009). Regarding the impact of epinephrine in memory consolidation, see, e.g., McGaugh & Roozendaal, "Role of adrenal stress hormones in forming lasting memories in the brain" (2002), and Cahill & Alkire, "Epinephrine enhancement of human memory consolidation: interaction with arousal at encoding" (2003).

light, infrared radiation, microwaves, or radio waves. Such capacities might be demonstrated, for example, by an implantable Wi-Fi radio transmitter, laser, or percutaneous LED display screen.

II. ELECTRIC CURRENT OR CHARGE

A neuroprosthetic device may generate an electrical current or charge. When done in a targeted fashion, this might allow a neuroprosthesis to transmit data through, for example, Ethernet or USB cables or proprietary channels to control peripherals such as an exoskeleton, vehicle, weapons system, 3D printer, medical device, or other equipment.

III. MAGNETIC FIELDS

A neuroprosthetic device may generate magnetic fields either for a particular functional purpose or as an unintentional side-effect of its internal electrical activity. A magnetic field might, for example, be purposefully generated by a subcutaneous component of a neuroprosthetic device in order to hold external components in place.[128]

b. Mechanical effectors

Some effectors are utilized primarily for the purpose of generating a particular type of motion or mechanical effect within the body of a device's human host or within the external environment.

I. SOUND WAVES

A neuroprosthesis might incorporate components such as a conventional audio loudspeaker, earphones, piezoelectric buzzer, or electrolarynx[129] that allow it to directly generate sound waves. Such technologies might allow a device's host to produce audible speech or to produce sounds that are not audible to the natural biological human ear but which might be detected by hosts of appropriate neuroprosthetic auditory implants or which might be used for other signaling or data-transmission purposes.[130]

---

[128] A common technique for holding the external portion of a neuroprosthetic device in place over the correct portion of its host's body is to insert one magnet in that external component and another magnet in the implantable portion of the device that may be percutaneous or hidden beneath the surface of its host's skin. The use of such mechanisms to secure, e.g., the external portion of middle ear implantable hearing devices (MEIHDs), cochlear implants, and auditory brainstem implants is discussed in Dormer, "Implantable electronic otologic devices for hearing rehabilitation" (2003).

[129] For an overview of such technologies, see Liu & Ng, "Electrolarynx in voice rehabilitation" (2007).

[130] Note that such technologies that directly produce sound waves are distinct from technologies that utilize, e.g., the microwave auditory effect; in the latter case, a device's immediate physical effect is the production of microwaves, not sound waves. See, e.g., Lin, "Hearing microwaves: The microwave auditory phenomenon" (2001).

II. PARTICLE KINETIC ENERGY

A neuroprosthesis may transfer energy to the body of its human host or to the external environment in the form of heat. A neuroprosthetic device's heating component might be used, for example, to regulate the body temperature of its host or (if attached to an external robotic arm) as a part of a welder or firelighting instrument.[131]

III. MOTION OF AN ACTUATOR

A neuroprosthetic device might incorporate or control the action of actuators that move in order to alter the position of some portion of its host's body or of some object in the external environment. Such actuators may include electric motors, hydraulic actuators, pneumatic actuators, electroactive polymers (or 'synthetic muscles'), or piezoelectric bimorphs. They may also incorporate servomechanisms that utilize feedback for automated error-correction.[132]

Such actuators may constitute key components of neuroprosthetic artificial limbs, artificial hearts, and other neuroprostheses involved with gestures or locomotion.

c. Chemical production and release

A neuroprosthetic device may produce or store chemicals which it releases either into a specific targeted substrate or into the environment more broadly under certain conditions.

I. RELEASE OF AIRBORNE CHEMICALS INTO ATMOSPHERE

A neuroprosthesis might release into the atmosphere external to its host's body substances such as perfumes, pheromones, inhalational anesthetics, or chemicals that generate a smoke screen; it might release into its host's lungs gases such as oxygen that can affect the host's physical performance.

II. RELEASE OF CHEMICALS INTO A LIQUID MIXTURE

A neuroprosthesis may alternatively release substances into a liquid mixture such as the bloodstream of its human host. Such substances might include drugs, hormones, chemical neurotransmitters, or substances added to the saliva in the host's mouth that will be experienced by the host through the sensory modality of taste.[133]

---

[131] For the use of welding electrodes or gas cutting torches as robotic end effectors, see, e.g., Saha, *Introduction to Robotics* (2008), pp. 18, 26, and Niku, *Introduction to Robotics: Analysis, Control, Applications* (2011), p. 7.

[132] See Saha (2008), pp. 32-50; Niku (2011), pp. 266-318; Shahinpoor et al., *Artificial Muscles: Applications of Advanced Polymeric Nanocomposites* (2007); and Kim et al., *Biomimetic Robotic Artificial Muscles* (2013).

[133] The use of implantable neuroprostheses as drug-delivery devices is discussed, e.g., in Merkel

d. Xenoactive phenomena

Many of the physical effects described above can be manifested in the form of effects that are already commonly produced by natural biological human beings. For example, the human body regularly produces sound waves, the motion of limbs, endocrine and exocrine secretions, exhaled gases, and electromagnetic fields.

However, some of the physical effects described above can also be manifested by a neuroprosthetic device in the form of 'xenoactive' effects that the natural biological human body is not capable of producing. For example, a neuroprosthesis that generates radio transmissions, microwave signals, visible light, or chemically synthesized pharmaceuticals is generating physical effects that exceed those that are normally possible for the human organism.[134]

F. Modalization of Motor Action

Sensory neuroprostheses allow a human host to experience reality through various *sensory modalities*. In that context, the word 'modality' does not directly describe the physical stimuli that give rise to a sensory experience nor the physical organ or mechanism by which the host's body detects the stimuli; instead, they reflect the distinct ways in which various types of sense data are experienced by the mind and include phenomena such as vision, hearing, taste, smell, and touch. By analogy, one can understand **motor modalities** as the distinct avenues by which a human being employs the motor capacities of his or her body in order to manifest his or her volitions within the world. A motor modality is not inherently identified with a particular motor organ or category of physical effect but with a desired means of expression. For example, using one's legs to walk across a room and using one's hands to play a musical instrument both involve similar kinds of motor action, if viewed at the microanatomical level of muscles being electrically stimulated by motor neurons that innervate them; however, from the perspective of the person performing these actions, each kind of action has a different purpose and constitutes a different sort of conscious experience. Several key types of motor modalities are discussed below.

---

et al. (2007); Bhunia et al., "Ultralow Power and Robust On-Chip Digital Signal Processing for Closed-Loop Neuro-Prosthesis" (2014); and Mercanzini & Renaud, *Microfabricated Cortical Neuroprostheses* (2010).

[134] In a sense, all fully implantable neuroprostheses that communicate with external systems using wireless (e.g., radio) transmission are xenoactive neuroprostheses. For an overview of such technologies, see, e.g., Wise et al., "Wireless implantable microsystems: high-density electronic interfaces to the nervous system" (2004); Wise et al., "Microelectrodes, microelectronics, and implantable neural microsystems" (2008); and *Implantable Biomedical Microsystems: Design Principles and Applications*, edited by Bhunia et al. (2015).

1. Symbolic Action

A neuroprosthetic device may participate in or control physical actions that generate or manipulate written or spoken language or other types of symbols that can be used for purposes of information storage and communication.

a. Language classified by its mechanism of production

Languages expressed through the use of a neuroprosthesis may be classified according to the mechanism of their production.

I. ORAL LANGUAGE

Some neuroprosthetic devices allow their hosts to communicate through the generation of oral language – for example, through the use of a natural biological voice box, prosthetic larynx, or mechanism for the production and transmission of spoken language within a virtual world.[135]

II. WRITTEN LANGUAGE

Some neuroprostheses allow their hosts to communicate through the production or manipulation of written language. This might be accomplished, for example, by a robotic prosthetic arm that allows its host to type on a keyboard or write within a stylus or pen or by an artificial eye that allows its host to 'type' words on a monitor by focusing his or her gaze on particular letters.[136]

III. SIGN LANGUAGE

A neuroprosthetic device may allow its host to communicate through the use of sign language, in which the movements of the host's biological, synthetic, or virtual limbs is not used to generate symbols within some other physical medium but instead become the symbols.

b. Language classified by its historical origin

Languages expressed through the use of a neuroprosthesis may be classified according to the historical process through which the languages were developed.

---

[135] See, e.g., Kshirsagar et al., "Personalized Face and Speech Communication over the Internet" (2001); Tang et al., "Humanoid audio–visual avatar with emotive text-to-speech synthesis" (2008); and *Computer Synthesized Speech Technologies: Tools for Aiding Impairment*, edited by Mullennix & Stern (2010).

[136] Various neuroprosthetic approaches to allowing those who are paralyzed to control a computer cursor are discussed, e.g., in Kostov & Polak, "Parallel man-machine training in development of EEG-based cursor control" (2000); Taylor et al., "Direct cortical control of 3D neuroprosthetic devices" (2002); Black et al., "Connecting brains with machines: the neural control of 2D cursor movement" (2003); and *Brain-Computer Interfaces: Principles and Practice*, edited by Wolpaw & Winter Wolpaw (2012).

I. NATURAL HUMAN LANGUAGE

A neuroprosthetic device may allow its host to communicate using a natural human language such as English, Polish, or Japanese that evolved organically through a gradual historical process of everyday use within one or more human cultures.[137]

II. SYNTHETIC LANGUAGE

A neuroprosthesis may allow its host to communicate with other human beings using a constructed language such as Esperanto or the 'artistic languages' developed within works of science fiction or fantasy to depict the language of intelligent nonhuman civilizations.[138] A host might use a constructed language such as the Robot Interaction Language (ROILA) to communicate with an array of compatible robotic entities using a form of linguistic social interaction.[139] A sufficiently sophisticated neuroprosthesis might allow its human host to directly control external computerized devices such as desktop computers, web servers, or mobile devices by 'thinking' directly in a programming language, command language, or even in machine code.

A synthetic language may be expressible through conventional speech or written text, or it may involve the use of icons or other forms (perhaps three-dimensional) that can only be expressed and represented with the aid of particular technologies or within a specific virtual world.

III. LANGUAGE OF THOUGHT

A neuroprosthetic device may allow its user to communicate or store information employing the modality of a hypothesized language of pure thought that is used internally by human minds to conceptualize ideas before translating them into natural human language.[140] It is unclear to what extent such a language of pure thought exists and, if it does exist, whether it takes essentially the same form for each human being or whether it differs in critical characteristics among human beings. It is also unclear to what extent it

---

[137] For an overview of the development of such human languages, see, e.g., Campbell, *Historical Linguistics* (2013), and *Evolutionary Linguistics*, edited by McMahon & McMahon (2013).

[138] See, e.g., Peterson, *The Art of Language Invention: From Horse-Lords to Dark Elves, The Words Behind World-Building* (2015).

[139] See Mubin et al., "Improving speech recognition with the robot interaction language" (2012), and Mubin et al., "Talk ROILA to your Robot" (2013).

[140] For just a small sampling of the many debates that touch on such issues, see, e.g., Smolensky, "The Constituent Structure of Connectionist Mental States: A Reply to Fodor and Pylyshyn" (1988); Braddon-Mitchell & Fitzpatrick, "Explanation and the Language of Thought" (1990); Clark, "Systematicity, Structured Representations and Cognitive Architecture: A Reply to Fodor and Pylyshyn" (1991); Cory, "Language, Brain, and Neuron" (2000); Taylor & Taylor, "The Neural Networks for Language in the Brain: Creating LAD" (2003); and Katz, "The Hypothesis of a Genetic Protolanguage: An Epistemological Investigation" (2008).

could be expressed using either synthetic or virtual effectors – given the fact that, by definition, it is not directly expressed in its normally occurring form through motor activity and does not otherwise naturally exist outside of a mind's internal cognitive processes. If neuroprostheses can indeed be constructed that are capable of accessing, interpreting, or producing such language of thought, they may thus be better classified as cognitive rather than motor neuroprostheses.

If each such language of thought is indeed unique to a particular person and can only be produced or accessed by a single human mind (or group of minds that are related in some way), it could potentially be understood and exploited as a means of encryption.[141]

2. Nonsymbolic Action

A neuroprosthetic device may participate in or control the generation of physical actions that do not involve the production of written or spoken words or other types of symbols.

a. Paralanguage

A neuroprosthetic device may allow its host to communicate using the modality of paralanguage (which includes elements such as tone of voice, voice volume and pitch, and throat-clearing sounds) as expressed through the motor organs of natural, synthetic, or virtual body components.[142]

b. Oculesics

A neuroprosthesis may allow its host to communicate using the modality of oculesics (which includes elements such as the direction of one's gaze, eye movement, pupil dilation, and use of eye contact) as expressed through the motor organs of natural, synthetic, or virtual eyes or other visual organs.[143]

c. Kinesics

A neuroprosthetic device may allow its host to communicate using the modality of kinesics (which includes elements such as facial expressions, gestures, and posture) as expressed through the motor organs of a natural, synthetic, or virtual body.[144]

---

[141] The idea of utilizing thoughts as a means of user authentication within the field of information security has already been raised; see, e.g., Thorpe et al., "Pass-thoughts: authenticating with our minds" (2005).

[142] For an overview of such phenomena, see Johar, *Emotion, Affect and Personality in Speech: The Bias of Language and Paralanguage* (2016).

[143] For an overview of oculesics in relation to other forms of nonverbal communication, see Andersen & Andersen, "Measures of Perceived Nonverbal Immediacy" (2005).

[144] For an overview of kinesics, see Andersen & Andersen (2005).

d. Haptic communication

A neuroprosthesis may allow its host to communicate using the modality of haptics (which includes specialized forms of touching such as handshakes, hugging, kissing, and tickling) and which is expressed through the fine manipulators and other motor organs of a natural, synthetic, or virtual body.[145]

e. Proxemics

A neuroprosthesis may allow its host to communicate using the modality of proxemics (which involves positioning the body in order to control space and interpersonal distance) and which is expressed through the natural, synthetic, or virtual motor organs that control locomotion and posture.[146]

f. Internal organ activity

A neuroprosthetic device may control or otherwise participate in the actions of internal biological or synthetic organs within its host's body. In this way, a neuroprosthesis may impact its host's cardiac activity, breathing, digestion, blood chemistry, and other phenomena.[147]

G. Mapping of Motor Routes

A motor neuroprosthetic device creates a 'mapping' by which a particular set of motor instructions is generated (e.g., within the brain or spinal cord or by an electronic computer), transmitted to a particular motor organ, and executed to produce some physical effect within the host's body or the external environment that displays a particular motor modality. Within the organism of a natural biological human being, a wide range of such motor routes or mappings can already be found: for example, a complex set of motor instructions generated in the brain may cause the lungs, throat, jaw, lips, and tongue to move in a way that produces a particular sequence of sound waves that allows an individual to impact his or her environment through the modality of *speech*;[148] if other human beings, voice-activated electronic devices, or audio recording devices are present to detect the speech, it may have a significant effect on the environment.

On the other hand, it is possible for a neuroprosthesis to participate in a 'non-standard' motor mapping. For example, some types of robotic prosthetic arms utilize the technique of targeted muscle reinnervation to allow an arm's movements to be controlled by its host's thoughts. In that procedure, nerves

---

[145] The significance of haptics as a form of communication is also discussed in Andersen & Andersen (2005).

[146] The phenomena constituting proxemics are also reviewed in Andersen & Andersen (2005).

[147] For discussion of such neuroprosthetic devices that affect the internal biological processes of their human host, see McGee (2008), p. 209, and Gasson (2012), pp. 12-16.

[148] For an overview of the brain regions and mechanisms involved with speech production, see Blank et al., "Speech production: Wernicke, Broca and beyond" (2002).

in the remaining portion of the biological limb are surgically relocated so that the innervate healthy muscle tissue within the host's chest: sensors are affixed to the relevant portion of the chest, and whenever the host attempts to move his or her arm, the motor instructions generated by the brain cause muscles in the chest to move; the data regarding such movement is detected by the sensors and transmitted to the neuroprosthetic arm's controller, which interprets the signals and causes the robotic arm to move in the way that its host desired.[149] In this case, the ultimate result of the host's motor instructions – the movement of the (artificial) arm – is typical, but the motor route used to effect it is not.

In the following sections we consider different ways in which a motor neuroprosthesis can link together motor instructions with a motor organ and motor modality to create a coherent avenue for motor expression by its human host.

1. Routing Natural Motor Instructions to Their Biologically Normal Effector

A neuroprosthesis may receive or generate motor instructions of a sort that are typically produced by a natural biological system (e.g., the brain or spinal cord) and convey them to a biological or artificial effector, where they produce some physical effect using the motor modality whose activity is naturally associated with that form of motor instruction. An example would be a robotic prosthetic leg that replaces a biological leg lost to illness or injury and which detects motor instructions that are produced in its host's motor cortex and conveyed through the sciatic nerve and converts those instructions into electrical signals that govern the movements of the robotic leg's actuators, allowing its host to act in the physical environment through the motor modality of locomotion.[150]

2. Routing Fabricated Motor Instructions to an Arbitrary Effector

Instead of detecting motor instructions generated within its host's brain or spinal cord and conveying them to a motor organ, a neuroprosthesis might instead generate such motor instructions itself, as determined by the programming of its internal computer or other built-in control mechanisms. An artificially intelligent neuroprosthesis might, for example, issue motor instructions in the form of electrochemical signals transmitted to natural biological motor neurons, muscle tissue, or gland cells in order to regulate its

---

[149] See *Targeted Muscle Reinnervation: A Neural Interface for Artificial Limbs* (2014).

[150] An overview of such a robotic prosthetic leg is provided in Hargrove et al., "Robotic leg control with EMG decoding in an amputee with nerve transfers" (2013).

host's heart rate, induce the secretion of hormones into the bloodstream, or disable or override the host's own voluntary control of skeletal muscles.[151]

3. Remapping Natural Motor Instructions to a Biologically Atypical but Natural Effector

It is possible for a neuroprosthesis to detect motor instructions that are generated naturally by the biological components of the brain or spinal cord and convey them to an effector which – while naturally occurring in human beings – is not the organ by which such motor instructions are typically expressed.

One variation of such a routing is for a neuroprosthetic device to map motor instructions that naturally govern the activity of a motor organ whose impact is largely felt outside of the host's body so that they instead control some effector within the host's body. An example would be a neuroprosthetic device that can detect motor instructions generated in the host's brain that cause muscles in the throat, tongue, jaw, and lips to produce a subvocalized sequence of sounds (e.g., a predetermined trigger word) and which – whenever the trigger word is subvocalized – stimulates the adrenal grands to release epinephrine into the bloodstream, thereby preparing the host's body for particular types of physical activity.[152] Conversely, a neuroprosthetic device may map motor instructions that typically regulate the activity of an effector within the body (such as cardiac muscle tissue) so that they instead control some effector whose impact is manifested in the environment external to the host's body. An example would be a neuroprosthesis that – upon detecting that preganglionic nerve fibers in the spinal cord are stimulating the adrenal gland to secrete epinephrine into the bloodstream – stimulates the secretion of sweat from sweat glands to cool the host's body in preparation for anticipated physical activity.

4. Routing Motor Instructions Using a Non-human Xenoexpressive Mapping

A neuroprosthetic device may incorporate into its motor mapping some form of non-natural 'xenoexpression' that is not manifested in the motor mappings of natural biological human beings. Through the use of such motor mappings, a neuroprosthetic device's human host may be able to produce physical effects that exceed the natural capacities of the human organism and

---

[151] Examples already in use or under development include some types of neuroprostheses designed to treat conditions such as epilepsy and Parkinson's disease. See Fountas & Smith, "A Novel Closed-Loop Stimulation System in the Control of Focal, Medically Refractory Epilepsy" (2007), and *Deep Brain Stimulation for Parkinson's Disease* (2007).

[152] Regarding the mechanisms of subvocalization, see, e.g., Pollatsek, "The Role of Sound in Silent Reading" (2015).

which – to unwitting observers – might appear inexplicable or even 'paranormal' in nature.[153]

a. Mapping motor instructions to a non-human xenergetic effector

One type of xenoexpressive mapping involves a neuroprosthetic device that is capable of expressing motor instructions through a non-human, 'xenergetic' effector of a type that is not present in the organism of a natural biological human being. Such effectors might potentially include electromechanical components such as wings or wheels, soft robotic limbs such as tentacles, electronic effectors such as radio transmitters or LED displays, or complete external vehicles or other systems.[154] By definition, such neuroprostheses utilize effectors for which the human mind has no corresponding natural default motor mapping; if such a device is to be controlled by its host's thoughts, then its designer must decide how to construct a cybernetic control pathway that allows such effectors' activity to be regulated by motor instructions generated by the brain.

b. Mapping motor instructions to manifest a non-human xenopractic motor modality

Another type of xenoexpressive mapping involves a neuroprosthetic device that expresses motor instructions using a non-human, 'xenopractic' motor modality that is not normally available to natural biological human beings. An example might include a neuroprosthesis that allows its host to inhabit and control a virtual body consisting of a swarm of spatially disjunct components existing in four-dimensional rather than three-dimensional space. It is not clear to what extent such fashioning of new motor modalities (or 'unlocking' of latent modalities) is theoretically and practically possible; the field of body schema engineering investigates such possibilities.[155]

Conclusion

In this text we have developed an ontology that attempts to envision, capture, and describe the full range of ways in which a neuroprosthesis may participate in the sensory, cognitive, and motor processes of its human host. By considering anticipated future developments in neuroprosthetics and adopting a generic biocybernetic approach, the ontology is able to account not only for therapeutic neuroprostheses already in use but also for future types of

---

[153] Here one may recall the potential indistinguishability of advanced technology and magic, as famously discussed in Clarke, "Hazards of Prophecy: The Failure of Imagination" (1973), p. 36.

[154] For example, neurotechnologies have been developed to allow a human being to successfully pilot an aerial drone in the primary physical world and drive a car within a virtual environment by means of his or her thoughts; see LaFleur et al., "Quadcopter control in three-dimensional space using a noninvasive motor imagery-based brain-computer interface" (2013), and Zhao et al., "EEG-Based Asynchronous BCI Control of a Car in 3D Virtual Reality Environments" (2009).

[155] See Gladden, "Cybershells, Shapeshifting, and Neuroprosthetics" (2015).

neuroprostheses that are expected to be developed and deployed for purposes of human enhancement. It is hoped that the use of such an ontology will allow biomedical engineers, ethicists, futurologists, and others to more easily, systematically, and robustly analyze, describe, and plan the biocybernetic role of neuroprostheses within their unique host-device systems.

Chapter Three

An Ontology of Neuroprostheses as Instruments of 'Cyborgization': Portals to the Experience of Posthumanized Digital-Physical Worlds

Abstract. The incorporation of a neuroprosthetic device into one's being at the physical, cognitive, and social levels constitutes a form of 'cyborgization' that imposes new constraints on one's existence while simultaneously opening a path to new forms of experience. This text explores the boundaries of this qualitatively novel form of being by formulating an ontology of the neuroprosthesis as an instrument that shapes the way in which its human host experiences and acts within emerging posthumanized digital-physical ecosystems.

The ontology addresses four main roles that a neuroprosthetic device may play in this context. First, a neuroprosthesis may serve as a means of human augmentation by altering the cognitive and physical capacities possessed by its host. Second, it may manipulate the contents of information produced or utilized by its human host. Third, a neuroprosthesis may shape the manner in which its host inhabits a digital-physical body and external environment. And finally, a neuroprosthesis may regulate the autonomous agency possessed and experienced by its host.

The development and use of such an ontology can allow researchers to better understand the psychological, social, and ethical ramifications of such technologies and can enable the architects of neuroprosthetic systems and the digital-physical ecosystems within which their human hosts operate to formulate principles of design and management that minimize the dangers and maximize benefits for the neuroprosthetically augmented inhabitants of such environments.

Introduction

The previous chapters have presented an ontology of the neuroprosthesis as a computing device and as a biocybernetic instrument that becomes integrated into the neural circuitry of a human organism in order to participate in processes of sensation, cognition, and motor action. In this chapter, we advance this exploration of the nature of neurocybernetic technologies by developing an ontology of the neuroprosthesis as a means for the 'cyborgization' of human beings that shapes how such individuals experience posthumanized digital-physical worlds.[1]

An Overview of Neuroprostheses

A neuroprosthesis may be defined as *an artificial device that is integrated into the neural circuitry of a human being* to create a neurocybernetic host-device system that possesses both human and computerized elements.[2] In principle, it is possible for neuroprostheses to be either 'invasive' (i.e., surgically implanted in the brain of a human host) or 'non-invasive' (e.g., consisting of an external device worn by a human host); however, it currently remains quite challenging to develop non-invasive technologies that can become fully integrated into the neural circuitry of a human being.[3] According to the definition employed in this text, contemporary neuroprostheses can thus typically be identified with invasive 'neural implants.' Devices involving non-invasive technologies such as EEG or fMRI are likely to be classified more generally as brain-computer interfaces (BCIs) or brain-machine interfaces (BMIs) rather than neuroprostheses.

---

[1] Here the term 'cyborgization' is used to describe the process by which a human host incorporates artificial biocybernetic components into his or her body, thereby becoming a cyborg. For use of the term in this context, see, e.g., Maguire & McGee, "Implantable Brain Chips? Time for Debate" (1999); Koltko-Rivera, "The Potential Social Impact of Virtual Reality" (2005); Novakovic et al., "Artificial Intelligence and Biorobotics: Is an Artificial Human Being Our Destiny?" (2009); and Nayar, *An Introduction to New Media and Cybercultures* (2010).

The term 'cyberization' is also sometimes used to describe the process of cyborgization. However, 'cyberization' is also used in a broader or alternative sense to refer to processes by which a human being becomes proficient in the use of (and perhaps psychologically and socially dependent on) – rather than physically integrated into – forms of electronic information and communications technology. When referring to the use of ICT such as email, social media, or computer gaming platforms, the term 'cyberization' does not imply that an individual has been subjected to physical biocybernetic augmentation; it is thus more appropriate to use the word 'cyborgization' when discussing the process of augmenting a host's body through the permanent incorporation of artificial biocybernetic components. For various uses of the term 'cyberization,' see, e.g., Miller, "Conclusion: Beyond the Human: Ontogenesis, Technology, and the Posthuman in Kubrick and Clarke's 2001" (2012); Baranyi et al., "Synergies Between CogInfoCom and Other Fields" (2015); and Ma et al., "Perspectives on Cyber Science and Technology for Cyberization and Cyber-Enabled Worlds" (2016).

[2] See Lebedev, "Brain-Machine Interfaces: An Overview" (2014), and Gladden, "Enterprise Architecture for Neurocybernetically Augmented Organizational Systems" (2016).

[3] See Gasson, "Human ICT Implants: From Restorative Application to Human Enhancement" (2012), p. 14, and Panoulas et al., "Brain-Computer Interface (BCI): Types, Processing Perspectives and Applications" (2010).

Neuroprosthetic devices are commonly classified as either sensory, motor, bidirectional sensorimotor, or cognitive neuroprostheses.[4] At present, such devices primarily fill therapeutic roles, as a means of restoring some capacity that is absent as a result of injury or illness: for example, auditory brainstem implants and retinal prostheses are used to restore sensory functionality to those who have lost the ability to hear or see, thought-controlled wheelchairs are used to restore some degree of mobility to those who are paralyzed, and experimental neural bridges are being developed to restore memory function in those who are unable to access their long-term memory due to hippocampal damage.[5] However, efforts are underway to develop and implement neuroprosthetic technologies whose purpose is not to restore some capacity typically found in human beings but to grant their human hosts sensory, cognitive, and motor capacities that greatly exceed those possible for natural biological human beings.[6]

The Emergence of Posthumanized Digital-Physical Ecosystems

The world within which the human hosts and users of neuroprosthetic devices exist is an increasingly rich and complex array of digital-physical ecosystems that reflect the ongoing 'technologization' of humankind.[7] The processes of technologization are manifested in phenomena such as the increasing physical integration of human beings with electronic computerized systems, our expanding interaction with and dependence on robots and artificial intelligences, our growing immersion in virtual worlds, and the use of genetic engineering to design human beings as if they were consumer products.[8]

---

[4] See Lebedev (2014).

[5] See, e.g., Cervera-Paz et al., "Auditory Brainstem Implants: Past, Present and Future Prospects" (2007); Weiland et al., "Retinal Prosthesis" (2005); Viola & Patrinos, "A Neuroprosthesis for Restoring Sight" (2007); and Soussou & Berger, "Cognitive and Emotional Neuroprostheses" (2008).

[6] Regarding the use of neuroprostheses for human enhancement, see, e.g., McGee, "Bioelectronics and Implanted Devices" (2008); Warwick & Gasson, "Implantable Computing" (2008); Gasson (2012); Gladden, "Neural Implants as Gateways to Digital-Physical Ecosystems and Posthuman Socioeconomic Interaction" (2016); and Gladden, "Enterprise Architecture for Neurocybernetically Augmented Organizational Systems" (2016).

[7] For a philosophical investigation (drawing on Actor-Network Theory) of ways in which human and nonhuman agents coexisting within digital-physical ecosystems might enter into 'symbioses' that are not simply metaphorical but are true symbioses at the physical, cognitive, and social levels, see Kowalewska, "Symbionts and Parasites – Digital Ecosystems" (2016).

[8] Processes of technologization are discussed as such in detail in Herbrechter, *Posthumanism: A Critical Analysis* (2013), and Gladden, *Sapient Circuits and Digitalized Flesh: The Organization as Locus of Technological Posthumanization* (2016). The relationship of posthumanism to the commercialization of the human entity is discussed in Herbrechter (2013), pp. 42, 150-52. For the analysis (and, in many ways, indictment) of technologization offered by critical posthumanism, see Herbrechter (2013), pp. 90, 19, and Gladden, *Sapient Circuits and Digitalized Flesh* (2016), p.

These processes of technologization are themselves among the most visible manifestations of larger forces of posthumanization that are at work within contemporary society. Such posthumanization can be understood as a process by which society comes to include at least some intelligent personal subjects that are *not* natural biological human beings and which leads to a nonanthropocentric understanding of reality. It is anticipated that our emerging future will include many different sources of intelligence and agency that create meaning in the universe through their networks and relations:[9] such entities might include 'natural' human beings, genetically engineered human beings, human beings with extensive neurocybernetic augmentation, human beings dwelling in virtual realities, social robots, artificially intelligent software, nanorobot swarms, sentient or sapient networks, and hive minds that link human and artificial intellects to create a unitary collective intelligence. Within the digital-physical ecosystems that constitute the functional infrastructure of that posthumanized world, the 'bioagency' possessed by traditional human beings will act alongside (and mutually influence) the 'cyberagency' of artificial beings and 'collective agency' of networks and hive minds.[10]

Within this context, neuroprosthetic devices are expected to increasingly become gateways that allow their human hosts to more deeply experience, control, and be controlled by the structures and dynamics of such digital-physical ecosystems.[11] The development of an ontology of the neuroprosthesis as a catalyst for cyborgization, technologization, and posthumanization would not only allow researchers to better understand the psychological, social, and ethical ramifications of such technologies; it would also allow the architects of neuroprostheses and the digital-physical ecosystems in which they and their hosts participate to formulate principles of design and management that minimize the dangers and maximize the beneficial outcomes for neuroprosthetically augmented individuals operating within such ecosystems.

46.

[9] See Ferrando, "Posthumanism, Transhumanism, Antihumanism, Metahumanism, and New Materialisms: Differences and Relations" (2013), for an excellent analysis of this and other aspects of posthumanization.

[10] See Fleischmann, "Sociotechnical Interaction and Cyborg–Cyborg Interaction: Transforming the Scale and Convergence of HCI" (2009).

[11] See Gladden, "Neural Implants as Gateways" (2016).

Figure 1: An Ontology of the Neuroprosthesis as Instrument of Cyborgization

I. The Neuroprosthesis as a Means of Human Augmentation

A. Capacity-adding
   1. Temporary addition
   2. Permanent addition
B. Capacity-enhancing
   1. Temporary enhancement
   2. Permanent enhancement
C. Capacity-restoring
D. Capacity-neutral
E. Capacity-suppressing
   1. Temporary suppression
   2. Permanent eradication

III. The Neuroprosthesis as a Means of Inhabitation

A. Embodiment: inhabitation of a material form
   1. The host as cyborg: reshaping the physical body
      a. Full-body cyborg
      b. Partial cyborg
      c. Extended cyborg
      d. Sessile cyborg
      e. Reverse cyborg
   2. The host as avatar: shaping others' perceptions of the host
      a. Cyberdouble
      b. Cybermorph
B. Embedding: inhabitation of an external environment
   1. The world viewed through augmented reality
      a. Overlaid information
      b. Refracted reality
   2. Occupancy of a virtual world
      a. Degree of immersion
         i. Maximal immersion
         ii. Partial immersion
      b. Period of immersion
         i. Temporary immersion
         ii. Long-term immersion
      c. Cognizance of immersion
         i. Recognized immersion
         ii. Unrecognized immersion

II. The Neuroprosthesis as a Manipulator of Informational Content

A. Content-providing
   1. Content-detecting
   2. Content-generating
   3. Content-importing ('downlink')
   4. Backup-restoring
B. Content-translating
   1. Content-encoding
   2. Content-decoding
C. Content-altering
   1. Content-editing
   2. Content-disrupting
   3. Content-deleting
D. Content-storing
E. Content-transmitting
   1. Content-exporting ('uplink')
      a. Content-outstreaming
      b. External archiving
   2. Production of external device backup

IV. The Neuroprosthesis as a Regulator of Agency

A. The neuroprosthesis as agent
   1. Neuroprostheses possessing agency
   2. Neuroprostheses lacking their own agency
B. The neuroprosthesis as instrument of external agency
C. The neuroprosthesis as modulator of a host's human agency
   1. Agency-realizing neuroprostheses
   2. Agency-extending neuroprostheses
   3. Agency-complementing neuroprostheses
   4. Agency-controlling neuroprostheses
   5. Agency-enhancing neuroprostheses
   6. Agency-impairing neuroprostheses
   7. Agency-eradicating neuroprostheses
D. The neuroprosthesis as hive mind infrastructure
   1. The hive mind as sensorimotor network
   2. The hive mind as facilitator of internal dialogue
   3. The hive mind as unitary collective intelligence

Developing an Ontology of the Neuroprosthesis as an Instrument of Cyborgization

The following sections develop an ontology of the neuroprosthesis as an instrument of cyborgization that shapes the manner in which its human host inhabits posthumanized digital-physical worlds. As delineated in Figure 1, the ontology addresses four main aspects of neuroprosthetic devices and the host-device systems that they form through structural and functional integration with their human hosts. First, a neuroprosthesis may serve as a means of human augmentation by altering the cognitive and physical capacities possessed by its host. Second, it may manipulate the contents of information produced or utilized by its human host. Third, a neuroprosthesis may shape the manner in which its host inhabits a digital-physical body and external environment. And finally, a neuroprosthesis may regulate the autonomous agency possessed and experienced by its host. These elements of the ontology are developed in detail below.

I. The Neuroprosthesis as a Means of Human Augmentation

Neuroprosthetic devices vary in the extent to which they enhance the naturally occurring capacities found within a typical biological human being. A device may add some new type of capacity that is not found in natural biological human beings; enhance or expand an existing capacity; restore some typical human capacity that is absent in a particular individual; suppress an existing capacity; or have no effect on the sensory, motor, and cognitive capacities possessed by a device's host.[12]

A. Capacity-adding

A neuroprosthetic device may grant its host some capacity that is not typically found in natural biological human beings. Examples might include a sensory neuroprosthesis gives its host the ability to perceive radio waves or a motor neuroprosthesis that allows its host to produce a particular pattern of visible light from an implanted photon emitter whose surface is exposed to the external environment (e.g., an LED display embedded in the host's arm). Such new capacities may be temporary or permanent.

---

[12] Various aspects of the use of neuroprostheses for human augmentation and enhancement is discussed, e.g., in *Converging Technologies for Improving Human Performance: Nanotechnology, Biotechnology, Information Technology and Cognitive Science*, edited by Bainbridge (2003); Merkel et al., "Central Neural Prostheses" (2007); McGee (2008); Gasson (2012); Warwick, "The Cyborg Revolution" (2014); and Gladden, "Neural Implants as Gateways" (2016).

1. Temporary Addition

In some cases it may not be practical or desirable for a neuroprosthesis to add a new capacity that is continuously manifested. For example, a cognitive neuroprosthesis that grants its host savant skills by electromagnetically disrupting the normal behavior of the left anterior temporal lobe[13] may do so at the cost of creating problems with other cognitive processes such as those involving shared attention, social cognition, and empathy. In that case, it may be most appropriate for the device to only be activated for brief periods of time when its functionality is required. Similarly, an artificial eye that provides its host with infrared vision or augmented reality displays may only activate these features when they are needed for a particular reason, so that the device's host can enjoy unimpeded normal visual perception at other times.[14] The use of neuroprostheses to provide only the sporadic and temporary addition of a capacity may be especially warranted in the case of devices that directly affect the functioning of the brain or other critical organs and whose long-term side-effects are not well understood.[15]

2. Permanent Addition

Some neuroprosthetic devices provide their host with a new capacity that is permanently active. For example, a sensory neuroprosthesis that records incoming visual sense data and wirelessly transmits it to an external system to create a backup copy may be running continuously, as it is not knowable in advance which visual experiences might later prove to be noteworthy and

---

[13] Damage to the left temporal lobe can cause adults to experience 'acquired savant syndrome,' in which they suddenly acquire savant skills; recent research suggests that it is possible to temporarily and artificially produce savant skills in individuals by temporarily disrupting the functioning of the left anterior temporal lobe by means of neurotechnologies such as transcranial magnetic stimulation (TMS). An implantable neuroprosthesis that is capable of briefly disrupting the behavior of the left temporal lobe might be able to temporarily provide its host with savant skills when needed, while at other times remaining inactive in order to avoid producing the deficits in shared attention, social cognition, and empathy that often accompany damage to the left temporal lobe. For a discussion of the possibility of inducing savant skills through the application of TMS, see Snyder et al., "Savant-like skills exposed in normal people by suppressing the left fronto-temporal lobe" (2003), and Snyder, "Explaining and inducing savant skills: privileged access to lower level, less-processed information" (2009).

[14] Regarding future neuroprosthetic devices that may grant such capacities, see, e.g., Warwick (2014); Gasson et al., "Human ICT Implants: From Invasive to Pervasive" (2012); and Merkel et al. (2007).

[15] Regarding the potential critical health impacts of implantable neuroprostheses, see *ISO 27799:2016, Health informatics – Information security management in health using ISO/IEC 27002* (2016); Ankarali et al., "A Comparative Review on the Wireless Implantable Medical Devices Privacy and Security" (2014); and Gladden, "Information Security Concerns as a Catalyst for the Development of Implantable Cognitive Neuroprostheses" (2016).

merit future replay or analysis.[16] The use of neuroprostheses to provide permanent, ongoing enhancement of a host's capacities may be especially warranted, for example, in cases where the repeated activation and deactivation of an enhancement might cause significant psychological or physical stress, create risks to a host's health, or cause other disruptions.[17]

B. Capacity-enhancing

A neuroprosthetic device may enhance some capacity in a qualitative or quantitative way to exceed what is typically possible for natural biological human beings but without granting its user an entirely new capacity. Examples might include an auditory prosthesis which allows its user to hear faint sounds whose volume falls just below the threshold of what the ear can normally detect. Such enhancement may be temporary or permanent in nature.[18]

1. Temporary Enhancement

Some enhancements may only be activated sporadically and temporarily. For example, an artificial eye might possess physical or digital mechanisms that amplify the available light and allow its user to discern environmental details in very low-light conditions; such an enhancement would be useful if it were nighttime and the user were attempting to navigate the environment, but it could be disruptive and dangerous if the user walked out into a bright sunlit environment with the enhancement still active and were blinded by its effects.[19]

---

[16] The use of neuroprostheses for sensory recording and playback is discussed, e.g., in Merkel et al. (2007); Robinett, "The consequences of fully understanding the brain" (2002); McGee (2008), p. 217; and Gladden, *Sapient Circuits and Digitalized Flesh* (2016).

[17] For the possibility that use of a neuroprosthesis may create dependencies that would result in psychological, physical, economic, or social harm to its human host if use of the device were to be discontinued, see Bostrom & Sandberg, "Cognitive Enhancement: Methods, Ethics, Regulatory Challenges" (2009), p. 323; McGee (2008), p. 213; Koops & Leenes, "Cheating with Implants: Implications of the Hidden Information Advantage of Bionic Ears and Eyes" (2012), p. 125; Gladden, "Neural Implants as Gateways" (2016); and Gladden, "Managing the Ethical Dimensions of Brain-Computer Interfaces in eHealth: An SDLC-based Approach" (2016).

[18] Note that as discussed here, 'enhancement' is defined in relation to the typical abilities of a natural biological human being, not in relation to the specific user who is receiving a device. For example, a motor neuroprosthesis that restores typical voluntary hand movement to an individual who has lost that ability due to injury or illness would more appropriately be seen as a restorative or therapeutic device rather than one that brings about human enhancement – although its host would experience it has having 'enhanced' his or her capacities beyond what existed prior to the device's activation.

[19] The use of neuroprostheses to grant abilities such as telescopic or zoom vision is discussed in Gasson et al. (2012); Merkel et al. (2007); and Gladden, "Enterprise Architecture for Neurocybernetically Augmented Organizational Systems" (2016).

2. Permanent Enhancement

Some enhancements may be permanently active. For example, an implantable cognitive neuroprosthesis that provides enhanced long-term memory functionality and which is deeply integrated into the brain structures of its host and powered by its host's internal biological processes may be continuously active following its implantation.[20]

C. Capacity-restoring

A neuroprosthetic device may restore or provide to its host some capacity that is typically found in natural biological human beings but which the host lacks (perhaps as a result of illness or injury). An example would include a sensorimotor prosthetic robotic arm that restores typical sensory and motor capacity to an individual who has lost one of his or her natural biological arms in an accident.[21]

D. Capacity-neutral

It is possible for a neuroprosthetic device to have no net impact on the sensory, motor, or cognitive capacities possessed by its host. For example, an individual might conceivably decide for purely aesthetic reasons to replace one of his or her natural biological body parts with an artificial neuroprosthesis that looks different but possesses the same functional capacities.[22]

E. Capacity-suppressing

A neuroprosthetic device may suppress capacities naturally possessed by its human host. For example, a cognitive neuroprosthesis that treats insomnia by inducing a sleeping state in its host could be understood as suppressing

---

[20] The 'memory prostheses' whose development is described in Soussou & Berger (2008) serve as a bridge between neurons that spans a damaged area within the hippocampus; future devices of this sort might be designed to operate continuously. For a discussion of technologies that might allow future implanted neuroprostheses to be powered by means of their hosts' own internal biological processes, see, e.g., Mitcheson, "Energy harvesting for human wearable and implantable bio-sensors" (2010); Zebda et al., "Single glucose biofuel cells implanted in rats power electronic devices" (2013); and MacVitte et al., "From 'cyborg' lobsters to a pacemaker powered by implantable biofuel cells" (2013).

[21] For an overview of the current state and anticipated future development of neuroprosthetic robotic limbs, see Farina & Aszmann, "Bionic limbs: clinical reality and academic promises" (2014), and Pazzaglia & Molinari, "The embodiment of assistive devices – from wheelchair to exoskeleton" (2016). For a broader discussion of therapeutic applications of neuroprosthetics, see, e.g., *Implantable Neuroprostheses for Restoring Function*, edited by Kilgore (2015), and Sanchez, *Neuroprosthetics: Principles and Applications* (2016).

[22] For cybernetic augmentation as a form of artistic expression, see, e.g., *The Cyborg Experiments: The Extensions of the Body in the Media Age*, edited by Zylinska (2002).

its host's ability for conscious awareness.[23] Such suppression may be temporary or permanent in nature.

1. Temporary Suppression

A neuroprosthesis may temporarily suppress some capacity within its human host. For example, a cognitive neuroprosthesis that suppresses its host's natural biological mechanisms for experiencing fear and anxiety may be activated when a soldier, aircraft pilot, or surgeon is about to perform some highly dangerous and sensitive maneuver, in order to allow him or her to act without any psychological and physical disruptions caused by nervousness – but at other times the device may be inactive, in order to allow the host's natural fear responses to prevent him or her from performing actions that are reckless and inappropriate in everyday life.

2. Permanent Eradication

A neuroprosthesis may permanently destroy some capacity previously possessed by its host. Note that such an outcome need not be an intentional effect desired by the device's designer, operator, or host. For example, a neuroprosthesis implanted in the brain might as a side-effect of its operation produce heat, electromagnetic radiation, or toxic chemical emissions that gradually destroy individual neurons or larger brain structures in a way that cannot be reversed or repaired and which permanently deprives its host of some sensory, motor, or cognitive capacity.[24]

II. The Neuroprosthesis as a Manipulator of Informational Content

A neuroprosthetic device may produce, receive, store, transmit, manipulate, or otherwise affect particular types of information that serve as the input or output of its host's sensory, cognitive, or motor processes. Such information might be found in sense data received from the environment, memories stored within the brain of the device's host, motor instructions that control the behavior of an effector, or in other components and contexts.[25] Below

---

[23] As noted earlier in the case of TMS used to artificially induce savant skills, the artificial suppression of some of the brain's capacities might simultaneously generate or enhance other capacities.

[24] The dangers of neuroprostheses that may be toxic or degrade over time in the body are noted in McGee (2008), pp. 213-16, and Gladden, "Enterprise Architecture for Neurocybernetically Augmented Organizational Systems" (2016).

[25] Indeed, one of the essential characteristics that distinguishes implantable neuroprostheses from other implantable medical devices (such as a prosthetic hip) is that the former are sophisticated pieces of *information and communications technology* (or ICT) – although even seemingly electronically 'inert' objects such as artificial hip joints, breast implants, or dental prostheses may

we discuss key roles that a neuroprosthetic device might play in shaping the nature and contents of such information.

A. Content-providing

There are a variety of ways in which a neuroprosthesis may provide informational content for a sensory, cognitive, or motor process.

1. Content-detecting

A neuroprosthesis may detect and capture information that is naturally existing within its environment and which did not need to be purposefully engineered or prepared in order to be detected by the device. A cochlear implant or retinal prosthesis that receives sense data in the form of environmental auditory or visual stimuli would be an example of such a neuroprosthesis.[26]

2. Content-generating

A neuroprosthetic device may autonomously generate content relating to a sensory, cognitive, or motor process – perhaps through the use of a software algorithm or the functioning of an artificially intelligent neural network. Such an approach might be used, for example, by a neuroprosthesis that is part of an immersive virtual reality system in order to generate the sense data corresponding to the virtual world to be experienced by the device's host.[27]

3. Content-importing ('Downlink')

A neuroprosthesis may import content in a specially prepared form that is immediately usable by the device from some external system that exists outside of the device and its human host. Such a neuroprosthesis would thus employ a 'downlink' by which information flows into itself from that outside source. Such importing may involve the receipt of an ongoing stream of real-

---

in the future increasingly include RFID chips used to facilitate device identification and diagnostics. Regarding RFID-enabled hip and breast implants, see Gasson, "Human ICT Implants" (2008), p. 22; for RFID-enabled dental implants, see Chang et al., "RFID applied in recognition and identification for dental prostheses" (2012). A key aspect of the nature of neuroprostheses as ICT is the need to maintain the information security of such devices and their host-device systems; that can be understood using InfoSec schemas such as the 'CIA Triad' relating to the confidentiality, integrity, and availability of information. See Gladden, *The Handbook of Information Security for Advanced Neuroprosthetics* (2015), and Gladden, "Information Security Concerns as a Catalyst for the Development of Implantable Cognitive Neuroprostheses" (2016).

[26] For a discussion of these types of sensory neuroprostheses, see Dormer, "Implantable electronic otologic devices for hearing rehabilitation" (2003); Gasson et al. (2012); Ochsner et al., "Human, non-human, and beyond: cochlear implants in socio-technological environments" (2015); Weiland et al. (2005); and Viola & Patrinos (2007).

[27] Regarding the potential use of implantable neuroprostheses as components of an augmented or virtual reality system, see, e.g., Sandor et al., "Breaking the Barriers to True Augmented Reality" (2015), pp. 5-6, and Gladden, *Sapient Circuits and Digitalized Flesh* (2016).

time data (i.e., 'content instreaming') or the periodic reception of a discrete file. The imported content might, for example, provide sense data to a sensory neuroprostheses or remote instructions to govern the actions of a motor neuroprosthesis.[28] Note that the neuroprosthetic device, its human host, and its operator may or may not recognize the fact that information is being imported from an external source; in the case of a neuroprosthesis whose information security has been compromised by an adversary, the existence of the downlink might be purposefully disguised.[29]

4. Backup-restoring

A neuroprosthesis may possess the capacity of reverting to an earlier functional state by loading a backup of stored data that is stored either remotely or within the device itself.[30]

B. Content-translating

A neuroprosthetic device may translate content from one form to another. For example, a sensory neuroprosthesis may perform a process of transduction by which some environmental stimulus (such as photons or sound waves) is converted into digital data for transmission to a computer for processing or into electrochemical signals for transmission to a biological neuron.[31]

1. Content-encoding

Particular examples of content translation by a neuroprosthetic device include the encoding of data for purposes of compression, encryption, storage,

---

[28] For the potential capacity of sensory neuroprostheses to receive live streams of sense data from a remote source, see Koops & Leenes (2012), pp. 115, 120, 126. Regarding the remote control of neuroprostheses (e.g., by a team of medical personnel controlling a device in order to deliver telemedicine), see Gasson (2012) and Gladden, *The Handbook of Information Security for Advanced Neuroprosthetics* (2015).

[29] For example, the possibility that false data might be supplied to a sensory neuroprostheses is raised in Koops & Leenes (2012); McGee (2008); and Gladden, *Sapient Circuits and Digitalized Flesh* (2016).

[30] Creating backup copies of information is a fundamental technique of information security; however, for neuroprostheses that store and process data in the form of a biological or biomimetic neural network, it may be impractical or even impossible to back up the devices' data in its entirety to a physically secure location in order to ensure its long-term availability. See *ISO 27799:2016* (2016) and Gladden, "Information Security Concerns as a Catalyst for the Development of Implantable Cognitive Neuroprostheses" (2016).

[31] Natural biological processes for the transduction of sense data are discussed in Smith, *Biology of Sensory Systems* (2008), pp. 1-30, and Møller, *Sensory Systems: Anatomy and Physiology* (2014), pp. 29-62.

or error correction to maintain integrity during transmission through a noisy channel.[32]

2. Content-decoding

An example of content decoding facilitated by a neuroprosthetic device would be the use of a mnemoprosthesis to retrieve and interpret particular memories stored in the brain's natural biological systems for short-term or long-term memory by detecting relevant neural structures and activity.[33]

C. Content-altering

A neuroprosthetic device may modify the contents of data at rest or in transit in a way that does not simply translate the data from one form or medium to another and which results in a loss of integrity of the original information.[34]

1. Content-editing

For example, a neuroprosthetic device may purposefully edit the content of some message, file, or signal in a purposeful and targeted way. An artificial eye might thus edit portions of the visual data presented to the mind of its human host in order to add specialized supplementary information through an augmented reality display.[35]

2. Content-disrupting

A neuroprosthetic device may alter the information present within a medium in a way that does not purposefully replace the information with some particularly meaningful targeted contents but simply disrupts it in a way that results in a permanent loss of the information.

3. Content-deleting

A neuroprosthetic device may delete information stored within itself or within connected biological systems, either as a normal part of its functioning or as an exceptional action. For example, a mnemoprosthesis may be capable

---

[32] For an overview of such practices, see, e.g., Neubauer et al., *Coding Theory: Algorithms, Architectures and Applications* (2007); Sayood, *Introduction to Data Compression* (2012); and Stallings, *Cryptography and Network Security: Principles and Practice* (2017).

[33] The brain's mechanisms for memory encoding and retrieval are discussed in Schwartz, *Memory: Foundations and Applications* (2014).

[34] The meaning of information 'integrity' within the context of information security is discussed in Parker, "Toward a New Framework for Information Security" (2002), p. 125.

[35] See, e.g., Sandor et al. (2015).

of erasing specific memories stored within the natural biological neural network of its human host's brain.[36]

D. Content-storing

A neuroprosthetic device may store information at rest. Such information may have been written to the device by its designers or operators prior to its activation or received or generated by the device during the time of its operation. Such contents may only be stored on the device temporarily (e.g., raw sense data that is stored momentarily before being processed) or permanently (e.g., operating system files). Information may be stored in the form of conventional binary digital files that can be accessed and interpreted by ordinary desktop computers, or they may be stored as connection and activation patterns within a physical neural network in a form that is difficult or impossible for external systems to access and interpret.[37]

E. Content-transmitting

A neuroprosthetic device may physically transmit information to external systems or components in the form of a digital or analogue signal.

1. Content-exporting ('Uplink')

A neuroprosthetic device may export content to some external system in a specialized form that is immediately usable by that system. Such a neuroprosthesis utilizes an 'uplink' by which information emanates from the device. Such exporting may involve the transmission of an ongoing stream of real-time data or the periodic generation and transmission of a discrete file. Such transmissions may or may not contain sufficient data to allow their recipients to restore the device to an earlier functional state in case of device failure. Note that the neuroprosthetic device, its human host, and its operator may or may not recognize the fact that information is being exported to an

---

[36] In some circumstances it might conceivably be desirable for a neuroprostheses to disrupt or delete undesirable memories stored within a brain's natural biological memory systems – e.g., because existence of the memories produces some unwanted psychological impact for the device's host or because the information is of a highly sensitive nature and was needed by the host only temporarily for the performance of a task that is now complete. The development of neuroprostheses capable of such actions might build on experimental technologies already used to successfully erase memories in mice. See Han et al., "Selective Erasure of a Fear Memory" (2009).

[37] Various approaches to binary digital data storage are discussed in *Information Storage and Management: Storing, Managing, and Protecting Digital Information in Classic, Virtualized, and Cloud Environments* (2012). The human brain's mechanisms for storage of long-term memories are discussed in Dudai, "The Neurobiology of Consolidations, Or, How Stable Is the Engram?" (2004), and Schwartz (2014).

external system; especially in the case of a neuroprosthesis whose information security has been compromised by an adversary, the existence of the uplink might be purposefully concealed by its creator.[38]

a. Content-outstreaming

A particular form of uplink is content-outstreaming, by which an ongoing stream of real-time data is transmitted to an external system. Such an uplink might, for example, allow online viewers around the world to vicariously experience reality 'through the eyes' of a human performance artist whose artificial eyes are continually broadcasting the sense data that they receive so that it can be experienced by others using virtual reality equipment.[39]

b. External Archiving

Another form of uplink involves content archiving, by which information received or generated by a device is periodically copied to an external system for potential future use by the device's human host or operator or by the device itself. For example, a cochlear implant might periodically transmit to its external support system an audio file containing the previous ten hours of auditory stimuli detected by the device. The device's host could later 'play back' particular conversations or other auditory experiences at will by downloading the correct archive file into the cochlear implant's internal computer.[40] Note that archived content does not necessarily constitute a backup file, as it may be fragmentary in nature and may not allow full restoration of a neuroprosthetic device to an earlier functional state.

2. Production of External Device Backup

A neuroprosthetic device may transmit information to an external system in the form of a single periodically generated file or an ongoing stream of data that can be used to restore the device to its current or an earlier functional state, should the device suffer a failure such as that caused by a power outage or physical damage. Note that the device itself may or may not possess the

---

[38] For the possibility that a hacker, computer virus, or other agent may be able to steal data contained in a neuroprosthesis or use the device to gather data (potentially including the contents of the thoughts, memories, or sensory experiences of the device's human host or others), see McGee (2008), p. 217; Koops & Leenes (2012), pp. 117, 130; Gasson (2012), p. 21; and Gladden, *The Handbook of Information Security for Advanced Neuroprosthetics* (2015).

[39] For a discussion of such possibilities, see Gladden, *The Handbook of Information Security for Advanced Neuroprosthetics* (2015), p. 291.

[40] The potential development of neuroprostheses that could allow 'playback' of recorded or previously experienced information is discussed in Merkel et al. (2007); Robinett (2002); and McGee (2008), p. 217.

capacity to autonomously retrieve remote backup files and restore itself; such actions may need to be manually performed by an external operator.[41]

III. The Neuroprosthesis as a Means of Inhabitation

Neuroprostheses can play powerful roles in transforming the way in which a host's mind is embodied within a particular corporeal form and the manner in which that body is embedded within an external environment.[42]

Distinguishing Primary ('Real') and Secondary ('Virtual') Physical Worlds

In order to appropriately analyze the potential involvement of neuroprostheses in their host's processes of embodiment and inhabitation of an environment, it is important to first develop a clear formulation of the difference between what are commonly referred to as 'real' and 'virtual' objects and phenomena. In everyday speech, a distinction is commonly made between the 'real world' in which human beings and their bodies, homes, automobiles, and computing devices exist and a 'virtual world' that exists only within a computer and which is experienced, for example, by staring at a computer monitor or wearing a VR headset. Implicit within this popular understanding is the notion that the 'real world' is one of tangible physical objects and a 'virtual world' is non-physical, a world whose apparent physicality is only the illusory product of a carefully arranged presentation of sense data.[43]

However, from the perspective of neuroprosthetic supersystems architecture, such a popular conception pitting 'the real' versus 'the virtual' is not only insufficient but even incorrect. A neuroprosthesis that immerses its human host in a virtual reality environment and provides its host with the experience of possessing a radically nonhuman body (e.g., a body in the form of a robotic octopus or a floating sphere of light) is not 'non-physical' in nature: after all, the neuroprosthetic device is made of physical components that are created

---

[41] The importance of regular creation of backup files is discussed in *NIST Special Publication 800-53, Revision 4: Security and Privacy Controls for Federal Information Systems and Organizations* (2013), p. F-87, and *ISO 27799:2016* (2016).

[42] There is a well-developed literature on the subject of embodied embedded cognition from perspectives such as human psychology, philosophy of mind, and robotics. See, e.g., Wilson, "Six views of embodied cognition" (2002); Anderson, "Embodied Cognition: A field guide" (2003); Sloman, "Some Requirements for Human-like Robots: Why the recent over-emphasis on embodiment has held up progress" (2009); and Garg, "Embodied Cognition, Human Computer Interaction, and Application Areas" (2012).

[43] Different approaches to defining virtual reality are discussed, e.g., in Heim, *The Metaphysics of Virtual Reality* (1993); *Communication in the Age of Virtual Reality*, edited by Biocca & Levy (1995); *Cybersociety 2.0: Revisiting Computer-Mediated Communication and Community*, edited by Jones (1998); Lyon, "Beyond Cyberspace: Digital Dreams and Social Bodies" (2001); Koltko-Rivera (2005); and Bainbridge, *The Virtual Future* (2011).

and maintained through physical processes, and it interacts physically with the biological components of the host's nervous system. Moreover, even the nonhuman virtual body that the device fashions for its host is not non-physical in character; the nature of the virtual body's structure and behaviors is stored as data (e.g., a set of binary digital files) that is contained within some physical substrate, such as a hard drive, RAM chip, or physical neural network. Damage to that physical substrate would result in the alteration or loss of the virtual body experienced by the host, just as damage to the host's natural biological body would result in the alteration or loss of that 'real' body. And the host does not sense and control his or her virtual body by means of some telepathic or psychokinetic powers that are non-physical in nature: sense data from the virtual world is provided by means of electrochemical signals that are an observable element of the physical world and which must physically affect neurons within the host's biological body, and in order to manipulate his or her body the host must generate or manipulate physical phenomena (such as electrical activity or chemical neurotransmitters in the brain) that can be detected by physical components of the neuroprosthetic device.

The distinguishing characteristic of virtual bodies and virtual worlds is thus not that they are 'non-physical' or 'unreal' but that they possess a special *type* of physicality. A human being's natural biological body is characterized by the fact that its physical components share an isomorphic and direct causal relationship with the components of the body that is experienced by the mind of that person. For example, a human being can see and feel that she possesses a leg whose components occupy a particular space and create a particular shape, and indeed the person's biological body includes cells and other physical components that are arranged in such a pattern. On the other hand, a virtual body belonging to a human being is characterized by the fact that its physical components do not share an isomorphic and direct causal relationship with the components of the body that is experienced by the mind of that person. For example, a human being might see and feel that she possesses a leg whose components occupy a particular space and create a particular shape, but the physical components determining the shape and nature of her leg are in fact a set of electrons stored within the capacitors of a RAM module's integrated circuit within her neuroprosthetic device.

Instead of counterposing terms such as 'real' versus 'virtual' or 'physical' versus 'digital' to distinguish these differing constellations of structures and activities, this text will utilize the phrase **'primary physical world'** to refer to the isomorphic physical world that includes a human being's natural biological body and surrounding environment and the phrase **'secondary physical world'** to refer to an anisomorphic physical world that determines the nature of a virtual body and virtual world to be experienced by a human being.[44] Having delineated these terms, we can consider in more detail the ways in which a neuroprosthesis may mediate a mind's situation in and interaction with the world through the processes of embodiment and embedding.

A. Embodiment: Inhabitation of a Material Form

Every neuroprosthetic device impacts the manner in which its host is embodied within a particular corporeal form or 'body.' Insofar as a neuroprosthetic device is integrated into the physical neural circuitry of its human host, the neuroprosthesis by definition affects the structure and behavior of its host's body. Some neuroprostheses only seek to support or restore the typical functioning of a host's natural biological body, while others provide their host with an enhanced or transformed (and potentially radically nonhuman) body. Such transformation may involve replacing or dramatically altering a significant portion of the natural physical components of a host's biological body, or it may involve leaving the host's biological body largely intact but providing the host – and others – with an *experience* of the host's possession of an enhanced or transformed body within some virtual world.[45] The difference between altering the host's body as it exists within the primary physical world (i.e., the 'real' world) and as it exists in a secondary physical world (i.e., a 'virtual' world) can be understood as the difference between the existence of the host as cyborg and the host as digital avatar.

1. The Host as Cyborg: Reshaping the Physical Body

A neuroprosthetic device that replaces or supplements part of its host's original biological physical body with new physical components that are designed to interact directly with the external physical environment (e.g., through physical touch, grasping, manipulation, gestures, locomotion, and audible speech) constitutes an isomorphic neuroprosthesis and can be understood as providing its host – to a greater or lesser extent – with a physical cyborg body. Such a process of 'cyborgization'[46] might involve the replacement of a severely damaged biological limb with a robotic prosthetic replica,

---

[44] There is only one primary physical world, while there is a limitless number and variety of secondary physical worlds in which a device's human host might become immersed – as well as tertiary or further worlds. We thus speak of 'the' primary physical world but 'a' secondary physical world.

[45] For the extent to which neuroprosthetic devices or other devices can become incorporated into a host's body schema, see Gladden, "Cybershells, Shapeshifting, and Neuroprosthetics: Video Games as Tools for Posthuman 'Body Schema (Re)Engineering'" (2015), and Pazzaglia & Molinari (2016).

[46] For use of the term 'cyborgization' to describe such a process, see, e.g., Maguire & McGee (1999); Koltko-Rivera (2005); Novakovic et al. (2009); and Nayar (2010). The term 'cyberization'

the augmentation of a healthy biological eye through the integration of additional sensors that allow a host to detect infrared light, or the implantation into the brain of a brain-computer interface that allows a host to wirelessly interact with and remotely control computerized systems.

Depending on the physical nature of a host's cybernetic augmentation, the manner in which it was installed in the host's body, and the way in which it interacts with the host's cognitive processes, a host may or may not realize that he or she has become a cyborg through the addition of such cybernetic components.[47]

a. Full-body cyborg

It is possible for a human host to replace at most a portion of his or her body with artificial cybernetic components; at least some critical components of the host's natural biological brain must remain intact.[48] Thus a 'full cyborg' should be understood not as a human being whose body has been wholly replaced with artificial biocybernetic components but one whose biological body parts have been replaced with artificial biocybernetic components *to the greatest extent possible* without causing the death of the host or loss of his or her personal identity. It is not clearly known to what extent a process of cyborgization can be safely applied to an individual before his or her limit for the maintenance of cognitive and biological integrity is exceeded, the host's personal identity is irrevocably lost, and death ensues.[49]

---

is also sometimes used to describe this process by which a human host incorporates artificial biocybernetic components into his or her body, thereby becoming a cyborg. However, the term 'cyberization' is also used in a broader or alternative sense to refer to processes by which a human being becomes psychologically (rather than physically) integrated into electronic information systems such as immersive virtual reality systems through long-term use and sensorimotor experience. This latter sense of 'cyberization' does not imply that an individual has been subjected to physical biocybernetic augmentation; it may be thus more appropriate to use the word 'cyborgization' when discussing the process of augmenting a host's body through the permanent incorporation of artificial biocybernetic components. For various meanings of the term 'cyberization,' see, e.g., Miller (2012); Baranyi et al. (2015); and Ma et al. (2016).

[47] For the possibility that a human host may not realize that he or she has been implanted with an invasive neuroprosthesis, see Gladden, *The Handbook of Information Security for Advanced Neuroprosthetics* (2015).

[48] At a minimum, some critical components of a host's natural biological brain must remain intact for a device to interface with; otherwise – by definition – the device is not a neuroprosthesis that is integrated into the neural circuitry of its human host. A technological system that replaces every one of a host's natural biological neurons with synthetic copies might thus be described as a 'neurotechnology' but not a 'neuroprosthesis.' Through the application of sufficiently sophisticated artificial intelligence such a device might even replicate much of the behavior of the human brain that has served as its template, but the device could not be understood as 'interfacing' with that brain.

[49] The notion that an excessive degree of cybernetic augmentation might result in an individual's

A full-body cyborg may be able to function safely and effectively in environments that are inhospitable or even fatal to unaugmented human beings (such as those lacking breathable air or possessing extremely high or low temperatures, pressure, acceleration, or levels of light or sound), if his or her organs are replaced with artificial substitutes or alternatives whose capacities differ significantly from those of the typical natural biological human body.

b. Partial cyborg

A 'partial cyborg' can be understood as a human being whose body has undergone a degree of permanent biocybernetic augmentation that is less than the maximum possible for that person. In cases of minimal cyborgization, the amount of biocybernetic augmentation may be trivial and it may be debatable whether an individual can appropriately be considered a 'cyborg,' depending on the precise definition of the term that is employed. For example, an individual who has received a dental bridge will likely not be considered a cyborg, due to the device's passive nature and lack of bioelectronic functionality. A person who has lost a hand due to injury and been given a conventional prosthetic hand whose fingers do not move will also likely not be considered a cyborg due to the device's easily detachable and nonpermanent nature, its lack of bioelectronic functionality, and its lack of interaction with the person's nervous system. An individual who has received an implantable RFID chip may be considered a cyborg by some experts, given the device's permanent incorporation into its host's body and its electronic functionality; however, he or she may not be considered a cyborg by others because of the device's lack of integration into its host's sensory, motor, and cognitive structures and processes. An individual who has received an artificial cardiac pacemaker may be considered a cyborg because of the device's long-term implantation in the body and integration into the functioning of the body's organs or may not because of the fact that the device supplements rather than replaces a biological component of the host's body, the fact that some of its parts (e.g., its battery) must be periodically replaced, and the fact that it is used for purely medical purposes rather than purposes of human enhancement.

---

loss of biological integrity, personal identity, or human 'essence' is discussed, e.g., in Miah, "A Critical History of Posthumanism" (2008), pp. 73-74; Fukuyama, *Our Posthuman Future: Consequences of the Biotechnology Revolution* (2002); and Gladden, *Sapient Circuits and Digitalized Flesh* (2016). It may be hypothesized that the threshold of maximum possible cyborgization may vary between individuals (e.g., depending on their age and health); the threshold for human beings as a whole may conceivably also increase over time, as new technologies allow the safer and more effective replacement of additional biological body components with artificial replicas or alternatives and human organisms are genetically engineered to become more amenable to such technologies.

c. Extended cyborg

An extended cyborg is one whose artificial biocybernetic components do not replace natural biological components with functionally equivalent replicas but which add new (and potentially radically non-human) ones. Such a cyborg might possess physical elements such as wheels, gills, additional limbs, or additional eyes providing 360° vision. The extent to which a cyborg can possess a non-human morphology is studied by the field of body schema engineering.[50]

d. Sessile cyborg

A neuroprosthesis may take the form of a biocybernetic housing or life-support system within which its host's brain (and perhaps additional body organs) is maintained and which is not designed to provide the host with a body that can be used to explore the world through locomotion and direct physical manipulation. Such a neuroprosthesis might instead allow its host's mind to inhabit, move within, and manipulate some virtual environment through a virtual reality system that creates biocybernetic sensorimotor feedback loops, even though such a host's neuroprosthetic shell (and thus its body) may be immobile within the primary physical world.[51]

e. Reverse cyborg

A 'reverse cyborg' is not truly a cyborg; it is a human being who has undergone a reversed process of cyborgization in which most of the person's body is maintained intact while critical components of the brain that are needed to preserve the individual's personal identity are replaced with artificial biocybernetic components that are capable of regulating the work of body organs and perhaps even replicating the person's patterns of social behavior and interaction by receiving and processing sense data and generating appropriate motor output.

Due to its use of similar biocybernetic technologies, it might superficially appear to non-specialists as though the process of reverse cyborgization is similar to that of creating conventional types of cyborgs. However, while the ethical and legal questions connected with the creation of regular full or partial cyborgs are already quite serious, the questions associated with the creation of reverse cyborgs are even more grave: an individual expecting to undergo a surgical procedure and awaken with newly augmented capacities

---

[50] See Gladden, "Cybershells, Shapeshifting, and Neuroprosthetics" (2015).

[51] Such an arrangement resembles the 'brain in a vat' scenario discussed by Harman and Putnam and many others since within the field of philosophy of mind, building on the thought experiment involving an 'evil demon' formulated by Descartes in his *Meditations on First Philosophy* (1641). See Harman, *Thought* (1973), p. 5, and Putnam, *Reason, Truth and History* (1981).

might instead have his or her brain destroyed and the remaining portion of his or her body artificially maintained and controlled by the operators of its neuroprosthetic interface as a sort of biological 'puppet' or 'zombie.'[52] From an ethical and legal perspective, intentionally creating such a being would appear to be highly illicit.

2. The Host as Avatar: Shaping Others' Perceptions of the Host

Some neuroprostheses are designed primarily to control, shape, or mediate the perceptions of a host's form and actions that other intelligent agents receive within a virtual environment. In other words, such neuroprostheses create an 'avatar' that constitutes or determines the host's body as it exists within some secondary physical world.

Even when the virtual body is designed to mimic as closely as possible its host's experience of his or her natural biological body, it actually provides its host with an anisomorphic body in the secondary physical world whose apparent size, shape, and construction do not correspond to the system's actual physical size, shape, and construction (comprising, for example, a set of electrons stored in the transistors of a RAM module and not a collection of biological cells) within the primary physical world.[53]

a. Cyberdouble

It is possible for a host's neuroprosthetically facilitated avatar to duplicate the host's actual physical appearance, features, and expressions in a way that is as authentic as possible, given the constraints of the virtual environment – thereby creating a virtual 'cyberdouble' of the host's body.

b. Cybermorph

Alternatively, a neuroprosthetically facilitated avatar may present to other inhabitants of a virtual environment an appearance that does not replicate its owner's actual physical appearance; such an avatar may be radically non-human in nature (e.g., appearing as a robotic spider or a floating ball of light), and inhabitants of the virtual environment may or may not be able to identify the avatar with its physical human owner.[54]

---

[52] The potential for such misuse of technologies for cybernetic augmentation is discussed in Gladden, *The Handbook of Information Security for Advanced Neuroprosthetics* (2015), pp. 98, 220.

[53] For a discussion of psychological, social, and political questions relating to repetitive long-term inhabitation of virtual worlds through a digital avatar, see, e.g., Castronova, "Theory of the Avatar" (2003).

[54] The extent to which a human being can make use of the (virtual) sensory and motor components and processes of a radically non-human avatar is limited by the adaptability of the individual's body schema. See Gladden, "Cybershells, Shapeshifting, and Neuroprosthetics" (2015).

B. Embedding: Inhabitation of an External Environment

A neuroprosthetic device may significantly alter the way in which its host senses, controls, and otherwise experiences the environment surrounding the host's body. For example, a neuroprosthesis may extend the region of space within which environmental objects and phenomena can be detected and manipulated; increase or decrease those aspects of the environment that can be sensed and manipulated; reduce the capacity of other agents or forces within the environment to detect or affect the host; increase or decrease the host's ability to understand the structures and dynamics of the environment; and increase or decrease the host's social, psychological, or physical dependence on elements within the environment. The difference between transforming the way in which a device's host experiences the primary physical world (i.e., the 'real world') and providing the host with the experience of a secondary physical world (i.e., a 'virtual world') can be understood as the difference between exposing the host to augmented reality and immersing him or her in a virtual reality.

1. The World Viewed Through Augmented Reality

A neuroprosthetic device may employ augmented reality to provide its host with information that is not available through the host's natural biological sense organs or cognitive processes.[55]

a. Overlaid information

One possibility is for a neuroprosthesis to 'overlay' fabricated sense data that conveys specialized information atop the natural sense data that the device's host is receiving from his or her body or the environment. For example, a retinal prosthesis might double as a clock by periodically displaying the time as a set of numerals hovering within its host's field of vision, or it might highlight streets and buildings to help the host navigate to a desired destination. An auditory prosthesis might periodically produce audible information about its host's blood glucose level or live transmission of a radio station's broadcast that the host can hear atop the natural sounds produced by the environment.

b. Refracted reality

Another possibility is for a neuroprosthetic device to temporarily or permanently present its host with specialized information that replaces rather than overlays the kind of sense data that would naturally be produced by the host's biological sense organs. For example, consider a retinal prosthesis

---

[55] Regarding the potential use of neuroprosthetic implants to provide an augmented reality experience, see Koops & Leenes (2012); Sandor et al. (2015), pp. 5-6; and Gladden, "Enterprise Architecture for Neurocybernetically Augmented Organizational Systems" (2016).

which – when activated – replaces its host's natural perception of visible light with infrared vision that only presents photons of infrared wavelengths that have been detected within the environment. Such a form of augmented reality can be understood as providing its user with a 'refracted' view of the world; it is as though the real world were being viewed through a particular type of filter or lens that both enhances and distorts. This is considered a form of augmented reality rather than virtual reality, because the world perceived by a device's host is isomorphic with the primary physical world; the device does not present a fabricated environment but a new way of experiencing the primary physical world.

2. Occupancy of a Virtual World

A neuroprosthesis may allow its host to inhabit a secondary physical world (or 'virtual world') by generating sense data that depicts the contents of the world and detecting motor instructions produced by the host's brain or spinal word which allow the host to manipulate the contents of the world.

a. Degree of immersion

Neuroprosthetic devices can be distinguished by the extent to which they immerse their hosts in a virtual environment.

I. MAXIMAL IMMERSION

In the strictest sense of the phrase, there is no such thing as a 'totally immersive' virtual reality system that can fully plunge its user into a virtual world, because the brain itself contains components that are sensitive to phenomena present within the primary physical world and whose functioning most likely cannot be wholly replaced or overridden by a neuroprosthesis. Even if a neuroprosthetic device were capable of completely blocking the sense data that a host's eyes, ears, nose, tongue, and skin normally receive from the primary physical environment and replacing them with fabricated sense data depicting a virtual environment,[56] the host's illusion of being present within that virtual environment would not be perfect or complete. For example, the virtual visual, auditory, and tactile data received by the host in a particular moment might give the impression that he or she is running

---

[56] It has been estimated that in principle, a virtual reality system would be capable of providing its user with a suite of visual, auditory, olfactory, gustatory, and tactile sense data whose quality equals that of sense data generated by the primary physical world, if the system were capable of presenting either roughly 200 Gbps of raw sense data to the host's natural biological sense organs (such as the retina, hair cells in the ear, and taste buds) through their external stimulation or roughly 250 Mbps of already-processed sense data in the form of direct electrochemical stimulation of the nerves (such as the optic and cochlear nerves) that carry such data to the brain or of the relevant brain regions themselves. See Berner, *Management in 20XX: What Will Be Important in the Future – A Holistic View* (2004), pp. 37-38, 45-47.

Chapter Three: An Ontology of Neuroprostheses as Instruments of 'Cyborgization' • 137

along a beach; but if within the primary physical world the host's body were actually lying motionless in a hospital bed, sense data relating to the senses of proprioception and balance would 'tell' the host that his or her body were in fact not moving at all.[57] A sufficiently well-trained host would notice such discrepancies, and they would diminish the experienced degree of immersion in the virtual world.

Even if a highly sophisticated future neuroprosthesis were somehow capable of providing 100% of the sense data experienced by its host, it could still not eliminate the reality that the host's brain exists within the primary physical world – not a virtual environment – and as such, the brain is subject to environmental phenomena present within the primary physical world like heat, electromagnetic radiation, acceleration, and the introduction of chemical substances into the brain that may directly affect the brain's functioning and create for a device's host experiences that are inconsistent with the characteristics of the virtual environment that is being fabricated by the neuroprosthesis.[58]

In ordinary everyday conversation, it may be convenient to speak of some neuroprosthesis as providing its host with 'full' immersion in a virtual world; however, the word 'full' should not be understood literally: it would be more appropriately taken to mean that such a neuroprosthesis immerses its host in a virtual world "to the fullest extent possible." It is thus more correct to speak of such a device as offering its host 'maximal' immersion in the virtual environment of a secondary physical world.

II. PARTIAL IMMERSION

A neuroprosthetic system creates 'partial immersion' if it provides its host with an experience of inhabiting a virtual environment that is less complete than that experienced with maximal immersion. Examples would include artificial eyes which, when activated, present their user with the visual experience of existing and moving within some virtual world – but which do not

---

[57] Regarding varying forms of 'cybersickness' that may be experienced by users of virtual reality systems, see Polcar & Horejsi, "Knowledge Acquisition and Cyber Sickness: A Comparison of VR Devices in Virtual Tours" (2013), and Davis et al., "A systematic review of cybersickness" (2014). Some forms of cybersickness may be generated or exacerbated when a device's host receives dissonant sense data through different sense modalities, some of which may be presenting authentic sense data from the primary physical world and others fabricated sense data from a virtual world.

[58] Such mechanisms of direct action upon the brain would not include phenomena such as the microwave auditory effect, which appears to act upon components of the ear rather than on the brain itself. See, e.g., Lin, "Hearing microwaves: The microwave auditory phenomenon" (2001).

affect the host's sense of hearing, which continues to present auditory sense data from the primary physical world.[59]

b. Period of immersion

Neuroprosthetic devices that allow their hosts to inhabit a virtual environment may differ by typically immersing their hosts for varying periods of time.

I. TEMPORARY IMMERSION

A neuroprosthetic device may provide its host with periodic and temporary immersion in a virtual environment. For example, institutional VR systems might be accessible to an organization's employees during designated working hours but inaccessible at other times, and VR gaming systems may be used for those relatively short and sporadic stretches during which a user can find time to play.

For a neuroprosthesis that periodically immerses its host in a virtual environment for a very brief period of time, less attention will need to be given by the device's designers and operators to the potential psychological, physical, or social effects that may result from inhabiting that virtual environment for an extended period of time – however, greater attention will need to be paid to any impacts that affect the host when he or she transitions into or out of the virtual environment, since those transitional effects may be experienced a large number of times and in close succession. In particular, any cumulative impacts produced by entering and existing the virtual environment must be carefully identified and studied.

II. LONG-TERM IMMERSION

A neuroprosthesis may provide its host with long-term or potentially even permanent immersion in a virtual environment. This might be the case, for example, with individuals whose physical bodies have been so severely injured that they can only be kept alive within a large, complex, immobile life-support system that uses a neuroprosthetic interface to allow a patient who can no longer move or sense the primary physical world through his or her natural physical organs to explore virtual environments and interact with other human beings or non-human intelligent agents within them.[60]

In the case of a neuroprosthesis that immerses its host in a virtual environment for an extended (or even indefinite) period of time, the device's designers and operators will be obliged to pay close attention to the potential

---

[59] Regarding the effects of varying degrees of immersion in virtual reality environments, see Cummings & Bailenson, "How immersive is enough? A meta-analysis of the effect of immersive technology on user presence" (2016).

[60] The ramifications of long-term immersion in virtual reality environments in discussed, e.g., in Heim (1993); Koltko-Rivera (2005); and Bainbridge (2011).

psychological, physical, or social effects that may result from long-term in-habitation of that virtual environment. It will also be necessary to study any impacts that affect the host when he or she transitions into or out of the vir-tual environment, as the rarity of such transitions may leave the host's mind and body ill-prepared for their effects. However, the cumulative impacts cre-ated by repeated transitions into and out of the virtual environment will be of less significance, as hosts are unlikely to encounter them.

c. Cognizance of immersion

Neuroprostheses may differ in the extent to which their hosts are aware of the fact that their devices are immersing them in a virtual environment.

I. RECOGNIZED IMMERSION

Many neuroprosthetic devices that immerse their hosts in a virtual envi-ronment do so in such a way that a host is consciously aware of when he or she is entering, leaving, or present within the virtual environment. It will be especially easy for hosts to gain and possess such knowledge when the virtual environment differs noticeably from the primary physical world and hosts periodically transition in and out of the virtual environment, rather than re-maining within it permanently.

II. UNRECOGNIZED IMMERSION

Some neuroprosthetic devices might conceivably immerse their hosts in a virtual environment in such a way that a host is not consciously aware of in-habiting a virtual environment when he or she is immersed in it.[61]

It may be difficult for a host to recognize that he or she is in a virtual en-vironment if, for example: a) the virtual environment strongly resembles the primary physical world with which the host was previously familiar; b) the transition into the virtual environment has occurred while the host was asleep or otherwise unable to consciously observe the transition; c) the neu-roprosthesis disrupts the host's memory functions that would allow the host to remember differences that have been experienced between the primary and secondary physical worlds;[62] or d) the host remains permanently im-mersed in the virtual environment, thereby being deprived of the possibility

---

[61] The fact that a host may not be consciously aware of the fact that he or she is immersed in a virtual environment is consistent with the fact – discussed earlier – that no VR system can create a state of 'full immersion' that severs the host from the influences of the primary physical world. As noted earlier, a host's physical brain will always be subject to the effects of phenomena such as cosmic rays, heat, or electromagnetic fields that exist in the primary physical world; however it is possible that such phenomena will not generate or influence sensory experiences in such a way that the host will become consciously aware of the phenomena's existence.

[62] The possible development of neuroprostheses that may alter or disrupt the memories of their human hosts is discussed in Gladden, *Sapient Circuits and Digitalized Flesh* (2016), and Gladden,

of noticing transitions into and out of the virtual environment and recognizing discrepancies between the virtual environment and primary physical world.

IV. The Neuroprosthesis as a Regulator of Agency

Neuroprostheses interact with the agency possessed and exercised by their human hosts in a range of ways. 'Weak' notions of agency define an agent as any entity that displays the externally observable characteristics of autonomy, reactivity, proactivity, and the ability for social interaction; 'strong' notions of agency insist that an agent also possess internal mental phenomena such as beliefs and desires (which, when joined, can constitute intentions).[63] The human beings who serve as hosts to neuroprosthetic devices are not only agents in the weak sense but also in the strong sense: as human beings, we experience our own beliefs, desires, and intentions, and we realize that any condition (such as illness or injury) that destroys our ability to experience beliefs, desires, and intentions would also eliminate our capacity to act as agents within the world.

Neuroprosthetic devices may themselves possess and exercise agency in the weak sense; the question of whether future neuroprostheses endowed with sufficiently sophisticated artificial intelligence might someday also possess agency in the strong sense is a contested issue. Below we consider neuroprostheses that manifest and interact with agency in different ways.

A. The Neuroprosthesis as Agent

A neuroprosthesis may or may not possess and exercise its own agency.

1. Neuroprostheses Possessing Agency

A neuroprosthesis may possesses and exercise its own autonomous agency within the context of its host-device system.[64] The agency of the neuropros-

---

"Enterprise Architecture for Neurocybernetically Augmented Organizational Systems" (2016). Such devices might potentially build on experimental technologies for the artificial generation, alteration, or deletion of memories currently being tested in mice. See, e.g., Han et al. (2009); Josselyn, "Continuing the Search for the Engram: Examining the Mechanism of Fear Memories" (2010); and Ramirez et al. (2013).

[63] For these definitions of agency, see Wooldridge & Jennings, "Intelligent agents: Theory and practice" (1995), and Lind, "Issues in agent-oriented software engineering" (2001). For more on the relationship of beliefs, desires, and intentions, see Calverley, "Imagining a non-biological machine as a legal person" (2008).

[64] For computerized devices such as neuroprostheses, autonomy can be understood as the state of being "capable of operating in the real-world environment without any form of external control for extended periods of time." See Bekey, *Autonomous Robots: From Biological Inspiration to*

thetic device may be generated and governed by, for example, a software program controlling the device or by the functioning of a physical neural network that controls the neuroprosthesis.

2. Neuroprostheses Lacking Their Own Agency

Some neuroprosthetic devices do not possess their own agency. Examples might be found in host-device systems in which only the human host possesses and exercises agency and the implanted device – while integrated into its host's neural circuitry – is passive in function.[65]

B. The Neuroprosthesis as Instrument of External Agency

While lacking its own autonomous agency, a neuroprosthetic device might act as a tool that extends the agency of some external agent into the organism of the device's host. An example would include a neuroprosthesis that is remotely controlled by medical personnel who use the device as a telepresence instrument to provide telemedicine services to the device's human host, who lives in a remote location where physicians are not available to administer medical services in person.[66] A neuroprosthesis that has been remotely hijacked by a hacker and whose operation is now being controlled by that adversary would be another example of a neuroprosthetic device functioning as an instrument of an external agent.[67]

C. The Neuroprosthesis as Modulator of a Host's Human Agency

Neuroprosthetic devices demonstrate a range of impacts on the autonomy and agency of their human hosts. While some devices may have no direct impact on their host's agency, other devices may enhance, impair, or eradicate their host's ability to possess and manifest agency.

1. Agency-realizing Neuroprostheses

A neuroprosthetic device may possess agency that is ultimately exercised not in the form of autonomy and independent action but through a purposefully designed subordination to the agency of the device's host and his or her

Implementation and Control (2005), p. 1.

[65] The existence of pieces of implantable information and communications technology (ICT) such as neuroprostheses that are passive in their functionality is discussed in Roosendaal, "Implants and Human Rights, in Particular Bodily Integrity" (2012), and Gladden, *The Handbook of Information Security for Advanced Neuroprosthetics* (2015).

[66] For the remote administration of implantable medical devices as a means of administering telemedicine, see Gasson (2012) and Gladden, "Managing the Ethical Dimensions of Brain-Computer Interfaces in eHealth" (2016).

[67] For such possibilities, see Denning et al., "Neurosecurity: Security and Privacy for Neural Devices" (2009); Krishnan, "From Psyops to Neurowar: What Are the Dangers?" (2014); and Gladden, *The Handbook of Information Security for Advanced Neuroprosthetics* (2015).

volitions; such a neuroprosthesis would exercise its agency by attempting to detect the volitions of its human host and faithfully implement and realize those volitions. Examples might include an artificially intelligent prosthetic robotic arm that replaces a natural biological arm that its host has lost due to injury and which employs its agency to detect motor instructions generated in its host's brain and move in such a way as to successfully enact the host's volitions.

2. Agency-extending Neuroprostheses

A neuroprosthesis might attempt to capture and preserve or replicate the natural agency that would have been exercised by its host, in order to manifest that agency in a remote location in which the host is not directly present or at a time when the host cannot directly exercise his or her normal agency (e.g., while the host is unconscious or asleep or after his or her death).

3. Agency-complementing Neuroprostheses

A neuroprosthetic device may possess agency which complements that of its human host and aids the host in his or her activities while ultimately maintaining independence as an agent and not being directly subject to the host's control. Such a device might employ a form of artificial intelligence to serve as an advisor, counsellor, companion, or friend to its human host.[68]

4. Agency-controlling Neuroprostheses

A neuroprosthetic device may possess some form of agency that it uses to directly or indirectly constrain or control its host's autonomous possession and exercise of agency. Such a neuroprosthesis might be employed as a means of medical treatment, surveillance, punishment, training, or workplace supervision.[69]

5. Agency-enhancing Neuroprostheses

A neuroprosthesis may enhance its host's ability to possess and exercise agency – perhaps by removing or inhibiting some obstacle that normally disrupts the host's agency. For example, some users of deep brain stimulation

---

[68] For discussions of robotic devices or artificially intelligent systems serving as colleagues and assistants to human workers, see, e.g., Ablett et al., "A Robotic Colleague for Facilitating Collaborative Software Development" (2006); Vänni and Korpela, "Role of Social Robotics in Supporting Employees and Advancing Productivity" (2015); and Gladden, "Leveraging the Cross-Cultural Capacities of Artificial Agents as Leaders of Human Virtual Teams" (2014). For robots that serve as charismatic leaders (and perhaps even spiritual guides) for human beings, see Gladden, "The Social Robot as 'Charismatic Leader': A Phenomenology of Human Submission to Nonhuman Power" (2014).

[69] For such possibilities, see Barfield, *Cyber-Humans: Our Future with Machines* (2015), p. 111, and Gladden, *The Handbook of Information Security for Advanced Neuroprosthetics* (2015).

devices employed to treat Parkinson's disease and other conditions have reported that their sense of autonomy and ability to exercise personal agency have been enhanced by use of such devices.[70]

6. Agency-impairing Neuroprostheses

A neuroprosthesis may temporarily or permanently impair its host's possession and exercise of agency without completely destroying that agency. Such devices might include an anesthetic neuroprosthesis that periodically induces a state of unconsciousness in its host or an emergency life support system that preserves a human brain intact but deprives it of the sensory and motor capacities that allow it to manifest its agency.[71]

7. Agency-eradicating Neuroprostheses

A neuroprosthetic device may permanently eradicate the ability of its host to possess and exercise agency within the world. This may occur if the presence or use of the device results in the death of its host's biological organism or if it damages or destroys neurons and brain structures to such an extent that the host – while still being maintained in a living state – can no longer exercise his or her natural agency.

D. The Neuroprosthesis as Hive Mind Infrastructure

A neuroprosthesis may link the mind of its human host with external intelligences (such as the minds of other neuroprosthetically augmented human beings or artificial intelligences) in such a way that the mind of the human host and the external agents form a sort of collective entity or 'hive mind.' The level at which and extent to which the cognitive processes of the hive mind's members are connected may vary.[72]

---

[70] See the discussion of such issues in Kraemer, "Me, Myself and My Brain Implant: Deep Brain Stimulation Raises Questions of Personal Authenticity and Alienation" (2011); Van den Berg, "Pieces of Me: On Identity and Information and Communications Technology Implants" (2012); McGee (2008); and Gladden, *Sapient Circuits and Digitalized Flesh* (2016).

[71] Such possibilities are discussed, e.g., in Gladden, *The Handbook of Information Security for Advanced Neuroprosthetics* (2015), and Gladden, "Neural Implants as Gateways to Digital-Physical Ecosystems and Posthuman Socioeconomic Interaction" (2016).

[72] The prospect of creating hive minds and neuroprosthetically facilitated collective intelligences is discussed, e.g., in McIntosh, "The Transhuman Security Dilemma" (2010); Roden, *Posthuman Life: Philosophy at the Edge of the Human* (2014), p. 39; and Gladden, "Utopias and Dystopias as Cybernetic Information Systems: Envisioning the Posthuman Neuropolity" (2015). For classifications of different kinds of potential hive minds, see Chapter 2, "Hive Mind," in Kelly, *Out of Control: The New Biology of Machines, Social Systems and the Economic World* (1994); Kelly, "A Taxonomy of Minds" (2007); Kelly, "The Landscape of Possible Intelligences" (2008); Yonck, "Toward a standard metric of machine intelligence" (2012); and Yampolskiy, "The Universe of Minds" (2014). For critical perspectives on the notion of hive minds, see, e.g., Maguire & McGee (1999); Bendle, "Teleportation, cyborgs and the posthuman ideology" (2002); and Heylighen, "The Global Brain as a New Utopia" (2002).

1. The Hive Mind as Sensorimotor Network

It is possible for a neuroprosthesis to link the mind of its host with other intelligent agents to form a collection of intelligences in which the host maintains his or her sense of personal identity and autonomy and experiences the thoughts and volitions of the other intelligences as phenomena whose origins are external to the host's own mind and which are perceived through the host's sense organs. The host's mind interacts with the hive mind's other member-intelligences, while still recognizing them as independent entities that are part of the external environment and not part of the host's own mind.

2. The Hive Mind as Facilitator of Internal Dialogue

A neuroprosthetic device may link its host's mind to external intelligences in such a way that the host becomes consciously aware of and experiences those intelligences' volitions and agency as 'voices' speaking to the host from within his or her own mind. Instead of experiencing the internal monologue that is a natural part of human mental life, the host would experience an internal dialogue in which his or her own inner voice converses with the voices of other members of the hive mind.

3. The Hive Mind as Unitary Collective Intelligence

A neuroprosthesis might conceivably link its host's mind to the cognitive processes of one or more other intelligent agents (either those of the device itself or of external artificial or human agents) in such a way that the host's mind ceases to directly experience its own volition or personal identity as such – and does not experience the cognitive processes of external agents as belonging to those agents – but instead shares with the external agents the experience of jointly creating a single mind and will. Over time, such a host may experience a loss of individuality and sense of self by becoming immersed in the collective hive mind whose thoughts and actions are determined jointly by the cognitive activity of its members.[73] Depending on the nature of the device and its long-term effects on the neural structures and cognitive processes of its human host, it may or may not be possible to restore the host's mind to its full experience of independent agency simply by terminating the device's operation.

Conclusion

In earlier chapters we considered an ontology of the neuroprosthesis as a computing device and as a biocybernetic instrument that becomes integrated

---

[73] A related network topology would be that of a 'quasi-hive-mind' or 'pseudo-hive-mind,' in which the host *experiences* the hive mind as though it were being created through the joint action of all its participants, while in reality the contents of the hive mind's cognitive processes are largely determined or controlled by the actions of one of its members or an external system.

into the neural circuitry of a human organism in order to participate in processes of sensation, cognition, and motor action. In this chapter, we completed our classification and analysis of neurocybernetic technologies by developing an ontology of the neuroprosthesis as a means for the 'cyborgization' of human beings that shapes how individuals possessing such devices experience posthumanized digital-physical worlds. The ontology addressed four roles that a neuroprosthesis may play in such processes. First, a neuroprosthesis can serve as a means of human augmentation by transforming the cognitive and physical capacities possessed by its host. Second, it can determine the contents of information generated or utilized by its human host. Third, a neuroprosthesis can affect the manner in which its host inhabits a digital-physical body and external environment. And finally, a neuroprosthesis can regulate the autonomous agency possessed and experienced by its host.

It is hoped that the development of such an ontology will enable researchers to better understand the psychological, social, and ethical implications of such technologies and will allow the architects of neuroprosthetic systems – and of the digital-physical ecosystems within which they are situated – to design and manage such systems in a way that safeguards and advances the well-being of devices' human hosts while maximizing the security, vibrancy, and efficiency of the digital-physical ecosystems within which they are situated.

Part II

Enterprise Architecture for Neuroprosthetic Supersystems

Chapter Four

The Organizational Deployment of Posthumanizing Neuroprostheses:
Motivating and Inhibiting Factors for Military Organizations and Other Early Adopters

Abstract. This text examines the types of organizations that are already working to intentionally deploy neuroprosthetic technologies for human enhancement among their workforce (or are expected to do so), factors that affect their adoption of such technologies, and the organizational roles that such neurotechnologies may play.

The current state of therapeutic neuroprosthetic device use is presented, along with an overview of posthumanizing neuroprostheses and the types of enhanced capacities that they offer human workers that may be relevant to organizations. A range of factors incentivizing or discouraging the organizational deployment of posthumanizing neuroprostheses is identified and discussed. The organizational roles of therapeutic and posthumanizing neuroprostheses are then analyzed. Many organizations already unknowingly incorporate workers possessing therapeutic neuroprostheses, while two key paths for the organizational deployment of posthumanizing neuroprostheses are highlighted. First is the 'transitional augmentation' of human workers as a stopgap measure on the path to eventual full automation of business processes through the use of AI. The second path involves retaining human workers in particular positions because exogenous factors (such as legal, ethical, or marketing requirements) mandate that human agents fill them, while augmenting the workers so that they can perform more competitively.

It is noted that military organizations play a key role among organizations likely to be early adopters of posthumanizing neuroprostheses. Known and hypothesized military programs for neuroprosthetic enhancement are discussed, along with characteristics of military organizations that remove obstacles that render the deployment of neuroprostheses impractical for most organizations. Other types of organizations are highlighted that share some traits as potential early adopters. Finally, enterprise architecture (EA) is discussed as a preferred management tool for many organizations that are likely to be early adopters; while EA does not directly address the serious ethical and legal questions raised by posthumanizing neuroprostheses, it can facilitate the functional aspects of integrating neuroprosthetically augmented workers into an organization's personnel structures, business processes, and IT systems.

I. Introduction

To date, neuroprosthetic technologies have been employed primarily for therapeutic medical purposes – to restore some sensory, cognitive, or motor capacity that is present in typical human beings but which is absent in a particular individual due to illness or injury. In that context, decisions about whether to implant neuroprosthetic devices have been driven largely by the medical needs and personal motivations of the individual patients to whom they are being provided – not the operational needs or strategic objectives of the businesses or other organizations for whom such individuals might work.

However, a growing range of neuroprosthetic technologies is being developed whose purpose is not therapy but human enhancement: such 'posthumanizing' neuroprostheses are expected to provide their human hosts with sensory, cognitive, or motor capacities that dramatically surpass or differ from those that are possible for unmodified human beings. For most contemporary organizations, the notion of intentionally deploying such advanced neuroprostheses among their human workers is a far-fetched one: the purposeful exploitation of such technologies would appear to be more appropriate as a plot device in a work of science fiction than as an element of a serious business model. And, indeed, the implantation and maintenance of such devices creates not insignificant dangers to the health and safety of the human beings who receive them, as well as generating high costs and raising complex ethical and legal issues – all while offering extremely limited business value in return. Thus, for the overwhelming majority of organizations, the possibility of actively deploying neuroprosthetic technologies among their workforce does not currently possess strategic, operational, or tactical relevance.

Nevertheless, a small but growing number of specialized organizations – primarily military research agencies and departments[1] – are actively seeking to develop posthumanizing neuroprostheses and deploy them among their personnel. For these organizations, the possibility of improving their employees' ability to operate safely and effectively in very dangerous and difficult situations and to successfully accomplish critical tasks is perceived to outweigh the high costs and medical risks involved with the use of such technologies. Moreover, the nature of neuroprosthetic technologies is evolving rapidly as their power and sophistication increases: as posthumanizing neuro-

---

[1] Chief among them is DARPA, the Defense Advanced Research Projects Agency of the US Department of Defense; this organization does not maintain extensive in-house research and development capacities but primarily commissions outside entities such as university research laboratories or commercial firms to develop strategically, operationally, or tactically important technologies. For an overview of the organization's work relating to neurocybernetic human enhancement, see Part IV of this text and its discussion of military organizations as early adopters of posthumanizing neuroprostheses.

prostheses become less dangerous (e.g., through the use of non-invasive technologies) and less costly and offer clearer business value, it is anticipated that the range of organizations actively deploying such technologies will grow.

In this text, we identify factors that are expected to incentivize or discourage the acquisition and use of posthumanizing neuroprostheses and investigate the types of organizations that are likely to be 'early adopters' of such technologies. Taking into account the characteristics typical of such organizations, we argue that the discipline of enterprise architecture is likely to provide them with a useful approach to managing the creation and maintenance of organizational units whose members include neuroprosthetically augmented human personnel.

II. The Current State of Neuroprosthetic Device Use

The overwhelming majority of neuroprosthetic devices have been implanted for therapeutic medical purposes. In this section, we consider the nature of such devices, the number of therapeutic neuroprostheses in use, and key factors that limit the adoption of such technologies.

Basic Types of Neuroprosthetic Devices

A neuroprosthesis can be defined as *an artificial device that is integrated into the neural circuitry of a human being.*[2] Such devices typically participate in or directly support the sensory, cognitive, or motor processes of their human host,[3] however, they may also be employed, for example, to gather real-time data about their host's biological processes and transmit it to a hospital to allow round-the-clock patient monitoring and the remote administration of health care.[4] In principle, neuroprostheses may be either 'invasive' (i.e., surgically implanted in the brain of a human host) or 'non-invasive' (e.g., consisting of an external device worn by a human host); however, a number of obstacles currently make it difficult for non-invasive technologies to become truly integrated into the neural circuitry of a human being.[5] Thus, as the term

---

[2] See Lebedev, "Brain-Machine Interfaces: An Overview" (2014), and Gladden, "Enterprise Architecture for Neurocybernetically Augmented Organizational Systems" (2016).

[3] See Lebedev (2014).

[4] See, e.g., Lorence et al., "Transaction-Neutral Implanted Data Collection Interface as EMR Driver: A Model for Emerging Distributed Medical Technologies" (2009), and Gladden, "Managing the Ethical Dimensions of Brain-Computer Interfaces in eHealth: An SDLC-based Approach" (2016).

[5] See Gasson, "Human ICT Implants: From Restorative Application to Human Enhancement" (2012), p. 14, and Panoulas et al., "Brain-Computer Interface (BCI): Types, Processing Perspectives and Applications" (2010).

is used in this text, contemporary 'neuroprostheses' can generally be identified with invasive 'neural implants.'

Given their nature, neuroprostheses are not presently provided to employees by their employers in the way that an employer might furnish a desktop computer or smartphone for use by an employee in effectively completing work-related tasks. Instead, neuroprostheses are generally provided by healthcare organizations to patients experiencing a particular medical condition that can be treated through the use of such devices.[6] In these cases, neuroprostheses are generally employed to restore some capacity that is found in typical human beings but which is absent in a particular patient as a result of injury or illness. For example, cochlear implants, auditory brainstem implants, and retinal prostheses can restore sensory functionality to those who have lost the ability to hear or see; robotic prosthetic limbs can replace natural biological limbs that have been amputated; robotic exoskeletons and thought-controlled wheelchairs can grant a degree of mobility to the paralyzed; and deep brain stimulation (DBS) devices can treat tremors in those suffering from Parkinson's disease.[7]

Total Global Installed Base of Neuroprosthetic Devices and Rate of Growth

The adoption of therapeutic neuroprosthetic technologies has thus far been slow but steady. The best estimate is that more than one million neuroprosthetic devices have been implanted worldwide, including 350,000 spinal cord stimulation (SCS) devices,[8] 324,000 registered cochlear implants,[9] more than 135,000 DBS devices,[10] and 1,000 auditory brainstem implants.[11] It can thus be estimated that at present, roughly one out of every 7,000 people

---

[6] For production neuroprostheses that are in general use, organizations involved in the provision of such devices may include hospitals, insurance companies, and government healthcare agencies. In the case of experimental devices, commercial or university research laboratories may play a key role in the process. For an overview of such institutional actors, see Gladden, "Information Security Concerns as a Catalyst for the Development of Implantable Cognitive Neuroprostheses" (2016).

[7] See, e.g., Gasson et al., "Human ICT Implants: From Invasive to Pervasive" (2012); Ochsner et al., "Human, non-human, and beyond: cochlear implants in socio-technological environments" (2015); Weiland et al., "Retinal Prosthesis" (2005); Viola & Patrinos, "A Neuroprosthesis for Restoring Sight" (2007); and *Deep Brain Stimulation for Parkinson's Disease*, edited by Baltuch & Stern (2007).

[8] See "Chronic Pain and Spinal Cord Stimulation (SCS): Frequently Asked Questions" (2013).

[9] The figures are as of 2012. See "Cochlear Implants" (2016).

[10] See "Products and Procedures" (2016) and Patoine, "Progress Report 2010: Deep Brain Stimulation – The 2010 Progress Report on Brain Research" (2010).

[11] See "Surgeons Publish Study on Auditory Brainstem Implant Procedure" (2015).

worldwide has received a neuroprosthesis. It is also worth noting that – while such devices are not neuroprostheses – more than 10,000 human beings worldwide are estimated to have received implantable RFID chips intended to allow access to secure facilities, to support the safe and accurate provision of medical care, or for other purposes.[12]

More than 100,000 new neuroprosthetic devices are implanted each year, including up to 50,000 SCS neurostimulators,[13] 50,000 cochlear implants (including 30,000 for children),[14] and around 6,000 DBS devices.[15] The annual rate of growth in the number of DBS implantation surgeries has recently been greater than 20% per year.[16]

Factors Limiting Adoption of Therapeutic Neuroprostheses

A variety of factors combine to limit the number of individuals who have thus far acquired neuroprosthetic devices for therapeutic medical purposes.

High Costs

A major constraint limiting the number of neuroprostheses that can be implanted worldwide each year is the significant cost of implantation surgery. For example, cochlear implant surgery can cost between $40,000-$100,000 per person,[17] while DBS implantation surgery can cost between $35,000-$100,000 per recipient.[18]

Required Surgical Expertise

Another factor limiting the rate of adoption is the number of doctors and facilities capable of performing neuroprosthetic implantations. For example, even in countries where DBS implantation surgery is funded by government health agencies, the lack of trained surgeons qualified to perform such complex and risky operations may result in waiting lists of 2-3 years for eligible device recipients.[19]

---

[12] See Min Neo & Fonsegrives, "For these 'cyborgs,' keys are so yesterday" (2015).

[13] See "Spinal Cord Stimulation" (2008).

[14] See Hochmair, "Cochlear Implants: Facts" (2013), and "Cochlear Implant Quick Facts."

[15] See Christen & Müller, "Current status and future challenges of deep brain stimulation in Switzerland" (2012), p. 2, and "Studies Find Disparities in Use of Deep Brain Stimulation" (2014).

[16] Christen & Müller (2012), p. 2.

[17] See "Cochlear Implant Quick Facts."

[18] See Okun, "Parkinson's Disease: Guide to Deep Brain Stimulation Therapy" (2014).

[19] For example, see Fayerman, "Funding, doctors needed if brain stimulation surgery to expand in B.C." (2013).

Limited Population of Potential Users

Finally, given the fact that current neuroprostheses are generally used to treat particular medical conditions, the number of individuals suffering from such conditions worldwide creates an upper limit on the number of neuroprostheses that are likely to ever be implanted for purposes of treating those conditions. For example, it has been estimated that the number of new individuals experiencing hearing loss who could benefit from implantation of a cochlear implant is roughly 134,000 persons per year;[20] while the current implantation rate of roughly 50,000 new cochlear implants sold and implanted per year does not yet match the level of potential demand, it can be seen that already the commercial, governmental, and societal mechanisms for manufacturing cochlear implants, implanting them surgically, and funding their use are capable of satisfying a significant portion of the total global demand for such technologies.

III. An Overview of Posthumanizing Neuroprostheses

Some organizations have a direct interest in providing therapeutic neuroprostheses to their employees, insofar as the types of injuries that makes such devices necessary may occur regularly as a result of employees' work for the organization. Thus military forces have an immediate interest – deriving from their social responsibility as employers – in providing neuroprosthetic limbs to their soldiers who have lost a natural biological limb in combat.[21] Other organizations have a less direct interest in supplying therapeutic neuroprostheses to employees; they may ensure that organizationally sponsored insurance plans subsidize the cost of employees' therapeutic neuroprostheses (and other medical care) as a general means of enhancing employees' wellbeing and retaining highly qualified workers, even if such devices do not improve the employees' job performance and do not treat an illness or injury acquired as a result of the employees' work for the employers.

---

[20] See Hochmair (2013).

[21] Regarding DARPA's coordination and funding of research programs to develop sophisticated robotic neuroprosthetic limbs that can restore motor and sensory capacity to, e.g., soldiers who have lost limbs as a result of wounds suffered during their military service, see Dellon & Matsuoka, "Prosthetics, exoskeletons, and rehabilitation" (2007); Ling et al., "Surgical innovations arising from the Iraq and Afghanistan wars" (2010); and Hutchinson, "The quest for the bionic arm" (2014). For a broader discussion of the philosophical and even theological implications of such research programs, see Keenan, "Enhancing Prosthetics for Soldiers Returning from Combat with Disabilities" (2013). The US Department of Veterans Affairs has also provided millions of dollars to fund the development of 'biohybrid' prosthetic limbs for amputees; see Evans-Pughe, "Smarter Prosthetics: Researchers Are Working to Develop Artificial Limbs with Almost Life-like Levels of Functionality and Control" (2006).

The relationship of employers to their workers' neuroprostheses that are used for therapeutic medical purposes is fairly straightforward; it resembles employers' interest in supporting the use of other long-established non-neuroprosthetic medical technologies. However, a range of emerging neurotechnologies designed for purposes of *human enhancement* is now transforming the ways in which employers assess the nature – and potential business value – of neuroprostheses. Efforts are underway to develop and deploy a broad range of 'posthumanizing' neuroprostheses[22] whose purpose is not to restore some capacity that is found in typical human beings but to grant their human hosts sensory, cognitive, and motor capacities that far exceed or differ from those that are possible for natural biological human beings.[23]

By providing knowledge, skills, and proficiencies that are otherwise absent in the human workforce, such posthumanizing neuroprostheses can create a powerful competitive advantage for their human hosts – and for the organizations that employ them. In this section we explore in more detail the new or enhanced capacities that posthumanizing neuroprostheses may grant to human workers and the forms that such devices might take.

Neuroprosthetically Enabled Capacities of Potential Relevance for Organizations

The human host of a neuroprosthesis can be viewed on three levels: 1) as a sapient metavolitional agent, a unitary mind that possesses its own conscious awareness, memory, volition, and conscience (or 'metavolitionality'); 2) as an embodied organism that inhabits and can sense and manipulate a particular environment through the use of its body; and 3) as a social and economic actor who interacts with others to form social relationships and to produce, exchange, and consume goods and services. A posthumanizing neuroprosthesis may generate new capacities for its human host at any or all of

---

[22] 'Posthumanizing' technologies can be understood as those that bring about an ecosystem in which entities other than natural biological human beings exist as intelligent agents and social actors that create meaning in the world. Technologies relating to artificial intelligence, artificial life, virtual reality, genetic engineering, and neuroprosthetic augmentation are a catalyst for processes of posthumanization; however, non-technological forces of posthumanization also exist. For more details, see Ferrando, "Posthumanism, Transhumanism, Antihumanism, Metahumanism, and New Materialisms: Differences and Relations" (2013); Herbrechter, *Posthumanism: A Critical Analysis* (2013); Miah, "A Critical History of Posthumanism" (2008); Birnbacher, "Posthumanity, Transhumanism and Human Nature" (2008); and "A Typology of Posthumanism: A Framework for Differentiating Analytic, Synthetic, Theoretical, and Practical Posthumanisms" in Gladden, *Sapient Circuits and Digitalized Flesh: The Organization as Locus of Technological Posthumanization* (2016).

[23] See, e.g., McGee, "Bioelectronics and Implanted Devices" (2008); Warwick & Gasson, "Implantable Computing" (2008); Gasson (2012); and Gladden, "Neural Implants as Gateways to Digital-Physical Ecosystems and Posthuman Socioeconomic Interaction" (2016).

these three levels that may create value for the individual as a worker and for his or her employer.[24]

Impacts on the Neuroprosthetically Augmented Worker as Sapient Metavolitional Agent

Below we describe potential new cognitive capacities that a neuroprosthesis might create for its human host when he or she is viewed as a sapient metavolitional agent and which positively affect that individual's ability to interface with organizational information systems, carry out work-related tasks, and participate in broader socioeconomic interaction.

Enhanced memory, skills, and knowledge stored within the mind (engrams)

Building on current technologies tested in mice, future neuroprostheses may offer human users the ability to create, erase, or otherwise modify memories stored within their brains' natural biological memory systems in the form of engrams.[25] This could potentially be used not only to affect a host's declarative knowledge but also to enhance motor skills or reduce learned fears. More speculative (but not yet clearly ruled out theoretically) is the possible use of a neuroprosthesis that is closely integrated with the brain's organic neural network to provide supplemental storage space for memories which the brain will be able to retrieve and experience as though they were engrams stored within the brain itself rather than treating them as exograms stored within an external system.

Enhanced creativity

A neuroprosthetic device may be able to enhance a mind's powers of imagination and creativity[26] by facilitating processes that contribute to creativ-

---

[24] Not considered here in detail is the fact that a neuroprosthesis may also create for its human host significant new impairments, including: a loss of agency; loss of conscious awareness; loss of information security for the user's internal cognitive processes; an inability to distinguish a real experience from an ongoing virtual one; an inability to distinguish true from false memories; other psychological side-effects; a loss of control over sensory organs, motor organs, or other bodily systems; other biological side-effects (such as poisoning); a loss of ownership of one's own body and intellectual property (including thoughts and memories); the creation of financial, technological, or social dependencies; subjugation of the user to external agency; social exclusion and employment discrimination; and vulnerability to crimes such as data theft, blackmail, and extortion. For more details regarding such potential impairments as well as possible advantages arising from the use of posthumanizing neuroprostheses, see Gladden, "Neural Implants as Gateways" (2016), from which this section draws heavily.

[25] Experimental memory-altering technologies currently being tested in mice are described in Han et al., "Selective Erasure of a Fear Memory" (2009), and Ramirez et al., "Creating a False Memory in the Hippocampus" (2013). For a broader perspective, see McGee (2008) and Warwick, "The Cyborg Revolution" (2014), p. 267.

[26] See Gasson (2012), pp. 23-24.

ity, such as stimulating mental associations between unrelated items. Anecdotal increases in creativity have been reported to result after the use of neuroprosthetic devices for deep brain stimulation.[27]

Enhanced emotion

A neuroprosthetic device might provide its host with more desirable emotional dynamics.[28] The ability to affect the emotions of their users have already been seen, for example, in neuroprosthetic devices used for DBS.[29]

Enhanced conscious awareness

Funded in large part by military research agencies,[30] efforts are being undertaken to develop neuroprosthetics that would allow the human mind to, for example, extend its periods of attentiveness and limit the need for periodic reductions in consciousness (i.e., sleep).[31]

Enhanced conscience

One's conscience can be understood as one's set of metavolitions, or desires about the kinds of volitions that one wishes to possess.[32] Insofar as a neuroprosthesis enhances processes of memory and emotion that allow for the development of the conscience, it may enhance one's ability to develop, discern, and follow one's conscience.

Impacts on the Neuroprosthetically Augmented Worker as Embodied Embedded Organism Interacting with an Environment

Neuroprostheses may affect the ways in which their hosts sense, manipulate, and occupy their environment through the interface of a physical or virtual body. Potential new capacities that a neuroprosthesis might create for its host when he or she is viewed as an embodied embedded organism are described below.

Sensory enhancement

A neuroprosthesis may allow its host to sense his or her physical or virtual environment in new ways, either by acquiring new kinds of raw sense data or

---

[27] See Cosgrove, "Session 6: Neuroscience, Brain, and Behavior V: Deep Brain Stimulation" (2004), and Gasson (2012).

[28] See, e.g., McGee (2008), p. 217.

[29] See Kraemer, "Me, Myself and My Brain Implant: Deep Brain Stimulation Raises Questions of Personal Authenticity and Alienation" (2011).

[30] Regarding DARPA's multimillion-dollar investment in its Continuous Assisted Performance program, see, e.g., Falconer, "Defense Research Agency Seeks to Create Supersoldiers" (2003).

[31] See Kourany, "Human Enhancement: Making the Debate More Productive" (2013), pp. 992-93.

[32] See Gladden, "Neural Implants as Gateways" (2016), and Calverley, "Imagining a Non-biological Machine as a Legal Person" (2008).

new modes or abilities for processing, manipulating, and interpreting the sense data provided by natural sensory organs.[33]

Motor enhancement

A neuroprosthesis may give its host new ways of manipulating physical or virtual environments through his or her body.[34] It may grant enhanced control over one's existing biological body, expand one's body to incorporate new devices (such as an exoskeleton or vehicle) through body schema engineering,[35] or allow the user to control external networked physical systems such as drones or 3D printers or virtual systems or phenomena within an immersive cyberworld.

Enhanced memory, skills, and knowledge accessible through sensory organs (exograms)

A neuroprosthesis may give its host access to external data-storage sites whose contents can be 'played back' to the host's conscious awareness through his or her sensory organs or access to real-time streams of sense data that augment or replace the natural sense data received through one's biological sensory organs.[36] The ability to record and play back all of the sense data that one has received could provide perfect audiovisual memory of one's experiences.[37]

Impacts on the Neuroprosthetically Augmented Worker as Social and Economic Actor

Neuroprostheses may affect the ways in which their hosts connect to, participate in, contribute to, and are influenced by the kinds of social relationships and structures that make workplace collaboration possible and the kinds of economic networks and exchange that facilitate and are generated by organizational business models. Potential new capacities that a neuroprosthesis might create for its host when he or she is viewed as a social and economic actor are described below.

Ability to participate in new kinds of social relations

A neuroprosthesis may grant the ability to participate in new kinds of technologically mediated social relations and structures that were previously

---

[33] See Warwick (2014), p. 267; McGee (2008), p. 214; and Koops & Leenes, "Cheating with Implants: Implications of the Hidden Information Advantage of Bionic Ears and Eyes" (2012), pp. 120, 126.

[34] See McGee (2008), p. 213, and Warwick (2014), p. 266.

[35] Such possibilities are explored in Gladden, "Cybershells, Shapeshifting, and Neuroprosthetics: Video Games as Tools for Posthuman 'Body Schema (Re)Engineering'" (2015).

[36] See Koops & Leenes (2012), pp. 115, 120, 126.

[37] See McGee (2008), p. 217.

impossible, perhaps including new forms of merged agency[38] or cybernetic networks with utopian (or dystopian) characteristics.[39]

Ability to share collective knowledge, skills, and wisdom

Neuroprostheses may link their human hosts in a way that forms communication and information systems[40] that can generate greater collective knowledge, skills, and wisdom than are possessed by any individual member of the system.[41]

Enhanced job flexibility and instant retraining

By facilitating the creation, alteration, and deletion of information stored in engrams or exograms, a neuroprosthesis may allow its host to download new knowledge or skills needed to perform some work-related task or to instantly establish relationships for use in a new job.[42]

Enhanced ability to manage complex technological systems

By providing a direct interface to external computers and mediating its host's interaction with them,[43] a neuroprosthesis may grant an enhanced ability to manage complex technological systems – for example, for use within an organization in the production or provisioning of goods or services.[44]

Enhanced business decision-making

By performing data mining to uncover novel knowledge, executing other forms of data analysis, offering recommendations, and alerting the host to potential cognitive biases, a neuroprosthesis may enhance its host's ability to execute rapid and effective business-related decisions and transactions.[45]

---

[38] See McGee (2008), p. 216, and Koops & Leenes (2012), pp. 125, 132.

[39] For a discussion of such possibilities, see Gladden, "Utopias and Dystopias as Cybernetic Information Systems: Envisioning the Posthuman Neuropolity" (2015).

[40] Regarding such possibilities, see McGee (2008), p. 214; Koops & Leenes (2012), pp. 128-29; and Gasson (2012), p. 24.

[41] This fact regarding the generation and storage of information by social systems was raised in Wiener, *Cybernetics: Or Control and Communication in the Animal and the Machine* (1961), loc. 3070ff., 3149ff., and is discussed further in Gladden, "Utopias and Dystopias as Cybernetic Information Systems" (2015).

[42] See Koops & Leenes (2012), p. 126.

[43] See McGee (2008), p. 210.

[44] Such possibilities are discussed in McGee (2008), pp. 214-15, and Gladden, "Enterprise Architecture for Neurocybernetically Augmented Organizational Systems" (2016).

[45] See Koops & Leenes (2012), p. 119, and Gladden, "Enterprise Architecture for Neurocybernetically Augmented Organizational Systems" (2016).

Storage mechanism for monetary value

By storing cryptocurrency keys, a neuroprosthesis may allow its host to store money directly within his or her brain for use on demand in financial transactions.[46]

Qualifications for specific professions and roles

Neuroprostheses may initially provide persons with abilities that enhance job performance in particular fields[47] such as computer programming, art, architecture, music, economics, medicine, information science, e-sports, information security, law enforcement, and the military; as expectations for employees' neural integration into workplace systems grow, possession of neuroprosthetic devices may become a requirement for employment in some professions.[48]

Examples of Posthumanizing Neuroprostheses

Posthumanizing neuroprostheses may take a number of forms. In this section we present examples of such devices whose development is already being pursued or which has been foreseen by researchers.[49]

An Artificial Ear with Full Recording, Playback, and Streaming Capacity

Researchers anticipate the development of future cochlear implants whose internal computers possess the ability to record everything that their human host hears and to play back this audio on demand, effectively providing the host with perfect auditory memory.[50] Such devices could also potentially download and play music, podcasts, or other audio content for their host or receive live audio through radio transmissions or the Internet.[51] By wirelessly transmitting a live stream of all the environmental audio input that

---

[46] Such applications of neuroprostheses are discussed in Gladden, "Cryptocurrency with a Conscience: Using Artificial Intelligence to Develop Money That Advances Human Ethical Values" (2015).

[47] See Koops & Leenes (2012), pp. 131-32.

[48] See McGee (2008), pp. 211, 214-15, and Warwick (2014), p. 269.

[49] For a discussion of such devices in the context of their potential organizational use, see Gladden, "Enterprise Architecture for Neurocybernetically Augmented Organizational Systems" (2016), from which this section draws much of its material.

[50] Such devices (or technologies that are expected to facilitate the development of such devices) are discussed, e.g., in Merkel et al., "Central Neural Prostheses" (2007); Robinett, "The consequences of fully understanding the brain" (2002); Koops & Leenes (2012); McGee (2008); and Gladden, "Enterprise Architecture for Neurocybernetically Augmented Organizational Systems" (2016).

[51] Koops & Leenes (2012), pp. 115, 120, 126; Gladden, "Enterprise Architecture for Neurocybernetically Augmented Organizational Systems" (2016).

it detects, such a device could allow others listening around the world to experience all that the host was hearing in real time.

An Artificial Ear with Superhuman Types of Perception

An artificial ear could also allow its host to perceive sounds whose frequencies fall outside of the range that is normally perceptible for human beings. For example, a cochlear implant could detect sounds that possess ultrasonic frequencies but utilize an algorithm to present such sounds as ones possessing frequencies that fall just within the range perceptible to human beings when transmitting signals to the cochlear nerve or brain of its human host.

Moreover, by employing a process of sensory substitution,[52] an implant could detect any number of environmental phenomena and convert them into particular patterns of auditory data, thereby allowing its host to 'hear' light, colors, smells, temperature, radiation levels, or other characteristics of the host's environment. Similarly, an implant could be employed to allow its host to 'hear' his or her current blood pressure, blood glucose level and other blood chemistry characteristics, body temperature, and the status of other internal biological processes.

An Artificial Eye with Full Recording, Playback, and Streaming Capacity

An advanced retinal prosthesis might be able to record, remotely store (via wireless transmission), and play back on demand everything that its host sees, thereby effectively granting the host perfect visual memory. Such an eye could potentially also allow its host to download and view films and other visual content without other people in the immediate vicinity realizing it, and by supplying live video from a remote camera the device's host could virtually 'inhabit' the environment in which the camera is located.[53] Conversely, by wirelessly transmitting a live stream of the video recorded by its camera, the implant could allow viewers around the world to experience the host's environment from his or her perspective.

An Artificial Eye with Superhuman Types of Perception

Future successors to contemporary retinal prostheses may give human beings the capacity to sense and interpret their environments in new ways, for

---

[52] As noted previously, technologies for sensory substitution were suggested as early as in Wiener (1961), loc. 2784ff. See also their discussion in the context of posthumanizing neuroprostheses in Gladden, "Cybershells, Shapeshifting, and Neuroprosthetics" (2015).

[53] For discussion of the InfoSec implications of the use of such technologies by an organization's senior personnel, see Gladden, "Implantable Computers and Information Security: A Managerial Perspective" (2016).

example through telescopic zoom or night vision functionality[54] or by using augmented reality displays to overlay the external visual data perceived by the implant's camera with supplemental information about the host's environment or internal biological processes, the contents of incoming text or video messages, or other information.

An Implantable Smartphone

An implantable neuroprosthesis with functionality similar to that of a miniaturized smartphone might offer its host wireless communication capacities including access to the Internet, GPS functions, and cloud-based software and data-storage services.[55] Depending on its level of sophistication, such a device might respond to voice or thought commands and present visual or auditory information to its host's mind by means of sensory neuroprosthetic components similar to those found in cochlear implants, auditory brainstem implants, or retinal prostheses.

An Implanted Controller for Vehicles, Games, and Other Systems

A neuroprosthesis might provide its host with a direct link to external systems such as desktop computers, vehicles, smart buildings, 3D printers and other manufacturing systems, domestic robots, microphones and speakers, cameras and displays, game systems, or systems within a virtual environment that allows such external devices to be controlled by the host's thoughts.[56]

A Thought-controlled Exoskeleton or Physical Cyborg Body

A neuroprosthesis might allow its host to control via his or her thoughts an external articulated exoskeleton or even an electromechanical cyborg body that largely (i.e., apart from the brain) replaces the host's natural biological body. Such a body could potentially be radically nonhuman in its form and functionality.[57]

An Implantable Memory Chip for Acquiring New Skills or Knowledge

Building on experimental technologies developed for creating and altering memories in mice,[58] future implantable mnemoprostheses for human beings

[54] See Gasson et al. (2012); Merkel et al. (2007); and Gladden, "Enterprise Architecture for Neurocybernetically Augmented Organizational Systems" (2016).
[55] Such a device is discussed in Gladden, "Enterprise Architecture for Neurocybernetically Augmented Organizational Systems" (2016).
[56] See, e.g., McGee (2008), pp. 213-15; Gladden, "Neural Implants as Gateways" (2016); and Warwick (2014), p. 266.
[57] The extent to which neuroprostheses may allow the adoption of nonhuman bodies is explored in Gladden, "Cybershells, Shapeshifting, and Neuroprosthetics" (2015).
[58] As noted previously, such technologies are described in Han et al. (2009) and Ramirez et al.

might take a number of forms – from that of a chip that allows its host to download content in order to acquire new knowledge or skills[59] to that of an ingestible 'knowledge pill' whose contents (perhaps a web-enabled nano-robot swarm[60]) travel to the brain, where they stimulate neurons to modify or create memories.[61]

Neural Scaffolding for Enhanced Cognitive Processing and Expanded Storage

Instead of producing or regulating specific behaviors, some neuroprosthetic devices might simply comprise a large set of artificial neurons that are connected to the natural biological neural circuitry of their host's brain in order to provide additional raw, general-purpose neural capacity that can be used, for example, for storing short- or long-term memories or processing information.[62]

A 'Savant Mode' Enabler

A neuroprosthesis may temporarily grant its host the ability to perform extraordinary mental feats of calculation or other savant skills by, for example, disrupting the functioning of the left anterior temporal lobe by means of magnetic stimulation in order to provide the host conscious access to and control over low-level cognitive processes that are normally hidden from the mind's conscious awareness.[63]

An Attention Booster

A neuroprosthesis may be able to enhance its host's conscious awareness and attentiveness by, for example, reducing the brain's need for sleep.[64] Such

---

(2013).

[59] Such possibilities are noted in McGee (2008).

[60] See Pearce, "The Biointelligence Explosion" (2012).

[61] See Spohrer, "NBICS (Nano-Bio-Info-Cogno-Socio) Convergence to Improve Human Performance: Opportunities and Challenges" (2002); and Gladden, "Enterprise Architecture for Neurocybernetically Augmented Organizational Systems" (2016).

[62] The extent to which such a device is theoretically possible is not yet clear and depends, e.g., on whether or not certain holographic models of memory are correct. For a discussion of such questions, see, e.g., Longuet-Higgins, "Holographic Model of Temporal Recall" (1968), and Pribram, "Prolegomenon for a Holonomic Brain Theory" (1990).

[63] The nature of savant skills is discussed in Treffert, "The savant syndrome: an extraordinary condition. A synopsis: past, present, future" (2009), and Rodriguez, *Autism Spectrum Disorders* (2011), pp. 36-39. The use of techniques such as transcranial magnetic stimulation (TMS) to artificially induce temporary savant skills is discussed in Snyder et al., "Savant-like skills exposed in normal people by suppressing the left fronto-temporal lobe" (2003), and Snyder, "Explaining and inducing savant skills: privileged access to lower level, less-processed information" (2009).

[64] See Kourany (2013), pp. 992-93.

technologies might potentially allow soldiers operating in hostile territory or the solo pilots of aircraft or watercraft to function for days without sleep.

An Emotional Regulator

A neuroprosthetic device might be able to regulate its host's moods, emotions, feelings, and desires in order to reduce or eliminate unwanted emotional phenomena and produce or strengthen desirable emotional behaviors.[65]

An Internal Advisor and Personal Consultant

A neuroprosthesis may be able to serve as a strategic consultant, business advisor, concierge, or even supplemental 'conscience' for its host – for example, by researching options and providing information through an augmented reality display to support decision-making or by warning its host when he or she is about to make a flawed decision that is influenced by some human cognitive bias.[66]

A Portal to Long-term Inhabitation of a Virtual World

A neuroprosthetic device could replace all of its host's sense data with real-time data depicting a virtual world (and the user's interactions with it), while the host's real physical body is maintained in a healthy state for an extended period of time in an artificial life-support system. Within such a fabricated virtual environment, the host might be given a radically nonhuman body and the laws of physics and biology may or may not apply, depending on the creative decisions made by the world's designer.[67]

A Hive Mind

Neuroprostheses may link the brains of multiple human beings in a manner that enables direct instantaneous communication between them,[68] the sharing of information, and the development of collective thoughts, desires, and decisions in a way that reduces the agency of each individual participating mind while simultaneously creating a form of combined agency and a

---

[65] Regarding such possibilities, see McGee (2008), p. 217; Kraemer (2011); and Gladden, "Neural Implants as Gateways" (2016).

[66] Such a device is discussed in Gladden, "Enterprise Architecture for Neurocybernetically Augmented Organizational Systems" (2016).

[67] The implications of long-term immersion in VR environments are discussed, e.g., in Bainbridge, *The Virtual Future* (2011); Heim, *The Metaphysics of Virtual Reality* (1993); Koltko-Rivera, "The potential societal impact of virtual reality" (2005); and Gladden, "Cybershells, Shapeshifting, and Neuroprosthetics" (2015).

[68] Such possibilities are discussed from different perspectives in Rao et al., "A direct brain-to-brain interface in humans" (2014), and Gladden, "Utopias and Dystopias as Cybernetic Information Systems" (2015).

collective mind whose knowledge, skills, and wisdom may exceed those of individual members of the system.[69]

IV. Factors Incentivizing Organizational Deployment of Posthumanizing Neuroprostheses

In addition to providing a competitive advantage to individual workers, there are a number of ways in which posthumanizing neuroprostheses might grant a distinct overall benefit to organizations that purposefully deploy such technologies. In this section we consider a few of these factors that may create an incentive for the organizational adoption of posthumanizing neuroprostheses.[70]

Neuroprostheses Provide Tools for Gathering Information

The information needed to successfully manage an organization includes not only easily quantifiable data relating to IT performance, financial transactions, production processes, or resource consumption but also more elusive qualitative data. In particular, the amount of cultural and social knowledge (much of which is hidden within the thoughts, beliefs, and behaviors of employees and other stakeholders) that must be captured in order to design and manage an enterprise is immense and is not always fully appreciated even by enterprise architects.[71] If an organization has access to the sensory, cognitive, and motor data recorded or generated by neuroprostheses implanted in its employees, each of those individuals becomes a suite of sensors that can be used to gather real-time data about such phenomena and allow live monitoring of ongoing organizational change.

For example, in-depth analysis of the social networks existing within an organization facilitates the development of an effective enterprise architecture for it.[72] An organization that includes neuroprosthetically augmented employees could gather information about its social networks by collecting and analyzing communications and other data from the employees' neuroprostheses; this may reveal actual network topologies and pathways of communication within the organization that differ greatly from the formal reporting relationships described in an official organizational chart. Indeed, it may

---

[69] For different aspects of this issue, see McGee (2008), p. 216; Koops & Leenes (2012), pp. 125, 132; Wiener (1961), loc. 3070ff., 3149ff.; and Gladden, "Utopias and Dystopias as Cybernetic Information Systems" (2015).

[70] For further analysis of these factors, see Gladden, "Enterprise Architecture for Neurocybernetically Augmented Organizational Systems" (2016), from which this section draws extensively.

[71] See Liu et al., "A Design of Business-Technology Alignment Consulting Framework" (2011).

[72] See Kazienko et al., "Social Network Analysis as a Tool for Improving Enterprise Architecture" (2011).

be possible to configure neuroprostheses to automatically perform the tracing and (re)construction of organizational social networks; some neuroprosthetic devices (such as those designed to enable massively multiuser virtual environments or hive minds) may include such functionality as a core feature.

Neuroprostheses Facilitate the Generation of Organizational Self-awareness

A particular aspect of information-gathering within an organization is its role in creating organizational self-awareness. Caetano et al. argue that in order to create and maintain alignment between an organization's electronic information systems, human resources, business processes, workplace culture, mission and strategy, and external ecosystem, the organization must be self-aware; this requires "knowledge to be shared and understood in such a way that the mismatch between the actual state of affairs and the state as perceived by its different organizational stakeholders is continuously minimized."[73] While organizations are not 'alive' in the same sense as biological organisms and are not conscious or sapient in the way that a typical adult human being is, an organization does display at least some degree of implicit functional self-awareness, which is necessary for the that entity to organize its internal structures and processes, regulate the 'metabolism' reflected in its component activities, and maintain homeostasis.[74] Such activities are supported by the organization's assimilation of real-time feedback and information-gathering to create and maintain an internal representation of the current and target state of all the organizational elements that must be managed and aligned;[75] the deployment of posthumanizing neuroprostheses could facilitate that work of gathering and processing information. Such devices might also give an organization's employees unlimited real-time access

---

[73] Caetano et al., "A Role-Based Enterprise Architecture Framework" (2009), p. 253.

[74] In formulating a definition of life that could be employed to determine whether artificial entities (such as robots and AIs) are 'alive,' Friedenberg (*Artificial Psychology: The Quest for What It Means to Be Human* (2008), pp. 201-03), draws on the criteria for biological life presented in Curtis, *Biology* (1983): namely, a living being manifests organization, metabolism, growth, homeostasis, adaptation, response to stimuli, and reproduction. See Gladden, "The Artificial Life-Form as Entrepreneur: Synthetic Organism-Enterprises and the Reconceptualization of Business" (2014), for a discussion of the application of such criteria to businesses and other human organizations understood as viable systems.

[75] This is one of the key practices of enterprise architecture. See Caetano et al. (2009), p. 253. The role of internal models and representations is highlighted in the field of cybernetics in Conant and Ashby's Good-Regulator Theorem. See Conant & Ashby, "Every Good Regulator of a System Must Be a Model of That System" (1970), and elaborations of such thought within the field of management cybernetics in texts such as Beer, *Brain of the Firm* (1981); Barile et al.. "An Introduction to the Viable Systems Approach and Its Contribution to Marketing" (2012); and Buckl et al., "A Viable System Perspective on Enterprise Architecture Management" (2009).

to the organization's centrally maintained (and continuously updated) internal representation of its full range of structures and dynamics, thereby allowing employees to more effectively identify opportunities for increasing organizational alignment or reducing misalignments and for carrying out actions to generate such improvements.

Neuroprostheses Allow the Creation of New Organizational Forms

During recent decades, new forms of information and communications technology (ICT) have enabled the development of novel organizational structures that were previously impossible or impractical. For example, email, videoconferencing, instant messaging, online commerce platforms, cloud computing (including tools for collaborative online document editing), and mobile apps have facilitated the creation of new types of organizational forms such as virtual organizations.[76] Insofar as posthumanizing neuroprostheses allow new forms of thought, physical action, communication, and collaboration to exist among an organization's workers, they are expected to enable the creation of new types of organizational forms and structures (such as hive minds capable of real-time, spatially dispersed, collective decision-making) that will both enable and require new types of organizational architectures that may provide competitive advantages over organizations employing more traditional architectures.

Neuroprostheses Can Enhance Information Security and Cyberwarfare Capacity

Few organizations require – and can legally and ethically justify – the possession of offensive cyberwarfare capacities used to proactively disable or destroy external computer systems perceived to pose a threat. National military agencies may desire such capacities in order to preemptively disrupt impending attacks by state or non-state actors that threaten to destroy critical infrastructure or imperil the health and safety of their nation's residents, while police agencies may desire such capacities in order to disrupt criminal conspiracies and apprehend suspected criminals.[77] Commercial organizations

---

[76] Various aspects of virtual organizations are discussed in Fairchild, *Technological Aspects of Virtual Organizations: Enabling the Intelligent Enterprise* (2004); *Virtual Organizations: Systems and Practices*, edited by Camarinha-Matos et al. (2005); and Shekhar, *Managing the Reality of Virtual Organizations* (2016).

[77] Regarding potential uses of offensive cybersecurity and cyberwarfare capacities by police and other security agencies in order to prevent various types of criminal activity (including cyberterrorism), see, e.g., Maitra, "Offensive cyber-weapons: technical, legal, and strategic aspects" (2015), and Leed, *Offensive Cyber Capabilities at the Operational Level: The Way Ahead* (2013).

such as large corporations may maintain competitive intelligence (CI) operations that exploit competitors' information security weaknesses to obtain information about their business operations and plans through lawful – if ethically questionable – means.[78] However, the legal ability of private organizations to actively compromise and disable threats such as botnets (which may involve thousands of computers belonging to entirely innocent third parties) is limited.[79] For an organization that does conduct legally and ethically permissible offensive cyberwarfare or cybersecurity operations, the deployment of posthumanizing neuroprostheses may significantly enhance the organization's capacity in those areas by merging in each host-device unit the speed, power, precision, and scope of automated cybersecurity technologies with the ability of a human agent possessing appropriate decision-making authority and legal responsibility to monitor, authorize, and guide such actions in real time.

For a broader range of organizations, the deployment of posthumanizing neuroprostheses might serve a legally and ethically permissible defensive role in enhancing information security by contributing to the prevention, detection, and mitigation of cybersecurity threats and vulnerabilities. The use of such technologies could potentially expand an organization's capacity for gathering real-time data about its processes and systems, analyzing that data to identify threats and vulnerabilities, and addressing risks by protecting personnel against social engineering, hardening electronic systems, establishing redundancy and failover capacity, creating honeynets, and executing other standard InfoSec practices.[80]

V. Factors Discouraging Organizational Deployment of Posthumanizing Neuroprostheses

While posthumanizing neuroprostheses have the potential to create competitive advantages for the individuals who use them and the organizations in which such persons are employed, such technologies also present serious

---

[78] For discussions of ethical issues inherent in the acquisition and use of competitive intelligence, see Collins & Schultz, "A review of ethics for competitive intelligence activities" (1996); Comai, "Global code of ethics and competitive intelligence purposes: an ethical perspective on competitors" (2003); Crane, "In the company of spies: When competitive intelligence gathering becomes industrial espionage" (2005); and Giustozzi et al., "The new competitive intelligence agents: 'Programming' competitive intelligence ethics into corporate cultures" (2011).

[79] For a discussion of such issues, see, e.g., Leder et al., "Proactive Botnet Countermeasures: An Offensive Approach" (2009).

[80] An overview of such information security practices can be found in *NIST Special Publication 800-53, Revision 4: Security and Privacy Controls for Federal Information Systems and Organizations* (2013) and Rao & Nayak, *The InfoSec Handbook* (2014).

disadvantages, problems, and dangers. Below are listed some factors that may discourage or preclude the organizational deployment of posthumanizing neuroprostheses.

Biomedical Factors

- The risk of a host's injury or death during implantation surgery and the risk of psychological or physical harm resulting from later use of the device.[81]
- A lack of the specialized biomedical expertise needed to safely operate and maintain such devices.
- An inability to monitor device hosts' medical status in real time (e.g., because the hosts operate in remote environments outside of communication range) in order to ensure ongoing device safety.

Business and Operational Factors

- The time and effort needed to train new hosts in the safe and effective use of their neuroprostheses, which may divert their attention from other work activities and result in a loss of related knowledge or skills.
- The failure of posthumanizing neuroprostheses to add substantial value to key organizational activities.
- The ability to successfully achieve business objectives and fulfill an organization's mission without the use of such complex neurotechnologies.
- A lack of the specialized neurocybernetic expertise needed to identify meaningful business applications for such technologies.
- Marketing and public relations concerns regarding the potential negative impact of neuroprosthetic deployment on an organization's brand image – for example, in cases in which a firm's use of neuroprostheses might be perceived as exploitative, unnatural, or otherwise 'sinister.'[82]

---

[81] Even relatively 'simple' implantation surgery for devices such as passive RFID chips involves risks; more complex invasive surgery for implantation of devices into the brain is even more dangerous. See, e.g., Rotter et al., "Passive Human ICT Implants: Risks and Possible Solutions" (2012), and Clausen, "Conceptual and Ethical Issues with Brain–hardware Interfaces" (2011), p. 499.

[82] In recent decades, the complex dynamics regarding public acceptance or active rejection of emerging technologies have been clearly witnessed and studied, for example, in the case of genetically modified organisms (GMOs) and their use as food for human beings; see, e.g., Frewer et al., "Societal aspects of genetically modified foods" (2004); Frewer et al., "Public perceptions

- The creation of new information security vulnerabilities arising from the use of such devices by organizational personnel.

Financial Factors

- The high initial costs of acquiring neuroprosthetic devices and surgically implanting them in their human hosts.[83]

- The high and uncertain long-term costs of maintaining implanted neuroprostheses and ensuring the health of their hosts (perhaps for decades, throughout the remainder of their natural lifespan).

Legal and Ethical Factors

- The potential existence of national or local laws and regulations or international treaties that explicitly ban the use of certain types of posthumanizing neurotechnologies or which create doubt regarding the legality of their use.

- Legal concerns regarding the propriety of an employer requiring, encouraging, or enabling employees to acquire neuroprostheses for work-related purposes.

- Legal uncertainties about whether intellectual property that is produced by employees (e.g., during their own free time) with the aid of employer-provided neuroprostheses is owned by the workers' employer, the workers themselves, or third-party firms that produced and maintain the devices.

- The creation of financial and legal liability for an organization for any future accidents or illicit behavior involving neuroprostheses that it has provided to its employees.

- Ethical concerns regarding the moral permissibility of implanting devices that alter basic human capacities and the creation of a workplace environment in which employees feel compelled to submit to

---

of agri-food applications of genetic modification – A systematic review and meta-analysis" (2013); and Patel et al., "Genetic engineering in agriculture and corporate engineering in public debate: risk, public relations, and public debate over genetically modified crops" (2005). More recently, such debates have also begun to emerge regarding technologies such as self-driving vehicles, caregiving social robots, and military robots. See, e.g., Woisetschläger, "Consumer Perceptions of Automated Driving Technologies: An Examination of Use Cases and Branding Strategies" (2016); De Graaf, & Ben Allouch, "Exploring influencing variables for the acceptance of social robots" (2013); and Royakkers & Van Est, "A literature review on new robotics: automation from love to war" (2015).

[83] As noted above in Part II's discussion of factors limiting the adoption of therapeutic neuroprostheses, the implantation surgery for such a device may cost up to $100,000.

such augmentation in order to preserve their jobs or receive promotions.

Tools such as a PESTLE analysis[84] may be employed to identify and weigh the risk that future changes to an organization's political, economic, social, technological, legal, and environmental context might render the use of implanted neuroprostheses undesirable or even untenable.

VI. Organizational Roles of Therapeutic vs. Posthumanizing Neuroprostheses

In this section, we consider the difference between the presence of therapeutic neuroprostheses among an organization's workforce and the organization's intentional deployment of posthumanizing neuroprostheses.

Organizations Already Unknowingly Include Neuroprosthetically Augmented Workers

The impacts of neuroprosthetic devices on organizations are most visible and significant when an organization intentionally deploys among its human workers neuroprostheses that are explicitly incorporated into the organization's enterprise architecture as elements of institutional information systems. However, such occurrences are not yet commonplace; far more frequent is the introduction of neuroprostheses into the workplace by individual workers who have acquired such devices of their own volition for their own personal reasons (e.g., as therapeutic devices to treat particular medical conditions). Today, such devices includes cochlear implants, auditory brainstem implants, retinal prostheses, or deep brain stimulation devices; in the future, they might also include new types of cognitive neuroprostheses such as implantable neural bridges intended to restore memory function in those suffering from hippocampal damage.[85]

Such personally acquired neuroprosthetic devices are already present in countless organizations, as it is estimated that more than one million neuroprosthetic devices have been implanted in human beings worldwide,[86] and some of the individuals possessing such devices are employees of organizations into whose workplace they bring their neuroprostheses on a daily basis. Thus many organizations already possess a workforce that includes neuro-

---

[84] See Cadle et al., *Business Analysis Techniques: 72 Essential Tools for Success* (2010), pp. 3-6, for a description of common variations on this analytic tool.

[85] Work on such a hippocampal mnemoprosthetic implant is described in Soussou & Berger, "Cognitive and Emotional Neuroprostheses" (2008).

[86] A more detailed breakdown of this estimate is found in Part II above, in the discussion of the total global installed base of neuroprosthetic devices and rate of growth.

prosthetically equipped individuals; however, the presence of such neuro-prostheses may not be known to those organizations. Indeed, medical privacy and employment discrimination laws may make it illegal for an organization to even attempt to discover which of its employees (if any) possess such neuroprostheses.

Therapeutic Neuroprostheses Do Not Directly Affect Organizational Architectures

While such therapeutic neuroprostheses can have a tremendous positive impact on the lives of the individuals who possess them, their existence does not directly affect the organizational architectures of the organizations within which such persons are employed, as there is typically little incentive (and few practical means) for such devices to become explicitly integrated into the organizations' personnel structures, business processes, and electronic information systems and IT infrastructure. There are indeed cases in which an employer is aware of the fact that its employees possess specific therapeutic neuroprostheses: for example, a military department that supplies robotic prosthetic limbs to its soldiers who have been injured during their tours of duty will be aware of the fact that they possess such therapeutic neuroprostheses, and the devices may – at some level – become incorporated into the organization's information systems (e.g., if the organization's own medical personnel remotely monitor the devices' functioning and provide ongoing maintenance services). To that extent, such neuroprostheses would indeed become components of an organization's enterprise architecture plan and might force changes to some of the elements governed by such plans (e.g., to safeguard the information security of such devices and their hosts). However, the possession of such therapeutic neuroprostheses does not provide their human users with sensory, cognitive, or motor capacities that significantly differ from those of typical employees, and thus the organization will generally not need to dramatically revise its enterprise architecture plans in order to account for the presence of such technologies.[87]

---

[87] Some changes to an enterprise architecture might be required, e.g., in the sphere of information security, to address the unique InfoSec vulnerabilities possessed by neuroprosthetic devices. Hoogervorst identifies information security as a component of the element of 'quality' within the information architecture domain. Other EA frameworks, such as the Siemens Framework described by Rohloff, do not explicitly identify information security as a domain or building block – not because it is unimportant, but because it underlies *all* of the architectural building blocks. See Hoogervorst, "Enterprise Architecture: Enabling Integration, Agility and Change" (2004), and Rohloff, "Framework and Reference for Architecture Design" (2008). For the unique InfoSec challenges presented by neuroprosthetic devices, see Denning et al., "Neurosecurity: Security and Privacy for Neural Devices" (2009), and Gladden, *The Handbook of Information Security for Advanced Neuroprosthetics* (2015).

Two Models for Organizations' Deployment of Posthumanizing Neuroprostheses

There are at least two general types of reasons why an organization might wish to intentionally deploy among its personnel posthumanizing neuroprostheses that enhance the capacities of such personnel or integrate them more intimately into the organization's business processes and IT infrastructure; each of these reasons is likely to yield its own model or path for deployment.[88]

'Transitional Augmentation' as a Stopgap Measure on the Path to Full Automation via AI

First, there may be cases in which posthumanizing neuroprostheses are deployed as a stopgap measure because an organization wishes to eventually remove human workers from the equation and fully automate some process. As soon as a robot or other artificially intelligent system can perform the job better than a human worker, the organization will replace the human worker with the robot; however, sufficiently sophisticated AI does not yet exist that can handle the particular task with the necessary degree of effectiveness, efficiency, and reliability and at a competitive cost. And yet, neither is it possible (or possible *any longer*) for an unmodified human worker to do the job that needs to be done in a satisfactory and competitive manner. In some such cases, augmenting a human worker with neuroprosthetically enhanced sensory, cognitive, and motor capacities may enable him or her to perform the work at a more competitive level. Through the process of neuroprosthetic augmentation, the core structures and functionality of the human worker's brain are being retained, because they provide the worker with the wisdom, good judgment, and experience needed to make complex work-related decisions – but the rest of the brain and body *surrounding* those critical neural structures is upgraded in order to enhance the person's capacities.

In essence, the core of the worker's brain would be used as the 'CPU' of a hybrid biological-electronic system while the organization waits for more capable AIs to be developed that can match the human brain's performance in the full range of necessary activities; neuroprostheses constitute new 'upgrades' or 'peripherals' that are added to the brain to expand or improve its range of capacities. In these sorts of organizational roles, the use of posthumanizing neuroprostheses is expected to be only temporary. They represent a form of 'transitional augmentation' that lies on the path to an eventual full

---

[88] Here we do not directly address the legal or ethical permissibility or propriety of these activities; indeed, such intentional deployment of posthumanizing neuroprostheses poses complex legal issues and grave moral and ethical problems that must be carefully considered. The discussion offered here highlights the fact that such legal and ethical analyses are urgently needed, insofar as operational factors exist that provide at least some organizations with a strong business incentive to consider exploiting such emerging neurotechnologies.

automation to be achieved through the use of robots and AIs. Yesterday, a natural biological human worker could adequately perform such tasks; tomorrow, an AI will perform them; in the meantime, a neuroprosthetically enhanced human worker is perceived to offer the most desirable mix of human rationality and flexibility and computerized speed, scope, and precision.

Augmentation of Personnel Where Exogenous Factors Mandate Involvement of Human Agency

Even if a robotic agent exists which – from a purely functional perspective – is able to perform a particular set of tasks better than a natural biological or neuroprosthetically augmented human being, there may be factors that effectively bar an organization from assigning a robotic agent to fill that position. For example, the laws and regulations governing corporations in a particular country may explicitly mandate (or implicitly assume) that corporate offices such as president and CEO be filled by particular human beings who must regularly submit forms to the government documenting their organization's legal, financial, and operational status and that those individuals bear legal responsibility for the accuracy and completeness of the documents that they sign and submit. Under currently existing legal regimes, robotic agents would be excluded from filling such organizational roles, insofar as they are not human legal persons who can, for example, be prosecuted and punished for perjury or fraud.

Similarly, there may be many jobs within professions such as law, medicine, accounting, health care, and education which – for reasons relating to regulation and licensing – can only be filled by human beings possessing particular professional qualifications. Many positions within military or police organizations may be restricted to human beings due to the fact that the agents holding such positions must make life-and-death decisions, and it is considered essential that human judgment – as flawed as it may be – bear ultimate responsibility for such choices. In other cases – such as work involving artistic creativity and performance, spirituality, or athletic competition – there may be no legal requirements that mandate the use of human agents in such roles; however, from a marketing perspective there may be a strong incentive for an organization to employ human rather than robotic agents, insofar as such work is perceived to involve a distinctly 'human' element, and the use of artificial agents in such roles would defeat the purpose for which the activities are undertaken.

In all of these situations, while legal, ethical, business, or cultural factors block organizations from employing *wholly* artificial robotic agents to perform such tasks, organizations might wish to exploit the gains in productivity and efficiency that can be achieved by employing at least *partially* artificial hybrid human-robotic agents. The neuroprosthetic incorporation of electronic, robotic, artificially intelligent elements into a human worker may represent a sort of compromise between competing interests: a robot cannot be handed the position in its own right, but by incorporating robotic capacities into the human worker who nominally holds the position, it allows the organization to ostensibly meet its legal and ethical obligations and employ humanistic marketing and advertising strategies, while simultaneously obtaining many or all of the perceived benefits of robotic and artificially intelligent workers. For natural biological human workers who already fill important roles within an organization, the acquisition of posthumanizing neuroprostheses may allow them to fill those roles in in a more effective, efficient, and reliable manner. In contrast to cases of 'transitional augmentation,' the use of neuroprosthetically augmented human workers to fill these positions might persist even after AIs exist that can perform the jobs better than human beings from a functional perspective. An eventual move to the use of robotic or artificially intelligent agents in such roles would depend on legal and cultural shifts that may occur only slowly, if at all.

VII. Military Organizations as Early Adopters of Posthumanizing Neuroprostheses

The intentional deployment of neuroprostheses within the workforce is not a pursuit that is necessary, feasible, or appropriate for the vast majority of contemporary organizations. As we have seen, it would raise complex (and potentially insurmountable) ethical and legal issues, typically requires invasive and risky surgeries, and oblige an organization to provide ongoing medical support and device maintenance – all for whatever limited organizational benefits, if any, would be gained in return. However, for a small range of specialized organizations, the deployment of posthumanizing neuroprostheses is an objective that is not only considered desirable but which is being energetically pursued. Chief among such organizations are military agencies and departments that envision the development of 'supersoldiers' who possess neuroprosthetically facilitated capacities such as the ability to carry heavy loads without fatigue, leap over walls, see in the dark and with telescopic vision, communicate silently with one another via their thoughts, and operate for days without sleep. Such organizations may find ready volunteers who are willing to undergo neurocybernetic augmentation in order to support an institutional mission that they value highly and for which they are willing to sacrifice their health, safety, and even – if necessary – their lives.[89]

---

[89] Regarding potential military applications of neurotechnologies for human enhancement, see, e.g., Falconer (2003); Moreno, "DARPA On Your Mind" (2004); Coker, "Biotechnology and War:

Known and Hypothesized Military Neuroprosthetic Programs

In some cases, current efforts by military organizations to develop such futuristic technologies can only be hypothesized, as the extreme levels of secrecy maintained within the research and development process make it impossible for outside observers to assemble an accurate account of the ongoing pursuit of such technologies through publically available information. However, a number of contemporary programs led by military agencies with the goal of developing posthuman neuroprostheses are publically known to exist. In some cases, agencies such as DARPA actively explain and promote their efforts to develop such technologies and organize public competitions seeking input and collaboration. Known initiatives created and funded by DARPA to develop technologies relating to neurocybernetic enhancement include:

- The Cognitive Technology Threat Warning System (CT2WS) program for developing computerized binoculars that detect potential threats in a soldier's field of vision by analyzing the soldier's own neural signals.[90]

- More than $100 million dedicated to the Revolutionizing Prosthetics program for developing bidirectional sensorimotor neuroprosthetic robotic limbs.[91]

- $20 million devoted to the Continuous Assisted Performance program designed to create technologies that allow soldiers stay awake and alert for up to seven days.[92]

- $40 million for the Exoskeletons for Human Performance Augmentation program that utilizes haptic interfaces to control an exoskeleton through the detection of minute muscle movements.[93]

- The multimillion-dollar Warrior Web program for developing a lightweight, flexible exoskeleton.[94]

The New Challenge" (2004); Clancy, "At Military's Behest, Darpa Uses Neuroscience to Harness Brain Power" (2006); Graham, "Imagining Urban Warfare: Urbanization and U.S. Military Technoscience" (2008), p. 36; Schermer, "The Mind and the Machine. On the Conceptual and Moral Implications of Brain-Machine Interaction" (2009); Brunner & Schalk, "Brain-Computer Interaction" (2009); Wolf-Meyer, "Fantasies of extremes: Sports, war and the science of sleep" (2009); Kourany (2013), pp. 992-93; and Krishnan, "Enhanced Warfighters as Private Military Contractors" (2015).

[90] See Weinberger, "Pentagon to Merge Next-Gen Binoculars with Soldiers' Brains" (2007), and Tennison & Moreno, "Neuroscience, Ethics, and National Security: The State of the Art" (2012).
[91] See Erikson, "Thought-Controlled Robotic Arm 'Makes a Big Negative a Whole Lot Better'" (2013), and "Prosthetics: Sponsor."
[92] See Falconer (2003).
[93] Falconer (2003).
[94] See Kusek, "The $3 Million Suit: Wyss Institute Wins DARPA Grant to Further Develop its Soft

- $19 million for the Brain Machine Interfaces program (which later become the Human Assisted Neural Devices program) with a goal of allowing thought-controlled manipulation of vehicles, weapons, and computer and downloading of information directly into a soldier's brain.[95]

- $4 million for the Silent Talk project to enable soldiers to exchange information silently on the battlefield through analysis of their neural signals and direct brain-to-brain communication.[96]

Apart from DARPA's efforts, $4 million has been allocated as part of the Multidisciplinary University Research Initiative program of the US Department of Defense to develop 'synthetic telepathy' facilitating communication among soldiers.[97]

Military Organizations Possess Unique Traits That Make Them Likely Early Adopters

Military departments and specialized military agencies are likely to be among the most notable 'early adopters' that proactively deploy posthumanizing neuroprostheses at an organizational level. While the unclear benefits and considerable obstacles involved with the use of such technologies are likely to dissuade most organizations from considering their use, military organizations (and especially those within more economically and technologically advanced nations) possess a range of characteristics that may cause the development and use of posthumanizing neuroprostheses to be viewed by national policymakers as something that is not only legally permissible but also strategically desirable and even a key part of the military's most essential mandate to counter all potential threats and safeguard a nation's security. Such characteristics displayed by military organizations that can facilitate the use of posthumanizing neuroprostheses include:

Biomedical Factors

- Access to premier medical facilities and biomedical and surgical experts needed for the successful implantation and maintenance of neuroprostheses.

- The ability to develop cutting-edge neurotechnologies based on in-house R&D expertise and longstanding relationships with leading external researchers and manufacturers.

Exosuit" (2014), and Cornwall, "In Pursuit of the Perfect Power Suit" (2015).

[95] See White, "Brave new world: Neurowarfare and the limits of international humanitarian law" (2008), and Falconer (2013).

[96] See Drummond, "Pentagon Preps Soldier Telepathy Push" (2009).

[97] See "Mind over Mouth? Study Could Lead to Communicating via Thoughts" (2008), and Bogue, Robert, "Brain-Computer Interfaces: Control by Thought" (2010).

Business and Operational Factors

- A mandate to protect national security and the well-being of an entire society that may ethically justify the governmental use of dangerous technologies whose use by private commercial interests would be considered impermissible.

- A wide range of organizational tasks for which the performance of human agents in hostile and stressful circumstances is critical to success and can be enhanced through the use of posthumanizing neuroprostheses.

- A critical need to anticipate, understand, and counteract the effects of posthumanizing neurotechnologies that could potentially be developed and deployed against a country by adversarial states or non-state actors.

- A demonstrated track record of successfully developing and employing innovative technologies.

Financial Factors

- Budgets sufficiently large to fund the development, deployment, and maintenance of such complex technologies.

- The ability to create and maintain mission-critical technology programs that do not demonstrate short-term 'profitability' in a commercial sense.

- Access to secret government-funded research relating to posthumanizing neuroprosthetics that is generally inaccessible to engineers at universities or private companies.

Legal and Ethical Factors

- Exemptions from many national legal and regulatory requirements that create obstacles for private commercial organizations attempting to develop, test, and deploy such technologies.

- A highly skilled workforce whose members are willing to take significant personal risks (including the risk of injury or death) in order to advance their organization's mission.

- The ability to legally conceal from public view the development and deployment of posthumanizing neurotechnologies that may be considered dangerous, overly expensive, or ethically repugnant by significant segments of society.

Other Types of Organizations Share Some Traits with Military Organizations

Other governmental organizations (such as specialized police agencies, space agencies, and health agencies) may share some but not all of these characteristics. While such organizations might deploy and benefit from such posthumanizing neuroprosthetic technologies after they have been developed, they are presumably less likely than military organizations to be the originators of such technologies. University research laboratories and private companies may participate in the development and production of posthumanizing neuroprostheses as contractors supporting military agencies, and as such they may occasionally attempt to incorporate advanced neuroprostheses into their own organizational structures, processes, and systems on a limited basis for experimental purposes – without, however, being the intended end users of such products.

VIII. Enterprise Architecture as a Preferred Management Tool for Early Adopter Organizations

Much has been written about the implications of posthumanizing neuroprostheses from the perspectives of biomedical engineering, philosophy, ethics, and law; however, almost no direct attention has been given to the management implications of intentionally integrating such devices into organizations. There are many lenses through which one might analyze the likely management impacts of posthuman neuroprosthetics on organizations, including those of marketing and sales, production management, HR management, finance and accounting, information security, and ethics and compliance. However, we would suggest that the management discipline of enterprise architecture provides an especially relevant and useful lens through which to analyze, plan, and manage the creation of posthumanizing neuroprosthetic supersystems within organizations.

There are two primary reasons for this – one of which relates to the nature and origins of EA and the other of which relates to the types of organizations that are likely to become early adopters of posthuman neuroprosthetic technologies. First, enterprise architecture provides a ready array of conceptual frameworks and tools through which to plan and administer the incorporation of advanced neuroprosthetic technologies into organizations, as it is a management discipline that was created explicitly to facilitate the successful integration of innovative IT into organizational structures, processes, and systems. The entire history and practice of EA has refined its capacity to be employed for such a purpose. Second, as large, complex, technology-intensive organizations, the military agencies and departments that are likely to be early adopters of posthumanizing neuroprostheses are just the types of organizations for which enterprise architecture is already a favored management approach – a key mechanism for integrating advanced IT seamlessly and effectively into the organizations' workforce.[98]

Managing the Integration of Neuroprosthetically Augmented Workers into Organizational IT Systems

Insofar as an organization such as a military department provides its employees with work-related neural implants, the organization's IT infrastructure will extend directly into the bodies and minds of those employees. Increasingly, biosensor networks and other kinds of remotely controlled or remotely accessible implantable devices are expanding the possibilities for external agents and systems to monitor and control the activities of those human beings in whom they are implanted;[99] on a functional level, the network of 'IT systems' subject to an organization's operational control might thus include not only the neuroprostheses implanted within organizational employees but the host-device systems that the employees form with their implants – and even the hosts themselves, in their role as biological organisms and information systems with which their neuroprostheses offer a convenient interface. While traditional EA approaches can aid organizations with the oversight of such phenomena, they do not in themselves offer an adequate framework: administering organizational IT will thus no longer be simply a matter of technology management but a matter of neuroscience, biomedical engineering, and healthcare. The fact that posthumanizing neuroprostheses may allow human persons to be incorporated into an organization's IT infrastructure in such ways raises profound philosophical, ethical, and legal questions that are not directly addressed by enterprise architecture.

IX. Conclusion

The world around us does not include a broad wave of organizations rushing to provide their human personnel with performance-enhancing neuroprosthetic devices, nor is such a phenomenon likely to be witnessed anytime soon. As has been discussed in this text, there exist a range of factors that make the deployment of posthumanizing neuroprostheses not only difficult but undesirable or impossible for most contemporary organizations. And yet there is an array of organizations which – having weighed the legal and ethical

---

[98] For example, in the 1990s the US Department of Defense developed the Command, Control, Communications, Intelligence, Surveillance, and Reconnaissance (C4ISR) Architecture Framework, which evolved in the early 2000s into the Department of Defense Architecture Framework (or DoDAF); Version 2.02 was released in 2010. See *C4ISR Architecture Framework Version 2.0* (1997); *DoD Architecture Framework Version 2.02* (2010); and Bergey et al., "U.S. Army Workshop on Exploring Enterprise, System of Systems, System, and Software Architectures" (2009).

[99] Gill, "Socio-Ethics of Interaction with Intelligent Interactive Technologies" (2007), p. 293.

considerations and evaluated the business value of such technologies – are moving forward energetically with the development and anticipated eventual deployment of posthumanizing neuroprostheses. Chief among these are military organizations for which the perceived benefits of such neurotechnologies are perceived to outweigh the dangers. Some such organizations may implement a 'transitional augmentation' of human personnel as a stopgap measure on the path to full automation of business processes through the use of AI; others may retain human workers in particular positions because factors such as legal, ethical, or marketing requirements mandate that human agents fill those roles, while simultaneously augmenting the workers so that they can perform more competitively. In many cases, the same characteristics that make such organizations likely to be early adopters of posthumanizing neurotechnologies also lead them to employ enterprise architecture as a key management practice. However, while EA can provide a valuable tool for integrating neuroprostheses into an organization's structures and dynamics at a technological level, in itself it does not directly address or resolve the serious ethical and cultural questions raised by the intentional organizational exploitation of technologies that promise to dramatically alter human workers' basic mental and physical capacities and relationship to the world. It is hoped that the introduction to such issues presented in this text can provide a foundation for further analyses of such questions both by the organizations that are contemplating the internal deployment of posthumanizing neuroprostheses as well as by device manufacturers, policymakers, philosophers of technology, individual consumers and citizens, and competing organizations that may be affected by the introduction of neuroprosthetically facilitated human enhancement into their sectors and industries.

Chapter Five

An Introduction to Enterprise Architecture in the Context of Technological Posthumanization[1]

Abstract. The discipline of enterprise architecture (EA) seeks to generate alignment between an organization's electronic information systems, human resources, business processes, workplace culture, mission and strategy, and external ecosystem in order to increase the organization's ability to manage complexity, resolve internal conflicts, and adapt proactively to environmental change. In this text, an introduction to the definition, history, organizational role, objectives, benefits, mechanics, and popular implementations of enterprise architecture is presented. The historical shift from IT-centric to business-centric definitions of EA is reviewed, along with the difference between 'hard' and 'soft' approaches to EA. The unique organizational role of EA is highlighted by comparing it with other management disciplines and practices.

The creation of alignment is explored as the core mechanism by which EA achieves advantageous effects. Different kinds of alignment are defined, the history of EA as a generator of alignment is investigated, and EA's relative effectiveness at creating different types of alignment is candidly assessed. Attention is given to the key dynamic by which alignment yields deeper integration of an organization's structures, processes, and systems, which in turn grants the organization greater agility – which itself enhances the organization's ability to implement rapid and strategically directed change. The types of tasks undertaken by enterprise architects are discussed, and a number of popular enterprise architecture frameworks are highlighted. A generic EA framework is then presented as a means of discussing elements such as architecture domains, building blocks, views, and landscapes that form the core of many EA frameworks. The role of modelling languages in documenting EA plans is also addressed.

In light of enterprise architecture's strengths as a tool for managing the deployment of innovative forms of IT, it is suggested that by adopting EA initiatives of the sort described here, organizations may better position themselves to address the new social, economic, and operational realities presented by emerging 'posthumanizing' technologies such as those relating to social robotics, nanorobotics, artificial life, genetic engineering, neuroprosthetic augmentation, and virtual reality.

---

[1] This text draws heavily on Part II B ("Fundamentals of Enterprise Architecture") of Gladden, "Enterprise Architecture for Neurocybernetically Augmented Organizational Systems" (2016).

I. Introduction

The management discipline of enterprise architecture (EA) is character-ized by its effort to engineer an enhanced 'alignment' between an organiza-tion's information systems, human resources, business processes, workplace culture, mission and strategy, and external ecosystem in order to strengthen the organization's ability to manage complexity, resolve internal conflicts, and adapt proactively to environmental change.

For small organizations – or those that are not especially dependent on sophisticated electronic information systems – the notion of employing the complex array of processes and tools offered by enterprise architecture may understandably appear both impractical and unnecessary; however, even small organizations can benefit from applying the basic principles of EA in a streamlined form. And for large organizations – and especially those whose competitive advantage depends on continually executing the rapid and suc-cessful adoption of innovative information technologies – EA in its fullest form can provide an essential management tool.

Enterprise architecture emerged in the 1980s and 1990s as a means for or-ganizations to bring greater coherence, competence, and strategic direction to their acquisition and use of computing technologies. From those earliest days, EA has cultivated a unique insight into the challenges and opportunities to be found in the integration of transformative technologies into organiza-tional life. Today, organizations are increasingly grappling with the strategic, operational, and tactical implications of a new wave of emerging technologies – such as those relating to social robotics, nanorobotics, artificial life, genetic engineering, neuroprosthetic augmentation, and virtual reality – that are cre-ating a technologized and 'posthumanized' world in which human beings are no longer the only intelligent agents who gather information from the envi-ronment, make strategic decisions, and act to transform the world.[2] The adoption of thoughtful EA initiatives can enable organizations to creatively and efficiently incorporate into themselves the most beneficial aspects of such innovative posthumanizing technologies while minimizing the organi-zational risks that they pose. In this way, enterprise architecture can position organizations to survive, compete, and adapt within an increasingly complex world of intensively networked digital-physical ecosystems that are evolving

---

[2] For a discussion of processes of technologization and its role in the larger phenomenon of posthumanization, see, e.g., Herbrechter, *Posthumanism: A Critical Analysis* (2013); Ferrando, "Posthumanism, Transhumanism, Antihumanism, Metahumanism, and New Materialisms: Dif-ferences and Relations" (2013); and Gladden, *Sapient Circuits and Digitalized Flesh: The Organi-zation as Locus of Technological Posthumanization* (2016).

with ever greater speed.[3] With these points in mind, the following sections provide an introduction to the definition, history, organizational role, objectives, benefits, mechanics, and popular implementations of enterprise architecture and present an overview of a generic enterprise architecture framework.

II. Definitions of Enterprise Architecture

Many management scholars and practitioners have formulated definitions of enterprise architecture. Although all definitions share similar elements, they possess nuances of emphasis that allow them to be divided into two broad groups of 'IT-centric' and 'business-centric' definitions.

IT-centric Definitions

'IT-centric' definitions of EA reflect the original vision of enterprise architecture as a discipline whose initial purpose was to aid organizations in designing and implementing IT systems. For example, Gammelgård et al. define enterprise architecture as "an established approach for the model-based and holistic management of IT,"[4] and Cane and McCarthy state that "Enterprise architecture frameworks provide a basis to systematically document and manage the information technology assets of an organization."[5] Similarly, Mezzanotte and Dehlinger contend that "Existing EA frameworks consider EA design solely from a techno-centric perspective focusing on the interaction of business goals, strategies, and technology." They argue that many EA programs fail as a result of this limiting and problematic focus on technological phenomena, which overlooks the social and human aspects that can make it impossible for key organizational stakeholders to effectively express what they wish to achieve from an enterprise architecture program.[6]

---

[3] Digital-physical ecosystems are discussed in numerous contexts, such as investigations of digital ecosystems, cyber-physical systems, virtual worlds, and the Internet of Things. For various perspectives on digital-physical ecosystems and their growing importance, see, e.g., Bainbridge, *The Virtual Future* (2011); Evans, "The Internet of Everything: How More Relevant and Valuable Connections Will Change the World" (2012); and *Digital Ecosystems: Society in the Digital Age*, edited by Jonak et al. (2016).

[4] Gammelgård et al., "An IT Management Assessment Framework: Evaluating Enterprise Architecture Scenarios" (2007).

[5] Cane & McCarthy, "Measuring the Impact of Enterprise Architecture" (2007).

[6] Mezzanotte & Dehlinger, "Enterprise Architecture: A Framework Based on Human Behavior Using the Theory of Structuration" (2012).

Business-centric Definitions

'Business-centric' definitions of enterprise architecture, on the other hand, reflect a newer and broader understanding of EA, emphasizing the discipline's role in shaping and optimizing an organization's structures, processes, and systems in order to best support the organization's business strategy. An example of such a definition is presented by Land et al.,[7] who after reviewing many established definitions of EA distill and synthesize those formulations to define enterprise architecture as:

> A coherent set of descriptions, covering a regulations-oriented, design-oriented and patterns-oriented perspective on an enterprise, which provides indicators and controls that enable the informed governance of the enterprise's evolution and success.

Such a definition does not make any specific reference to an organization's technological infrastructure or IT systems.

The Shift from IT-centric to Business-centric Approaches

Turner et al.[8] have charted the evolution of enterprise architecture from a field primarily concerned with IT management and dominated by technologists to a more mature and holistic discipline that informs and shapes strategic, operational, and tactical decision-making throughout an organization and which calls for enterprise architects to possess a broader range of experience and expertise than simply familiarity with IT management. Similarly, Bean acknowledges that the origins of enterprise architecture lay explicitly in IT management but explores ways in which the field has been extended to guide the engineering of all aspects of an organization, rather than merely its IT systems.[9] Nevertheless, the vision of enterprise architecture as a discipline that is not merely a robust form of IT management but instead a general managerial discipline whose proper object comprises all of an organization's components and activities is belied by the fact that, in practice, EA personnel typically report to an organization's chief information officer rather than to the chief strategy officer or CEO.[10]

---

[7] Land et al., "Positioning Enterprise Architecture" (2009).

[8] Turner et al., "Architecting the Firm – Coherency and Consistency in Managing the Enterprise" (2009).

[9] Bean, "Re-Thinking Enterprise Architecture Using Systems and Complexity Approaches" (2010).

[10] Lindström et al., "A Survey on CIO Concerns – Do Enterprise Architecture Frameworks Support Them?" (2006).

'Hard' vs. 'Soft' Approaches to Enterprise Architecture

The distinction between IT-centric and business-centric definitions of enterprise architecture is also sometimes understood as a difference between 'hard' and 'soft' approaches to EA. Magoulas et al. classify elements of enterprise architecture frameworks as either 'hard' approaches that focus on the engineering of organizational IT systems or 'soft' approaches that focus primarily on the human aspects of implementing architectural changes and which resemble many of the techniques employed in more humanistic and 'anthropocentric' disciplines such as organizational design and, especially, organization development.[11]

Historically, enterprise architects have tended to focus on hard approaches. In particular, early practitioners of enterprise architecture largely underappreciated the value (and even necessity) of explicitly incorporating into their work 'soft' elements such as the political and cultural knowledge, interpersonal communication, leadership, and incentivization that are needed to reduce the doubts and resistance of stakeholders and successfully implement an enterprise architecture program within an organization.[12] However, even from the beginning, some EA frameworks (such as that developed by Xerox) have done a more effective job than others of incorporating complementary hard and soft aspects.[13] And over time, the field of EA management has broadened its focus to include such concerns – to the extent of even attempting to make enterprise architecture 'fun' for those stakeholders whom it impacts.[14]

Comparing Enterprise Architecture to Other Management Disciplines

Enterprise architecture is often confused with or assumed to be identical to a number of other management disciplines and techniques[15] that similarly involve the optimization of organizational components and dynamics, the

---

[11] Regarding the humanistic origins and nature of organization development, see Bradford & Burke, *Reinventing Organization Development* (2005), and Gladden, "Organization Development and the Robotic-Cybernetic-Human Workforce: Humanistic Values for a Posthuman Future?" (2016).

[12] Ahlemann et al., "People, Adoption and Introduction of EAM" (2012).

[13] See Magoulas et al., "Alignment in Enterprise Architecture: A Comparative Analysis of Four Architectural Approaches" (2012), p. 98, and its discussion of Howard, "The CEO as Organizational Architect: An Interview with Xerox's Paul Allaire" (1992).

[14] Ahlemann et al. (2012).

[15] See Stelzer, "Enterprise Architecture Principles: Literature Review and Research Directions" (2010).

administration of IT resources and processes, or the management of organizational change. Below we consider the relationship of EA to some of these practices.

Enterprise Architecture vs. Organizational Architecture

In its focus on creating 'alignment,' enterprise architecture shares many similarities with the field of organizational architecture that emerged at roughly the same time and which strives to generate a high degree of 'fit' or 'congruence' between an organization's components.[16] However, while enterprise architecture is rooted in the highly technical perspectives of IT management and its methodologies for installing and operating concrete hardware and software platforms, organizational architecture is grounded more broadly in philosophies of design, employing principles from the world of architecture in a more metaphorical and less literal fashion.

The disciplines of organizational architecture and enterprise architecture might be understood as mirror images or isomorphs of one another. Both attempt to design and implement an optimized target state for an entire organization; however, while organizational architecture sees the successful design of IT systems as one small component of overall organizational design, enterprise architecture tends to see designing an organization's broader strategies, personnel structures, and processes as one constituent task that must be performed in order to achieve the successful design and implementation of critical IT systems.[17]

Enterprise Architecture vs. Business Models

Iacob et al. suggest that an organization's enterprise architecture is linked to its business model, and thus changes to an organization's IT systems that affect an organization's EA will also affect its business model. In particular, they demonstrate how organizational IT changes reflected in the ArchiMate enterprise architecture model can be correlated with changes reflected in Osterwalder's Business Model Canvas.[18]

Enterprise Architecture vs. Service-Oriented Architecture (SOA)

Service-oriented architecture (SOA) – which "takes an architectural approach to designing and implementing IT solutions" – is best understood not

---

[16] See Nadler & Tushman, *Competing by Design: The Power of Organizational Architecture* (1997), and Hoogervorst, "Enterprise Architecture: Enabling Integration, Agility and Change" (2004), pp. 4-5.

[17] See, e.g., Rohloff, "Framework and Reference for Architecture Design" (2008), pp. 11-12.

[18] See Iacob et al., "From Enterprise Architecture to Business Models and Back" (2012), and Osterwalder & Pigneur, *Business Model Generation: A Handbook for Visionaries, Game Changers, and Challengers* (2010).

as an alternative to enterprise architecture but as a particular approach to managing the design and implementation of enterprise architectures.[19]

Enterprise Architecture vs. Management Cybernetics

Enterprise architecture can also be compared to management cybernetics and the Viable Systems Approach. Early iterations of management cybernetics (e.g., those of Stafford Beer) were explicitly developed by studying the processes of communication and control within intelligent biological organisms and developing general principles that could be applied to systems such as businesses and other organizations; however, more recent formulations of management cybernetics draw extensively on more generalized systems theory.[20]

Enterprise architecture is overwhelmingly a practical management discipline whose goal is to generate improved performance within real-world organizations in the immediate to near future. As such, EA frameworks implicitly incorporate a raft of assumptions about the nature and capacities of contemporary organizations, their human workforce, and their IT resources that may be valid in the present moment but which are not guaranteed to hold true in the future as human societies and technologies evolve. That does not typically pose a problem, insofar as EA is valued primarily for its practical value in enhancing the productivity of existing organizations and not for its theoretical insights or long-term predictive capacity. Management cybernetics, on the other hand, attempts to formulate more generalized principles regarding the behavior of organizations that are applicable not only to businesses as they exist today but to any types of organizations that might appear in the future. If enterprise architecture is understood primarily as a technology-oriented approach to managing contemporary organizations that is informed by various theories of organizational behavior, management cybernetics can conversely be understood largely as a theory of systems structure and dynamics that possesses useful applications for the management of organizations.[21]

---

[19] See MacLennan & Van Belle, "Factors Affecting the Organizational Adoption of Service-Oriented Architecture (SOA)" (2013).

[20] See, e.g., Beer, *Brain of the Firm* (1981); Barile et al., "An Introduction to the Viable Systems Approach and Its Contribution to Marketing" (2012); Pérez Ríos, "Systems Thinking, Organisational Cybernetics and the Viable System Model" (2012); and Gladden, "The Artificial Life-Form as Entrepreneur: Synthetic Organism-Enterprises and the Reconceptualization of Business" (2014).

[21] For a discussion of connections between EA and management cybernetics, see, e.g., Buckl et al., "A Viable System Perspective on Enterprise Architecture Management" (2009).

Enterprise Architecture as a Discipline, Process, or State

Caetano et al. explicitly use the phrase 'enterprise architecture' to describe the optimized target state that is designed for a particular organization by an enterprise architect. They thus explain that:

> An enterprise architecture is the result of the continuous process of representing, integrating and keeping consistently aligned the elements that are required for managing and understanding the organization.[22]

However, this is by no means the only way of conceptualizing enterprise architecture. The phrase 'enterprise architecture' can be used in at least four ways:

- To refer to the **discipline and body of knowledge** that describe (and seek to continually improve) the array of processes and techniques that constitute enterprise architecture.

- To refer to a **process** of enterprise architecture that is intentionally employed by a particular organization at a specific point in time in order to design and implement a new organizational state.

- To refer to the **optimized target state** that has been knowingly designed for a particular organization by an enterprise architect and which is detailed in an EA plan.[23]

- To refer to the *de facto* **arrangement** of an organization's components and dynamics at a particular point in time – i.e., the actual 'architecture' of that 'enterprise' – regardless of whether or not that state was intentionally designed by an enterprise architect.

If the phrase is employed in the final sense, it could be said that *every* functioning organization possesses an 'enterprise architecture,' even if it is inefficient and suboptimal and the organization has never intentionally employed the *process* of enterprise architecture. In this text, the phrase 'enterprise architecture' is used in all four senses, with the exact meaning depending on the context. Thus one might speak of "drawing on the principles of enterprise architecture to carry out a process of enterprise architecture that replaces an organization's current, suboptimal enterprise architecture with a new enterprise architecture that creates greater alignment."

Within the field's literature, the phrase 'enterprise architecture management' is often employed to specify the discipline of EA or the process of designing, implementing, and maintaining a particular EA within a given organization. Similarly, the phrase 'enterprise architecture framework' refers to a specific approach (often given a proper name and marketed by a particular

---

[22] Caetano et al., "A Role-Based Enterprise Architecture Framework" (2009), p. 253.
[23] See Iacob et al. (2012).

author or institution) to conducting an EA process and preparing the documents that depict an organization's target EA.

III. Goals and Benefits of Enterprise Architecture

Perhaps reflecting their origins in the world of IT management, the many EA frameworks in use today generally do a robust and rigorous job of defining technical tools and processes (such as UML-based schematics) for analyzing or designing an organization's enterprise architecture. However, there is a distinct lack of agreement regarding the goals that such activities are meant to achieve. Moreover, there is a lack of clarity regarding the success factors that contribute to a particular enterprise architecture effort being able to deliver beneficial results.[24]

A comprehensive analysis by Boucharas et al. suggests that EA is capable of delivering a hundred different benefits that are produced through the three key value-generating mechanisms of standards, models, and IS/IT governance frameworks.[25] However, individual frameworks often focus on generating a smaller number of more targeted benefits. A typical account is that presented by Magoulas et al., who assert that the goals of enterprise architecture include reducing an organization's complexity, increasing its changeability, and providing a clearer basis for evaluation, in order to enhance the organization's competitiveness.[26] Similarly, Rohloff[27] notes that the objectives of enterprise architecture include promoting an organization's effectiveness, efficiency, continuity and structural stability, ability to adapt to change, transparency in communication, and ability to plan and control activities in order to implement the organization's strategy and business model.

For the kinds of organizations (such as military agencies and departments) that are likely to be early deployers of posthumanizing technologies such as those used to neuroprosthetically enhance human personnel, we can identify at least three goals of enterprise architecture that are likely to be relevant; these are managing complexity, resolving internal conflicts, and generating integration, agility, and the ability to manage change. Below we consider these goals and benefits of enterprise architecture in more detail.

---

[24] Nakakawa & Proper, "Quality Enhancement in Creating Enterprise Architecture: Relevance of Academic Models in Practice" (2009).

[25] Boucharas et al., "The Contribution of Enterprise Architecture to the Achievement of Organizational Goals: A Review of the Evidence" (2010), pp. 6-7.

[26] Magoulas et al. (2012), p. 88.

[27] See Rohloff (2008).

Managing Complexity

Managing change and complexity are the two greatest challenges reported by contemporary business leaders, and the discipline of enterprise architecture has in large part been developed to aid organizations in confronting these challenges.[28] Enterprise architecture makes it easier for organizations to grapple with tremendous internal complexity by creating a conceptual framework that reduces such intricacies to a more manageable array of core components and dynamics whose behavior and interactions can be easily analyzed, predicted, and controlled.[29] Historically, enterprise architecture has been especially valuable in aiding organizations to deal with internal technological complexity: in many contemporary organizations, key processes of data analysis, decision-making, and control are supported or managed by ineffective and inefficient IT systems that comprise a vast and bewildering collection of hardware and software platforms – old and new, off-the-shelf and custom-built – that are held together by a makeshift jumble of mechanisms and interfaces.[30] EA can be employed to make sense of such environments.

Resolving Internal Conflicts

The implementation of a thoughtfully designed enterprise architecture can also aid in avoiding or resolving internal conflicts[31] such as differences in prioritizing organizational objectives; competing claims on organizational resources; or disagreements over the timing, scope, or tactics of organizational activities being planned.[32]

Generating Integration, Agility, and the Ability to Manage Change

Enterprise architecture seeks to optimize an organization's ability to successfully manage change by achieving a high degree of agility, which in turn results from effectively integrating the organization's components and behaviors. We can explore these dynamics in more detail.

Organizations Must Be Able to Implement Rapid and Directed Change

Phenomena such as the growing complexity and dynamism of business environments, intensive globalized competition, and increasingly fickle and knowledgeable consumers mean that a "traditional modus operandi based on command and control" no longer has the ability to generate and maintain a

---

[28] Højsgaard, "Market-Driven Enterprise Architecture" (2011).

[29] Rohloff (2008).

[30] Sundberg, "Building the Enterprise Architecture: A Bottom-Up Evolution?" (2007).

[31] See Fritz, *Corporate Tides* (1996), and its discussion in Hoogervorst (2004), pp. 4-5.

[32] In this way, EA supports the conflict-resolving role of 'System 2' as described in the Viable System Model formulated within management cybernetics. See, e.g., Pérez Ríos (2012).

competitive advantage for a contemporary business; in response to such environmental challenges, enterprise architecture seeks to provide a 'blueprint' that allows a business to better understand its internal structure, processes, and systems and to deftly and efficiently adjust those features as needed in order to achieve a desired future state.[33]

Indeed, an organization's ability to successfully design and implement a process of transformational change in order to allow it to execute its business strategies is arguably even more important than the development of particular strategies that possess high-quality content; a business with a highly adaptable and well-integrated architecture but suboptimal strategies might be expected to outperform one with good strategies but a brittle, inelastic, unresponsive, labyrinthine architecture.[34]

In Order to Change Rapidly, Organizational Agility Is Needed

Not all organizations are in a position to implement rapid and directed change on an ongoing basis. Hoogervorst notes that an ever-greater degree of organizational agility is needed in order for businesses to regularly introduce new kinds of goods and services and to modify their existing offerings with increasing frequency – activities that are required by the quickening pace of societal and industrial change.[35] However, large, complex, and technology-intensive enterprises face especially great challenges in generating and maintaining the degree of organizational agility that is needed in order to create and control such dynamic processes of change. Proactive steps must be taken in order to achieve such agility.

In Order to Develop Agility, Organizational Integration is Needed

Proponents of enterprise architecture argue that the best way to develop organizational agility is by achieving a high degree of organizational integration on both the functional and conceptual levels. Attempting to implement major changes in one element of an organization without first understanding all of that element's interdependencies with other parts of the organization can easily end in disaster.[36] If the changes taking place throughout a large organization are not consistent with one another, they risk pulling the organization away from its strategies or even tearing the organization apart. The more radical the change that an organization is attempting to implement, the

---

[33] Magoulas et al. (2012), p. 88.

[34] See Kaplan & Norton, *The Strategy-Focused Organization: How Balanced Scorecard Companies Thrive in the New Business Environment* (2001), and the discussion of that text in Hoogervorst (2004), pp. 4-5.

[35] Hoogervorst (2004), p. 5.

[36] Hoogervorst (2004), p. 2.

stronger and deeper the integration that the organization must possess.[37] An organization lacking such integration might be compared to an animal that lacks a central nervous system: although the body parts that constitute such an animal might change in response to environmental stimuli, the changes are not internally coordinated or purposefully directed by the organism.

Hoogervorst notes that as a business (or any other system) grows more 'extended' in its geographical scope, complexity and diversity of activities, number of internal subsystems, and range of interaction with its external ecosystem, it becomes more essential for the business to possess a single coherent architecture that can integrate all of the organization's components and processes and allow them to function seamlessly in unison with one another.[38] Similarly, Caetano et al. argue that in order to be able to adapt to change in an agile manner, an organization must possess an integrated representation that contains all relevant knowledge about its personnel, business processes, IT systems, and technology[39]; the process of enterprise architecture attempts to formulate such a knowledge representation for a particular organization and the system that it constitutes.

EA as Facilitator of an Ongoing Process of Change Management

To summarize, Hoogervorst argues that enterprise architecture has a threefold purpose: by creating a thorough *integration* of an organization's components and behaviors, EA gives the organization a high degree of *agility*, which allows the organization to *change* quickly and intelligently in order to adapt to rapidly evolving environmental conditions.[40] Likewise, Van der Raadt and Van Vliet emphasize the role of enterprise architecture in facilitating successful change management, noting that:

> Enterprise Architecture (EA) is an increasingly important instrument to better manage enterprise transformations. EA provides a means for getting insight into the current state landscape, creating a target blueprint, and setting out a roadmap to achieve that target state.[41]

Enterprise architecture not only supports the implementation of planned change; it also prepares organizations to better handle unplanned change.[42]

---

[37] Hoogervorst (2004), p. 17.

[38] Hoogervorst (2004), p. 4.

[39] Caetano et al. (2009).

[40] Hoogervorst (2004), p. 5.

[41] Van der Raadt & Van Vliet, "Assessing the Efficiency of the Enterprise Architecture Function" (2009).

[42] Hoogervorst (2004), p. 6; Kandjani et al., "Enterprise Architecture Cybernetics for Complex Global Software Development: Reducing the Complexity of Global Software Development Using Extended Axiomatic Design Theory" (2012).

In this way, it positions contemporary organizations to continuously adapt in the face of disruptive new technologies, dynamic global markets, and rapidly evolving regulatory environments.[43]

IV. Creating Alignment: The Methodology of Enterprise Architecture

Regardless of which specific benefits an organization is seeking to obtain through the use of enterprise architecture, the means by which EA achieves those benefits is typically through the creation and maintenance of *alignment* within the organization. Within enterprise architecture, the notion of alignment has been used to refer to such diverse concepts as 'integration,' 'fusion,' 'compatibility,' 'harmony,' 'balance,' 'conformity,' and 'fit.'[44]

The History of Enterprise Architecture as a Generator of Alignment

The path toward the development of mechanisms for generating such alignment began in the 1980s, when many organizations made large investments in IT systems but were then surprised to discover that the systems did not generate the desired productivity benefits. Such failures often resulted from the fact that an organization's IT purchases were not adequately 'aligned' with its business needs and strategies:[45] often the IT systems provided business capacities that were not needed, failed to provide business capacities that *were* needed, could not be managed effectively by the organization's personnel, or otherwise could not be integrated successfully into the organization's existing business processes. The Strategic Alignment Model (SAM) was developed in the early 1990s by Henderson and Venkatraman to facilitate the alignment and integration of an organization's IT systems with its business processes and needs.[46]

Scholars later noted the challenges that arise when attempting to integrate IT systems and business strategy directly, given the fact that these are two very different spheres with different conceptual frameworks, vocabularies, objectives, and methodologies; various proposals were developed for formulating intermediary structures or disciplines that could serve as a neutral and effective interface between IT and business strategy.[47]

---

[43] Buckl et al., "A Situated Approach to Enterprise Architecture Management" (2010).

[44] Magoulas et al. (2012), pp. 92-93.

[45] Magoulas et al. (2012), p. 89; see also Hoogervorst (2004), p. 16.

[46] See Henderson & Venkatraman, "Strategic alignment: Leveraging information technology for transforming organizations" (1993), and Magoulas et al. (2012), p. 89.

[47] Magoulas et al. (2012), p. 89.

Types of Alignment

Chan and Reich[48] distinguish four kinds of alignment: strategic and intellectual, structural, social, and cultural. In principle, an effective EA framework should be capable of facilitating all four types of alignment. However, in practice EA frameworks focus primarily on creating strategic alignment between business and IT strategies, and the success or failure of an enterprise architecture's implementation is evaluated primarily on the basis of whether the architecture generates quantifiable 'extrinsic' gains in performance, not whether it yields 'intrinsic' cultural and social gains such as an enhanced understanding and acceptance of an organization's strategies by its workforce.[49]

Magoulas et al., meanwhile, identify five key kinds of alignment that can be generated by enterprise architecture: socio-cultural alignment (which creates harmonious relationships between information systems and the values, goals, and objectives of an organization's personnel), functional alignment (which creates harmonious relationships between information systems and an organization's activities and processes), structural alignment (which integrates information systems with the sites of power that serve as sources of authority and responsibility within an organization), infological alignment (which creates harmonious relationships between information systems and an organization's individual stakeholders), and contextual alignment (which creates harmonious relationships between an organization's information systems, the organization as a whole, and the external environment).[50]

EA's Ability to Create Different Types of Alignment

The analysis conducted by Magoulas et al.[51] has found that while a range of popular EA frameworks typically possess robust tools and guidance for creating functional, structural, and contextual alignment, they generally lack adequate (or even any) mechanisms or resources for creating infological or socio-cultural alignment; Magoulas et al. attribute this deficiency in part to the fact that some popular EA frameworks "presuppose that information resources should be treated as independent of organization and culture."[52] Such presuppositions underestimate the extent to which social and cultural factors impact every aspect of an organization's functioning, including elements as apparently 'impersonal' as IT systems.

---

[48] See Chan & Reich, "IT alignment: what have we learned?" (2007), and its discussion in Magoulas et al. (2012), p. 98.

[49] Magoulas et al. (2012), pp. 88-90.

[50] Magoulas et al. (2012), pp. 93-95.

[51] Magoulas et al. (2012), pp. 93-95, 98.

[52] Magoulas et al. (2012), p. 95.

How Enterprise Architects Engineer Alignment

Enterprise architecture utilizes explicit design principles to generate alignment and integration between domains such as those of *business, organization, information,* and *technology.*[53] Insofar as it seeks to intentionally design the form and functioning of such domains rather than to simply manage the behavior of existing structures, enterprise architecture requires a 'white-box' constructional perspective that understands all of the internal components and dynamics of such domains rather than a more limited 'black-box' functional perspective that is capable of manipulating the elements of these domains but not creating them.[54] Enterprise architects utilize their analytical and design skills in order to:

- Gather the vast quantities of data needed to map out the characteristics of an organization's current enterprise architecture.

- Identify missing elements, sources of conflict, and duplication of efforts within the current EA.

- Design a new EA that reflects an optimized 'target state' toward which the organization can move by means of change management processes.

In more mature organizations that have a track record of consciously employing the techniques of enterprise architecture, an enterprise architect's job may involve maintaining an EA that was designed at an earlier point in time, evaluating the ongoing effectiveness of that EA, and developing and implementing changes to the EA in response to internal or external change.

V. Popular Enterprise Architecture Frameworks

In their extensive review of EA frameworks, Magoulas et al. highlight the Zachman Framework, The Open Group Architecture Framework (TOGAF), the Extended Enterprise Architecture Framework (E2AF), and the Generalised Enterprise Reference Architecture and Methodology (GERAM) as leading examples of EA frameworks.[55] Meanwhile, Rohloff's comprehensive analysis of EA frameworks includes TOGAF, GERAM, and the Zachman Framework, as well as the Federal Enterprise Architecture Framework (FEAF), the Gartner and META Group Enterprise Architecture Frameworks, and the Siemens Framework that Rohloff formulates in his text.[56] Other EA frameworks

---

[53] Hoogervorst (2004), p. 18.
[54] Hoogervorst (2004), pp. 16-18.
[55] Magoulas et al. (2012), p. 88.
[56] Rohloff (2008), p. 3.

include those described in ISO 15704, ISO 19439, and ISO/IEC 15288.[57] Because different enterprise architecture frameworks all address the same underlying organizational dynamics (such as personnel structures and IT systems), it is generally possible to map one EA framework to another, regardless of how different their terminology or conceptual foundations might appear.[58]

VI. Key Elements of Enterprise Architecture: Presenting a Generic EA Framework

By analyzing the basic structures and dynamics that underlie a range of popular EA frameworks, it is possible for us to synthesize a high-level EA framework that provides a schematic overview of the types of elements commonly found in particular EA frameworks.[59] Such a generic EA framework is presented in the following sections; its key elements are constituent architectures (or domains), building blocks, views, and landscapes.[60] We can consider each of these in turn.

The Four Constituent Architectures or 'Domains'

The largest and most general features distinguished in an EA framework may be referred to as its constituent architectures or 'domains.' Our generic EA framework includes the four architecture domains of *business, organization, informatics,* and *infrastructure.*[61] The business architecture domain encompasses those elements that define the enterprise as a business that must

---

[57] See *ISO 15704:2000, Industrial automation systems – Requirements for enterprise-reference architectures and methodologies* (2000); *ISO 19439:2006, Enterprise integration – Framework for enterprise modelling* (2006); *ISO/IEC 15288:2002, Systems engineering – System life cycle processes* (2002); and Martin & Robertson, "A Comparison of Frameworks for Enterprise Architecture Modeling" (2003).

[58] For examples of such mappings, see Noran, "A Mapping of Individual Architecture Frameworks (GRAI, PERA, C4ISR, CIMOSA, ZACHMAN, ARIS) onto GERAM" (2003), and Williams & Li, "PERA and GERAM – Enterprise Reference Architectures in Enterprise Integration" (1999).

[59] Our generic framework draws on elements of EA frameworks such as TOGAF, FEAF, and, especially, the Siemens EA Framework described by Rohloff. See *TOGAF® Version 9.1* (2011); *Federal Enterprise Architecture Framework* (2013); and Rohloff (2008).

[60] These elements are commonly found across the broad range of EA frameworks; however, the vocabulary used to describe them in different frameworks varies in ways that can potentially create confusion. For example, Caetano et al. (2009) use the word 'view' to refer to what Rohloff (2008) and Hoogervorst (2004) call a 'domain'; Rohloff in turn uses the word 'view' to refer to what Caetano et al. might call a 'concept.' Nevertheless, as previously noted, it is generally easy to correlate such divergent sets of terminology, insofar as they address the same underlying enterprise structures and dynamics.

[61] TOGAF includes the four architectures of *business, data, applications,* and *technology*; see Part II (8) through Part II (12) of *TOGAF® Version 9.1* (2011). This can be compared, e.g., with the EA framework described by Hoogervorst (2004, p. 5), which includes the four domains of *business,*

secure resources from the environment, transform them into goods or services, and release those finished products into the environment to be acquired by consumers. The organization architecture domain encompasses those elements that define the structure and dynamics of the collection of intelligent social agents (who have historically been human beings but increasingly might instead include artificial agents) who undertake the enterprise under consideration. The informatics architecture domain encompasses all of an organization's information, the conceptual frameworks by which it organizes and classifies information, and the applications and processes (such as conventional software programs or artificial neural networks[62]) by which it manipulates information and makes it available to end users. The infrastructure architecture domain encompasses all of the physical IT systems and components that enable the execution of computing processes of all sorts; it also includes low-level electronic communication functions such as those described in the Open Systems Interconnection (OSI) model.[63]

The Building Blocks that Constitute Domains

In our framework, each of the four domains is in turn composed of three separate 'building blocks,' as depicted in Figure 1.[64] The business architecture

---

information, organization, and *technology;* with the EA framework formulated by Caetano et al. (2009, p. 254), which includes the four domains or 'concerns' of *organization, business, information,* and *system;* and with the Siemens EA Framework described by Rohloff (2008), which includes the three constituent architectures or 'domains' of *business, application,* and *infrastructure* that are tied together by an organization's *strategy.*

[62] A 'virtualized' neural network that is run as a software program on a conventional computer with a CPU-based Von Neumann architecture would be an application falling within the informatics architecture domain; a physical artificial neural network (such as one comprising an array of memristors) would constitute both a neural computing process falling within the informatics architecture domain as well as a physical IT system that falls within the infrastructure architecture domain. For a review of the standard Von Neumann architecture, see, e.g., Dumas, *Computer Architecture: Fundamentals and Principles of Computer Design* (2006). Regarding artificially intelligent systems that utilize physical ANNs, see, e.g., Snider, "Cortical Computing with Memristive Nanodevices" (2008); Versace & Chandler, "The Brain of a New Machine" (2010); *Advances in Neuromorphic Memristor Science and Applications,* edited by Kozma et al. (2012); and Lohn et al., "Memristors as Synapses in Artificial Neural Networks: Biomimicry Beyond Weight Change" (2014).

[63] See *ISO/IEC 7498-1:1994, Information technology – Open Systems Interconnection – Basic Reference Model: The Basic Model* (1994).

[64] For use of the term 'building blocks' to describe the elements of an enterprise architecture, see Parts I (2) and IV (37) of *TOGAF® Version 9.1* (2011), where building blocks may include such discrete elements as a single employee, business process, or software application. In the Siemens EA Framework described by Rohloff (2008), a 'building block' is a larger and more abstract element that is a constituent part of an architecture domain; each of the three domains is composed of four building blocks. The nature of the building blocks found in our generic EA framework is closer to Rohloff's usage, and there is some overlap between the building blocks defined in the

domain is constructed from the building blocks of: (1) an organization's *mission*, which represents the purpose for which the organization exists and provides the ultimate criterion by which it chooses which business avenues to pursue; (2) the *business model* that reflects the complex dynamics by which resources are transformed and products marketed in order to generate value for the organization; and (3) the *business processes* that are carried out in order to realize the dynamics represented by the business model.

The organization architecture domain is constructed from the building blocks of: (1) an *organization design* that defines intended roles, activities, and reporting relationships for all of an organization's human agents (or, potentially, artificial agents such as social robots or artificial general intelligences); (2) the *management practice* comprising the approaches by which managers actually direct activities on a daily basis[65] (and which may differ from the idealized management practices implicit in the organization design); and (3) the *organizational culture* that shapes and reflects workers' ethical standards, degree of motivation and commitment, and expectations for one another and which is only partly subject to the organization's centralized control.

The informatics architecture domain is constructed from the building blocks of: (1) the *information model* that provides a schema for identifying, capturing, and classifying all information relevant to the organization, such as that relating to business processes, products, physical resources, employees, suppliers, customers, and competitors; (2) *business applications* in the form of software that allows workers or artificial agents to perform business processes; and (3) *support applications* (such as generic web browsers or building automation system software that regulates a facility's heating and lighting) that are not designed or optimized to directly execute business processes but which facilitate other necessary activities.

Finally, the infrastructure architecture domain is constructed from the building blocks of: (1) *internal and external networks* that enable communication among all of the organization's stakeholders and systems and the gathering of intelligence from the external environment; (2) *servers and data storage systems* that enable the secure and efficient storage and retrieval of data; and (3) *productivity systems* including desktop and mobile systems that provide basic computing resources to workers as well as industrial systems that perform a specialized function.

---

two frameworks.

[65] Building on the classic management framework of Henri Fayol, Daft identifies the four essential functions that must be performed by a manager as *planning, organizing, leading,* and *controlling* the activities of those workers and systems that fall within the manager's purview. See Daft, *Management* (2011), p. 8.

Business architecture domain

- *Mission* that represents the purpose for which an organization exists and the ultimate criterion by which it selects which businesses to pursue
- *Business model* that reflects the dynamics by which resources are acquired, transformed, and exchanged to generate value
- *Business processes* that are performed to produce goods and services and exchange them within the external ecosystem

Organization architecture domain

- *Organization design* that defines intended roles, activities, and reporting relationships for the organization's human (or artificial) agents
- *Management practice* reflecting the approaches by which managers actually plan, lead, organize, and control organizational activities
- *Organizational culture* reflecting and shaping workers' ethics, motivation, commitment, and expectations for one another

Informatics architecture domain

- *Information model* that provides a schema for identifying, capturing, and classifying all information relevant to the enterprise
- *Business applications* that allow workers or artificial agents to perform business processes
- *Support applications* that do not directly execute business processes but facilitate other necessary activities

Infrastructure architecture domain

- *Internal and external networks* that enable communication among stakeholders and systems and the gathering of intelligence
- *Servers and data storage systems* that enable the storage and retrieval of data
- *Productivity systems* including desktop and mobile systems that provide computing resources and specialized industrial systems

S
T
R
A
T
E
G
Y

Fig. 1: A high-level schema for a generic enterprise architecture framework. Illustrated here are the four architecture 'domains' of business, organization, informatics, and infrastructure, along with the three 'building blocks' that constitute each domain and strategy as a unifying force that shapes all four domains. Different EA frameworks merge, subdivide, rename, and relate these basic elements in various ways.

Elements that Link or Transcend Domains

Researchers also identify various elements of EA frameworks that are said to link or transcend individual architecture domains. For example, Rohloff notes that the manifold interrelationships between the domains are themselves constituent elements of an enterprise architecture; he also contends that strategy and information security are elements that cannot be contained within a single domain but which instead link, infuse, and support all of the domains.[66] The Federal Enterprise Architecture Framework similarly depicts security as a concern that impacts and interconnects the five other domains (or 'reference models') of *performance, business, data, application,* and *infrastructure.*[67] Hoogervorst, on the other hand, is willing to localize information security more narrowly within the EA framework that he formulates, positioning InfoSec as a component of the element of 'quality' within the information architecture domain.[68]

Three Basic 'Views' for Describing Domains

It is possible to analyze an architecture domain from different perspectives, each of which reveals unique insights. For example, consider an employee of an organization's accounting department who is responsible for tracking the organization's acquisition and disposal of desktop computers so that he or she can perform annual asset depreciation calculations. That employee would conceptualize the organization's infrastructure architecture domain primarily in terms of information such as the quantity, purchase price, and serial numbers of desktop computers and their current location within the organization; he or she would have no particular interest in knowing the computers' performance characteristics or learning which business processes they facilitate. On the other hand, an employee of the organization's training and development department would analyze the same infrastructure architecture domain in terms of how workers might best be trained to successfully execute business processes using the types of computers possessed by the organization; he or she would not have a need to track the current state of repair or financial value of every particular computer. Both employees' perspectives on the infrastructure architecture domain are important and should be represented within an enterprise architecture plan; however, they are best captured using different lenses or 'views.'

The use of formalized 'views' as a means of analyzing architecture from different perspectives has been facilitated by the development of a number of models and standards. For example, the IEEE 1471 standard describes at an

---

[66] Rohloff (2008).

[67] *Federal Enterprise Architecture Framework* (2013), pp. 20-21.

[68] Hoogervorst (2004), p. 15.

202 • Neuroprosthetic Supersystems Architecture

abstract level the characteristics that architectural views should possess.[69] A particular array of views is defined by Kruchten's '4+1' model, which includes the *logical view* (employing the perspective of end-user functionality), *process view* (focusing on communication, integration, and other system behaviors and dynamics), *development view* (prepared from the perspective of the design, implementation, and management of software), and *physical view* (focusing on the deployment and interconnection of hardware components).[70] Similarly, the Siemens EA Framework described by Rohloff explicitly formulates three perspectives: the *component, communication*, and *distribution* views.[71] Meanwhile, The Open Group's ArchiMate modelling language for enterprise architecture defines eighteen standard 'viewpoints' or perspectives that can serve as the basis of views.[72]

For purposes of developing our generic EA framework, the streamlined approach taken by Kruchten and Siemens offers an appropriate guide. Our framework incorporates three primary perspectives on architecture domains: the *component, interaction*, and *membership* views.[73] The component view highlights all of the entities that together constitute the enterprise, including employees, physical facilities, computing devices, vehicles, products, and financial, material, and informational resources, as well as the capacities and internal processes that these entities possess. The interaction view highlights the network topology of the ways in which these entities are connected to

---

[69] See *ANSI/IEEE 1471-2000, IEEE Recommended Practice for Architectural Description for Software-Intensive Systems* (2000), later superseded by *ISO/IEC/IEEE 42010:2010, Systems and software engineering – Architecture description* (2011).

[70] See Kruchten, "The 4+1 view model of architecture" (1995).

[71] See Rohloff (2008), pp. 5-6.

[72] The eighteen standard 'viewpoints' defined by ArchiMate 2.1 are the *Introductory, Organization, Actor Co-operation, Business Function, Business Process, Business Process Co-operation, Product, Application Behavior, Application Co-operation, Application Structure, Application Usage, Infrastructure, Infrastructure Usage, Implementation and Deployment, Information Structure, Service Realization, Layered*, and *Landscape Map* viewpoints. See Section 8.4 of *ArchiMate® 2.1 Specification* (2013).

[73] Our component view is roughly analogous to Rohloff's component view and similar to Kruchten's physical view; our interaction view is similar to Rohloff's communication view and Kruchten's process view; and our membership view is comparable to Rohloff's distribution view and incorporates aspects of Kruchten's development and physical views. However, in contrast with Rohloff's model, we consider the 'relationships' among entities primarily as a part of their dynamic interaction rather than as static aspects of the components themselves (i.e., they are highlighted in the interaction rather than component view); we also suggest that VR technologies may be able to fashion stable and resilient virtual worlds that can be occupied by workers and other organizational elements rather than simply facilitating communication among organizational elements (i.e., worlds that can be fruitfully analyzed using the membership view rather than solely via the communication or interaction views). In contrast with Kruchten's model, we do not include a logical view dedicated to end-user functionality, insofar as our EA framework encompasses more diverse organizational elements than simply the software systems that Kruchten's model addresses.

one another and the processes (such as those of communication and control) by which they interact. The membership view highlights the boundaries and occupants of those spatiotemporal regions (such as physical buildings, countries, time zones, or virtual environments) within which organizational elements are located or operate and of those functional or conceptual groupings (such as corporate departments or project teams) to which elements belong or to whose authority they are subject. Figure 2 presents a simplified schematic that demonstrates the ways in which different aspects of a hypothetical enterprise become apparent depending on whether one analyzes it using the component, interaction, or membership views.

Virtual Teams, Virtual Reality, and EA Views

Both the interaction and membership views highlight connections between workers, but in different ways. For example, two employees might work in the same building but have no interaction with one another, while two other employees may work at facilities in different countries but interact on a daily basis via email, telephone, and instant messaging as part of a virtual project team.[74] The emergence of new immersive virtual reality technologies adds a new dimension to membership views, as VR systems can be employed to create persistent multiuser environments in which human agents who are geographically dispersed in the 'real' world may inhabit a shared virtual world in which they interact with one another[75] – and perhaps also with virtual agents whose behavior may be governed by computers housed in yet other

---

[74] For more about such virtual teams, see Zofi, *A Manager's Guide to Virtual Teams* (2012), and Settle-Murphy, *Leading Effective Virtual Teams: Overcoming Time and Distance to Achieve Exceptional Results* (2012).

[75] In this context, what is commonly referred to as the 'real' world might more precisely be called the 'primary physical world,' to contrast it with 'secondary physical worlds' (or 'virtual' worlds). The contents of the primary physical world possess an objective existence and an underlying physical form that is isomorphic with the form in which the world is experienced by its human inhabitants. Secondary physical worlds are virtual worlds whose contents are determined by the computational processes of a computerized VR system; the contents of such virtual worlds are arbitrary, insofar as they are not constrained by the organization of the primary physical world (i.e., the shape of a virtual desk or virtual tree is determined by a digital file whose string of binary data does not in itself possess the 'shape' of a desk or a tree) and can be dramatically altered at will by a virtual world's human designer or world-management algorithms. However, even secondary physical worlds are still 'real' and 'physical' insofar as the organization of their contents is maintained within real physical objects (e.g., the hard drives or ROM chips of a VR computer system) and is experienced by their human inhabitants through the mediation of real physical stimuli (such as electrons or chemical neurotransmitters used to stimulate neurons in a host's sensory system or brain).

For the ramifications of long-term immersion in VR environments, see, e.g., Koltko-Rivera, "The potential societal impact of virtual reality" (2005), and Bainbridge (2011). Regarding psychological, social, and political questions arising from the repetitive long-term inhabitation of virtual worlds by means of a digital avatar, see, e.g., Castronova, "Theory of the Avatar" (2003).

geographical regions or whose control processes are not geographically localizable.[76] An organization's elements might thus be grouped according to the virtual facilities, regions, or worlds within which they operate. A membership view designed to reflect the elements of an organization's virtual environments may require a complex multilayered representation: while a single employee typically inhabits just a single geographical region, he or she might simultaneously enjoy access to numerous virtual environments.

Landscapes: From Current to Target States

Many enterprise architecture approaches use a term such as 'blueprint' or 'landscape' to describe a document (or 'artifact') that depicts the relationships and dependencies that a single domain building block possesses with all of an enterprise's architecture domains, as analyzed according to multiple views.[77] Some researchers restrict use the term 'blueprint' to refer exclusively to a target architecture that is to be implemented in the future, employing a different term to refer to a document that describes a (potentially suboptimal) architecture that currently exists.[78]

Our generic EA framework facilitates the creation of similar types of landscapes. The landscape for a particular organizational element (such as an individual staff position, business process, smart building, or piece of enterprise software) can be developed by starting with a blank template of the sort shown in Figure 3 and filling in the various fields to reflect the ways in which the element being considered relates to the other elements of all four architecture domains when analyzed through the component, interaction, and membership views.

---

[76] Regarding the participation of artificial agent in virtual teams, see, e.g., Gladden, "Leveraging the Cross-Cultural Capacities of Artificial Agents as Leaders of Human Virtual Teams" (2014). For the possibility of networked artificial agents whose structure and behaviors cannot be traced or assigned to a particular physical location, see Gladden, "The Diffuse Intelligent Other: An Ontology of Nonlocalizable Robots as Moral and Legal Actors" (2016).

[77] For example, in the Siemens EA Framework presented by Rohloff, three commonly used blueprints or landscapes are the *application landscape* (which highlights the ways in which a particular business process is supported by various applications), *data repository landscape* (which illustrates the ways in which databases are deployed to create 'information clusters' within the organization's information architecture), and *service landscape* (which depicts the ways in which infrastructure services are deployed to support the organization's applications). See Rohloff (2008). TOGAF, on the other hand, uses the term 'Architecture Landscape' to refer to an organized collection of the diverse assortment of architectures at use within an organization at any given moment; see Part III (20) of *TOGAF 9.1* (2011).

[78] See, e.g., Bischoff et al., "Use It or Lose It? The Role of Pressure for Use and Utility of Enterprise Architecture Artifacts" (2014), which uses the term 'blueprint' to describe a target architecture and 'literature' to refer to artifacts that documenting an existing architecture.

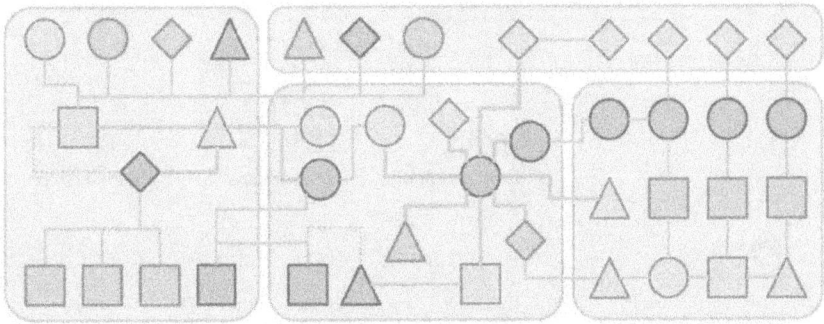

(A) Component view

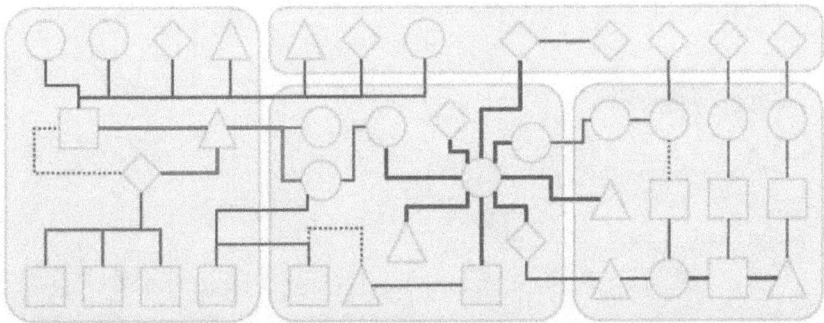

(B) Interaction view

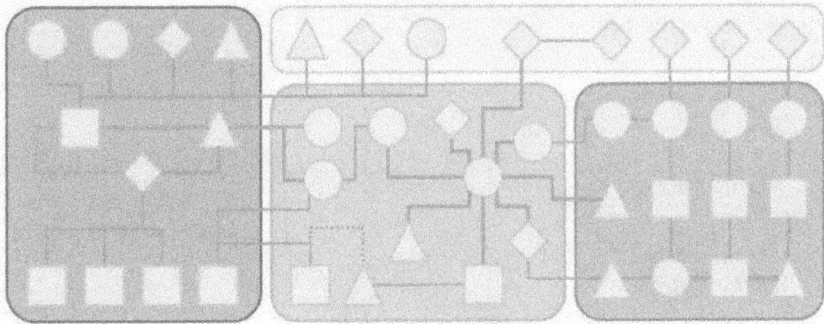

(C) Membership view

Fig. 2: A generic depiction of the role of different 'views' in an EA framework. Illustrated here are the structure and dynamics of a single enterprise as analyzed according to three views. The (A) component view highlights the entities and objects that constitute the organization; the (B) interaction view highlights the network topologies and processes of interaction between those entities; and the (C) membership view highlights the way in which the components operate and interactions occur within regions determined by spatiotemporal, functional, or conceptual boundaries.

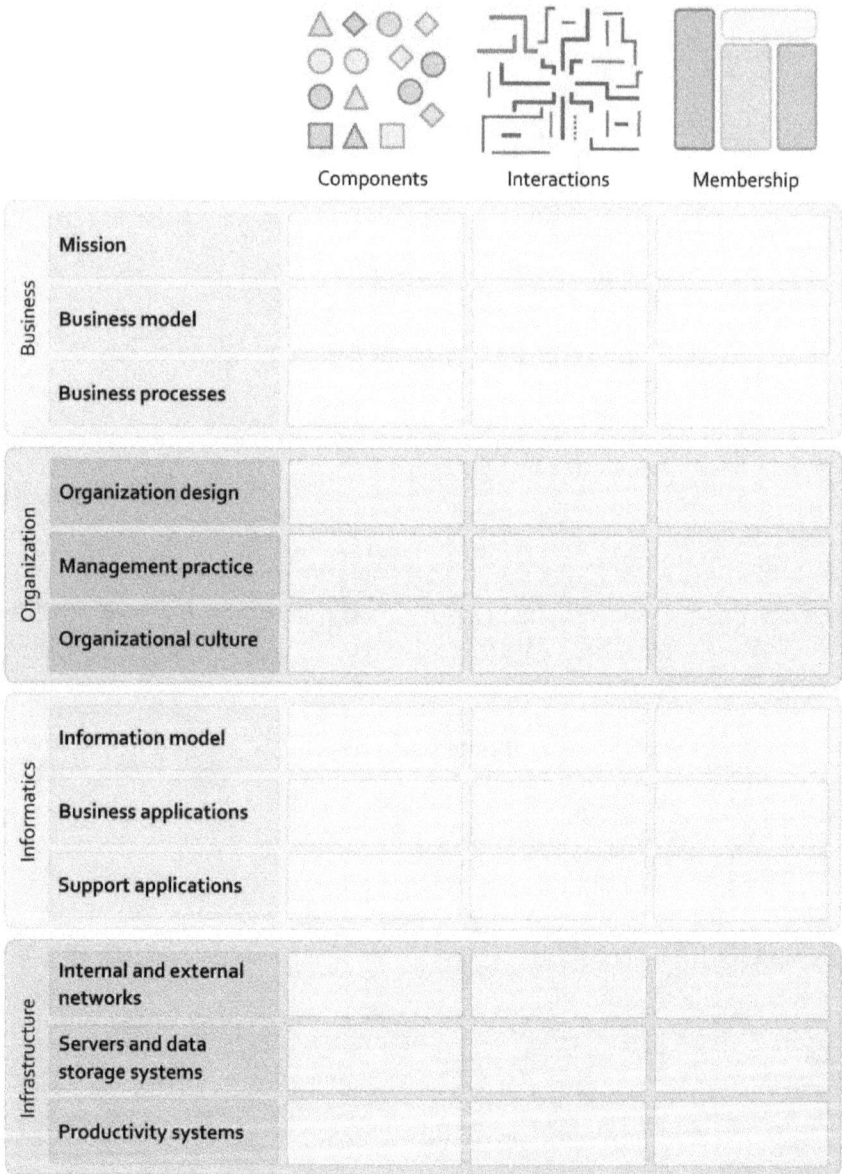

		Components	Interactions	Membership
Business	Mission			
	Business model			
	Business processes			
Organization	Organization design			
	Management practice			
	Organizational culture			
Informatics	Information model			
	Business applications			
	Support applications			
Infrastructure	Internal and external networks			
	Servers and data storage systems			
	Productivity systems			

Fig. 3: A generic template for creating an enterprise architecture 'landscape' or 'blueprint.' A particular organizational element (such as a specific business process or IT system) is chosen as the focus, and its relationship to the building blocks and domains is analyzed via the component, interaction, and membership views. Depending on the element chosen, some (less relevant) fields may be omitted or considered only generally, while (more relevant) others will be expanded into detailed sub-diagrams.

A common EA practice is to first prepare a landscape showing how, for example, a particular piece of scheduling software is currently being used within the organization; in the process of preparing the landscape, an enterprise architect might discover that one corporate division is utilizing the software to great advantage while other divisions that could potentially benefit from use of the software are underutilizing it or are not even aware of its existence. Alternatively, the enterprise architect might realize that the software is inadequate and should likely be replaced, by observing the fact that superior alternative software is already being used to greater effect within other parts of the organization. The work of preparing the landscape might also highlight business processes that are currently performed manually throughout the organization but which the scheduling software could easily automate for increased efficiency, or it might reveal ways in which the scheduling software is being used in an *ad hoc* and unsecure manner to handle business processes for which it was never intended.[79] Once a landscape has been prepared that documents the ways in which such an organizational element is currently being used, a new target landscape (or 'blueprint') can be designed that depicts an improved state of affairs in which the element is used more efficiently, securely, and productively to advance the organization's strategy and mission. A change management process can then be designed and implemented to move the element from its current state to its target state.

The landscape template shown in Figure 3 contains 36 fields to be taken into consideration; however, it is not necessary (or, typically, appropriate) to dedicate equal attention to all of the fields. EA frameworks should be freely adapted to suit the circumstances at hand; they are meant to provide a useful tool, not to create a burden.[80] Thus, depending on the organizational element that is the chosen focus of the landscape, an enterprise architect might reasonably decide to consider some of the fields only very generally or to omit them altogether – while other fields may be so significant that it is necessary to address them in detail, perhaps even expanding them to create new subdiagrams or additional documents. While detailed landscapes may be necessary for frontline personnel who implement or manage technological systems, blueprints displaying a greater degree of abstraction may be especially useful for explaining a target architecture to organizational stakeholders such

---

[79] The use of landscapes for, e.g., identifying systemic deficiencies or redundancies in which business processes are supported by either an insufficient number or superfluity of applications is discussed in Rohloff (2008), p. 10.

[80] The importance of adapting an EA framework to the unique needs and circumstances of the organization employing it is discussed in Haki et al., "Beyond EA Frameworks: Towards an Understanding of the Adoption of Enterprise Architecture Management" (2012), and Magoulas et al. (2012), pp. 90-91.

as senior executives who require a high-level overview of the architecture but do not need access to all of its underlying details.[81]

Standards and Modelling Languages for EA Notation

Enterprise architecture facilitates the implementation of standards across an organization's architectural building blocks, allowing greater interoperability and ease of communication between systems.[82] Modelling languages such as ArchiMate and UML are often employed as visualization and notation systems to aid in the implementation of such standards.[83] However, while UML is well-suited for some architecture-related tasks such as the deployment of new software systems, its conceptual and terminological focus on information technology means that UML is – at least in its raw form – not appropriate for notating enterprise architecture, which involves a much broader range of organizational components and phenomena, not all of which can be understood as providing a 'service' to end users in the way that software does.[84] Even a language like ArchiMate that has been optimized for use in enterprise architecture cannot easily represent 'soft' building blocks such as organizational culture or the leadership aspect of management practice, which are critical for the success of an EA plan.[85]

VII. Conclusion

As we have explored in this text, the discipline of enterprise architecture is defined by its effort to engineer alignment between an organization's information systems, human resources, business processes, workplace culture, mission and strategy, and external ecosystem in order to enhance the organization's ability to manage complexity, resolve internal conflicts, and adapt proactively to environmental change.

The field of enterprise architecture is a rich (and potentially bewildering) one: while EA's original IT-centric focus has generally evolved to take on a broader business-centric outlook, contrasting 'hard' and 'soft' approaches to EA exist and are reflected in many competing EA frameworks. What all EA frameworks share is a commitment to the creation of alignment as the core

---

[81] See Van der Torre et al., "Landscape Maps for Enterprise Architectures" (2006).

[82] Rohloff (2008), p. 6.

[83] See Caetano et al. (2009), p. 253, and *ArchiMate® 2.1 Specification* (2013).

[84] Rohloff (2008), p. 6.

[85] The importance of psychological, social, and cultural factors for successful EA implementation is discussed, e.g., in Magoulas et al. (2012); Weiss & Winter, "Development of Measurement Items for the Institutionalization of Enterprise Architecture Management in Organizations" (2012); and Stephan Aier, "The Role of Organizational Culture for Grounding, Management, Guidance and Effectiveness of Enterprise Architecture Principles" (2014).

mechanism for fashioning a deeper integration of an organization's structures, processes, and systems – which provides the organization with greater agility, which in turn strengthens the organization's ability to implement rapid and strategically directed change.

While enterprise architecture cannot by itself provide clear direction and objectives for an organization, it can beneficially complement, coordinate, and bring additional rigor to the work of other management disciplines such as strategic planning, organizational architecture, organization development, management cybernetics, and the development of business models. Moreover, EA's unique historical grounding in the field of information technology will render the discipline even more critical to businesses' success, as organizations increasingly grapple with the potential – and challenge – of harnessing emerging posthumanizing technologies such as those relating to social robotics, nanorobotics, artificial life, genetic engineering, neuroprosthetic augmentation, and virtual reality. Through the development and implementation of thoughtful enterprise architecture initiatives, organizations may better position themselves to encounter, withstand, and – where appropriate – exploit such dynamics of technologization and posthumanization that promise to transform the complex digital-physical ecosystems within which all contemporary organizations dwell.

Chapter Six

The Deepening Fusion of Human Personnel and Electronic Information Systems:
Implications of Neuroprosthetic Augmentation for Enterprise Architecture[1]

Abstract. When designing target architectures for organizations, the discipline of enterprise architecture has historically relied a set of assumptions regarding the physical, cognitive, and social capacities of the human beings serving as organizational members. In this text we explore the fact that for those organizations that intentionally deploy posthumanizing neuroprosthetic technologies among their personnel, such traditional assumptions no longer hold true: the use of advanced neuroprostheses intensifies the ongoing structural, systemic, and procedural fusion of human personnel and electronic information systems in a way that provides workers with new capacities and limitations and transforms the roles available to them.

Such use of neuroprostheses has the potential to affect an organization's workers in three main areas. First, the use of neuroprostheses may affect workers' physical form, as reflected in the physical components of their bodies, the role of design in their physical form, their length of tenure as workers, the developmental cycles that they experience, their spatial extension and locality, the permanence of their physical substrates, and the nature of their personal identity. Second, neuroprostheses may affect the information processing and cognition of neurocybernetically augmented workers, as manifested in their degree of sapience, autonomy, and volitionality; their forms of knowledge acquisition; their locus of information processing and data storage; their emotionality and cognitive biases; and their fidelity of data storage, predictability of behavior, and information security vulnerabilities. Third, the deployment of neuroprostheses can affect workers' social engagement, as reflected in their degree of sociality;

---

[1] This text draws heavily on "The Posthuman Management Matrix: Understanding the Organizational Impact of Radical Biotechnological Convergence" in Gladden, *Sapient Circuits and Digitalized Flesh: The Organization as Locus of Technological Posthumanization* (2016), pp. 133-201, which explores a broader range of phenomena (such as the emerging organizational significance of social robotics and artificial life) than is considered here.

relationship to organizational culture; economic relationship with their employers; and rights, responsibilities, and legal status.

While ethical, legal, economic, and functional factors will prevent most organizations from deploying advanced neuroprostheses among their personnel for the foreseeable future, a select number of specialized organizations (such as military departments) are already working to develop such technologies and implement them among their personnel. The enterprise architectures of such organizations will be forced to evolve to accommodate the new realities of human-computer integration brought about by the posthumanizing neuroprosthetic technologies described in this text.

Introduction

The discipline of enterprise architecture (EA) seeks to generate alignment between an organization's electronic information systems, human resources, business processes, workplace culture, mission and strategy, and external environment in order to increase the organization's agility and enhance its ability to manage complexity, resolve internal conflicts, and adapt proactively to environmental change.[2] As part of its work of designing and implementing target architectures for organizations, enterprise architecture has historically made a series of implicit assumptions regarding the physical, cognitive, and social capacities of the human beings that serve as organizational members. These assumptions presume that an organization's human workers differ fundamentally in their nature from the electronic information systems that the organization employs. For example, EA has always been able to take for granted the fact that:[3]

- The physical components of human workers are biological, while those of computers are electronic.

- The basic physical capacities of human workers are inherited from their parents through randomized biological processes, while those of electronic computers can be purposefully designed by engineers.

---

[2] For definitions of enterprise architecture, see Gammelgård et al., "An IT Management Assessment Framework: Evaluating Enterprise Architecture Scenarios" (2007); Cane & McCarthy, "Measuring the Impact of Enterprise Architecture" (2007); and Land et al., "Positioning Enterprise Architecture" (2009). Regarding EA's goals and benefits, see Rohloff, "Framework and Reference for Architecture Design" (2008); Boucharas et al., "The Contribution of Enterprise Architecture to the Achievement of Organizational Goals: A Review of the Evidence" (2010); Højsgaard, "Market-Driven Enterprise Architecture" (2011); Rohloff (2008); Hoogervorst, "Enterprise Architecture: Enabling Integration, Agility and Change" (2004); Buckl et al., "A Situated Approach to Enterprise Architecture Management" (2010); and Caetano et al., "A Role-Based Enterprise Architecture Framework" (2009).

[3] For more about these assumptions, see Gladden, "Enterprise Architecture for Neurocybernetically Augmented Organizational Systems" (2016).

- The means of 'upgrading' the capacities of human workers include techniques such as education, experience, physical exercise, and training, while electronic information systems are upgraded through techniques such as software updates, file downloads, the installation of additional memory chips, or the attachment of new peripherals.

- The locus of information processing and data storage for a human worker is the biological neural network of his or her brain, while for computers it includes components like CPUs, RAM chips, and non-volatile digital media.

- The main information security threats that directly target human workers include social engineering techniques, while those directly targeting computers include electronic hacking and malware.

Under these traditional assumptions, it would be grossly inappropriate, for example, for an enterprise architecture plan to assign a human worker the role of serving as a database that is responsible for accurately storing and manipulating the client records of millions of customers – a task that a computerized system could perform with ease. It would be similarly incoherent to attempt to design digital software applications that could be 'run' on the biological computing platforms comprising workers' minds and brains.

However, a broad range of posthumanizing neuroprosthetic technologies is now being developed that has the potential to reshape the sensory, cognitive, and motor capacities of human agents by integrating artificial computing devices directly into the neural circuitry of their natural biological organisms. For those organizations that choose to deploy such technologies among their personnel, traditional assumptions regarding the capacities and limitations of human workers – and their relationship to organizational information systems – will become increasingly obsolete. Such use of advanced neuroprostheses for purposes of human enhancement intensifies the ongoing structural, systemic, and procedural fusion of human personnel with electronic information systems in a way that transforms the roles and activities that enterprise architectures can assign to human workers.

Scope and Limitations of Our Analysis

In this text, we will explore the fact that posthumanizing neuroprostheses have the potential to affect workers especially in the three areas of their *physical form, information processing and cognition*, and *social engagement*. First, neuroprostheses may affect workers' physical form, as manifested in their bodies' physical components, the extent to which their physical form is the subject of organizational design, their length of tenure as workers, the developmental and operational cycles that they experience, their spatial extension and locality, the permanence of their physical substrates, and the nature of their personal identity. Second, neuroprostheses can affect the intellects of neuroprosthetically augmented workers, as reflected in their degree of sapience, autonomy, and volitionality; their means of knowledge acquisition; their locus of information processing and data storage; their emotionality and cognitive biases; and their fidelity of data storage, predictability of behavior, and information security vulnerabilities. Finally, the deployment of neuroprostheses may affect workers' social engagement, as manifested in their degree of sociality; relationship to organizational culture; economic relationship with their employers; and rights, responsibilities, and legal status. An overview of these impacts is presented in Figure 1.

Figure 1: Impacts of Posthumanizing Neuroprostheses on Human Personnel That Are Relevant for Enterprise Architecture

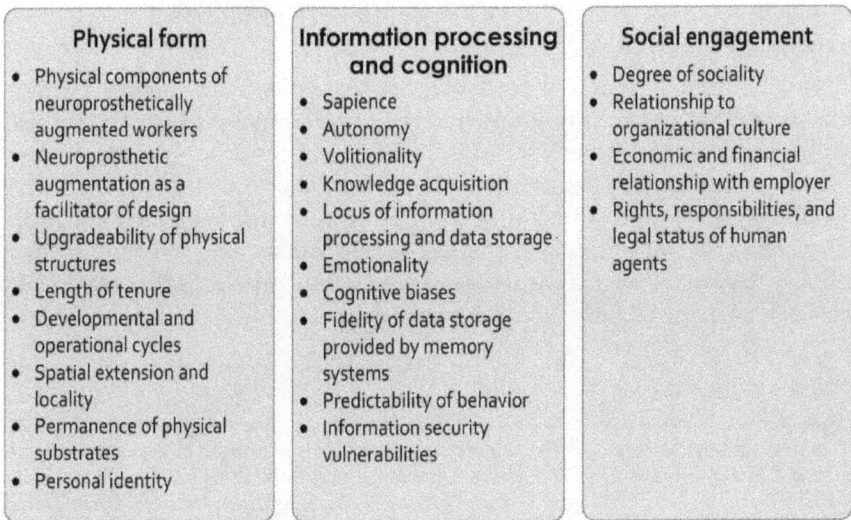

Physical form	Information processing and cognition	Social engagement
• Physical components of neuroprosthetically augmented workers • Neuroprosthetic augmentation as a facilitator of design • Upgradeability of physical structures • Length of tenure • Developmental and operational cycles • Spatial extension and locality • Permanence of physical substrates • Personal identity	• Sapience • Autonomy • Volitionality • Knowledge acquisition • Locus of information processing and data storage • Emotionality • Cognitive biases • Fidelity of data storage provided by memory systems • Predictability of behavior • Information security vulnerabilities	• Degree of sociality • Relationship to organizational culture • Economic and financial relationship with employer • Rights, responsibilities, and legal status of human agents

It is not claimed or expected that the types of neuroprosthetic augmentation discussed here will soon be purposefully exploited by a broad range of organizations: indeed, a complex array of ethical, legal, political, economic, and functional factors will prevent most organizations from deploying advanced neuroprostheses among their personnel for the foreseeable future. However, there already exists a small and specialized array of organizations (largely comprising military agencies and departments) that are actively seeking to develop and deploy among their personnel 'posthumanizing' neuroprosthetic technologies that grant some sensory, cognitive, or motor capacities which exceed or differ from those that are possessed by natural, unaugmented human beings and which can assist their human hosts in the performance of particular organizational roles.[4] The immediate relevance of this text is primarily for such organizations, which will be forced to transform their enterprise architecture practices and plans in order to address the deepening human-computer hybridization brought about by posthumanizing neuroprosthetic technologies.[5] The kinds of impacts discussed in this text will not be so strongly felt by organizations that include members who have personally acquired neuroprostheses (e.g., cochlear implants or deep brain stimulation devices) for therapeutic medical purposes, nor by organizations that deploy among their personnel neuroprostheses for therapeutic medical purposes (e.g., robotic prosthetic limbs provided by a military department to soldiers who have been injured during a tour of duty), insofar as the neuroprosthetic devices that are present within the workplace in such circumstances do not become integrated into an organization's institutional information systems and enterprise architecture.[6]

Finally, the discussion of particular neuroprosthetic technologies in this text should not be taken to imply that their intentional deployment and exploitation by organizations would be ethical or – necessarily – even legal. This text does not explicitly analyze the ethical or legal propriety of the organizational use of neuroprosthetic technologies; instead it investigates at a functional level the ways in which the use of posthumanizing neuroprostheses impacts the organizational structures, systems, and processes that are the objects of enterprise architecture. The problem of defining licit and illicit organ-

---

[4] For potential military use of neurotechnologies for human enhancement, see, e.g., Schermer, "The Mind and the Machine. On the Conceptual and Moral Implications of Brain-Machine Interaction" (2009); Brunner & Schalk, "Brain-Computer Interaction" (2009); Coker, "Biotechnology and War: The New Challenge" (2004); Graham, "Imagining Urban Warfare: Urbanization and U.S. Military Technoscience" (2008), p. 36; Krishnan, "Enhanced Warfighters as Private Military Contractors" (2015); Falconer, "Defense Research Agency Seeks to Create Supersoldiers" (2003); Moreno, "DARPA On Your Mind" (2004); Clancy, "At Military's Behest, Darpa Uses Neuroscience to Harness Brain Power" (2006); and Kourany, "Human Enhancement: Making the Debate More Productive" (2013), pp. 992-93.

[5] Similarly, our investigation of these issues may be of value to those policymakers, ethicists, futurists, and others who seek to regulate, evaluate, or anticipate the behavior of such organizations.

[6] Even in these cases, some changes to an organization's enterprise architecture might be required – e.g., to address the unique information security vulnerabilities possessed by neuroprostheses. Various enterprise architecture frameworks identify information security either as a discrete component within an architectural domain (see Hoogervorst (2004)) or as an element that underlies all architectural building blocks (see Rohloff (2008)). For the unique InfoSec challenges presented by neuroprosthetic devices, see Denning et al., "Neurosecurity: Security and Privacy for Neural Devices" (2009), and Gladden, *The Handbook of Information Security for Advanced Neuroprosthetics* (2015).

izational uses of such technologies – or determining whether such technologies are, in themselves, ethically or legally permissible – is one that requires thoughtful expert analysis. It is hoped that the questions explored in this text may support such analysis by robustly envisioning the ways in which organizations may attempt to exploit emerging neuroprosthetic technologies for strategic, operational, or tactical ends by incorporating them into their core enterprise architectures.

Impacts of Posthumanizing Neuroprosthetic Augmentation on Organizational Personnel

A neuroprosthesis can be defined as *an artificial device that is integrated into the neural circuitry of a human being.*[7] Historically, such devices have been used primarily for therapeutic medical purposes, to restore some capacity that is absent to due illness or injury; however, a growing range of neuroprostheses are being developed for purposes of human enhancement, to provide their human hosts with sensory, cognitive, or motor capacities that exceed what is possible for unmodified human beings.[8] Such 'posthumanizing' neuroprostheses help create a world in which natural biological human beings are no longer the only intelligent social agents who create meaning through their imagination, industry, and interaction; instead, artificial entities (such as social robots) and hybrid biological-electronic entities (such as neurocybernetically augmented human beings) will also contribute to the creation of meaning within the world's digital-physical ecosystems.[9]

In the following sections, we consider in detail the ways in which the deployment of posthumanizing neuroprostheses among an organization's human personnel may reshape such workers' physical form, information processing and cognition, and social engagement in ways that affect the organization's practice of enterprise architecture.

---

[7] See Lebedev, "Brain-Machine Interfaces: An Overview" (2014), and Gladden, "Enterprise Architecture for Neurocybernetically Augmented Organizational Systems" (2016), for more regarding the definition of a 'neuroprosthesis' and the basic division of such devices into sensory, cognitive, and motor neuroprostheses.

[8] For the distinction between 'traditional' neuroprostheses that are utilized for therapeutic medical purposes to restore some capacity that is absent to due illness or injury and more 'futuristic' neuroprostheses that are designed for purposes of human enhancement, see Gladden, *Sapient Circuits and Digitalized Flesh* (2016); Merkel et al., "Central Neural Prostheses" (2007); Gasson, "Human ICT Implants: From Restorative Applications to Human Enhancement" (2008); Gasson, "ICT implants" (2008); and McGee, "Bioelectronics and Implanted Devices" (2008).

[9] For a discussion of processes of posthumanization and the role of advanced neuroprostheses in processes of technologization and posthumanization, see, e.g., Herbrechter, *Posthumanism: A Critical Analysis* (2013); Ferrando, "Posthumanism, Transhumanism, Antihumanism, Metahumanism, and New Materialisms: Differences and Relations" (2013); and Gladden, *Sapient Circuits and Digitalized Flesh* (2016).

Physical Form

By deploying posthumanizing neuroprostheses, organizations expand the range of physical forms and capacities available to their human personnel – such as through the use of artificial limbs or sensory organs possessing non-biological materials and non-human shapes and dynamics. Below we consider a number of physical aspects of human agents and the ways in which the assumptions regarding the nature of those characteristics that have traditionally been relied upon by enterprise architects will no longer hold for organizations utilizing posthumanizing neuroprostheses.

Physical Components of Neuroprosthetically Augmented Workers

▶ **Historical assumption:** Human beings are composed of biological components that significantly limit the roles they can fill.

▶ **Impact of neuroprosthetic augmentation:** Human beings may incorporate electronic components that expand or further constrain the organizational roles they can fill.

The nature of the human body's physical composition means that the roles which an enterprise architecture plan assigns to a human agent have traditionally differed greatly from those that can be assigned to computerized systems. For example, a conventional computer is typically composed of mass-produced electronic components that are durable and readily repairable and whose behavior can easily be analyzed and predicted.[10] Such components are often able to operate in conditions of extreme heat, cold, pressure, or radiation in which biological material would not be able to survive and function. Electronic components can also be designed to incorporate significant functional redundancy, making them more reliable in the face of adverse operating conditions. Moreover, they can be manufactured either with large physical dimensions and structural reinforcement (potentially increasing durability and ease of inspection and repair) or in miniaturized form (thereby increasing mobility and undectability), depending on the intended purpose of a particular computer. The ability to manufacture electronic components to precise specifications with little variation means that millions of copies of a single artificial agent can be produced that are functionally identical.

On the other hand, enterprise architecture has traditionally been required to assume that the physical forms of human workers fall within a narrow

---

[10] For an in-depth review of the historical use of electronic components in computers as well as an overview of emerging possibilities for (non-electronic) biological, optical, and quantum computing, see Null & Lobur, *The Essentials of Computer Organization and Architecture* (2006). Regarding the degree to which the failure of electronic components can be predicted, see Băjenescu & Bâzu, *Reliability of Electronic Components: A Practical Guide to Electronic Systems Manufacturing* (1999).

range of possibilities and can be modified only to a very slight degree. Historically, the body of a human being is composed of biological material and not mechanical or electronic components; the qualities of such biological material place limits on the kinds of work that human employees can perform. Thus it is not possible for human beings to work in areas of extreme heat, cold, or radiation without extensive protection; nor is it possible for a human employee to work for hundreds of consecutive hours without taking breaks for sleep or meals or to use the restroom.

For organizations that deploy posthumanizing neuroprostheses, it is anticipated that the bodies of human agents will increasingly include electronic components in the form of artificial limbs and exoskeletons, artificial sense organs, memory implants, and other kinds of neuroprosthetic devices.[11] Such changes to the physical components or capacities of a human body may dramatically expand or further constrain the roles that human agents can be assigned within an enterprise architecture plan.

The Role of Neuroprosthetic Augmentation as a Facilitator of Design

▶ **Historical assumption:** The basic physical characteristics of human workers result largely from randomized processes of biological inheritance.

▶ **Impact of neuroprosthetic augmentation:** The basic physical characteristics of human workers may be intentionally engineered using neurotechnologies.

Enterprise architecture is, in many ways, a discipline based around effective *design* – of structures, processes, and systems. A major focus of EA's work involves designing the ways in which human workers interact with computerized systems, the ways in which computers interact with one another, and the internal components and dynamics of individual computing devices. The nature of electronic computers facilitates these kinds of design efforts: historically, almost all aspects of a conventional computer's physical form and basic functionality have been intentionally planned and constructed by human scientists, engineers, manufacturers, and programmers in order to enable the computer to successfully perform particular tasks.[12]

In that regard, the nature of computers differs greatly from that of contemporary human workers. Traditionally, enterprise architecture has been constrained by the fact that the basic physical form of a particular human

---

[11] See Gasson (2008); Gasson et al., "Human ICT Implants: From Invasive to Pervasive" (2012); McGee (2008); Merkel et al. (2007); and Gladden, "Enterprise Architecture for Neurocybernetically Augmented Organizational Systems" (2016).

[12] See, e.g., Dumas, *Computer Architecture: Fundamentals and Principles of Computer Design* (2006).

being is determined largely by genotypic factors resulting from the random-ized inheritance of genetic material from the individual's biological parents; the individual's particular biological components cannot be intentionally en-gineered.[13]

However, for those organizations that deploy posthumanizing neuropros-theses, human beings' basic physical and cognitive capacities will no longer simply be inherited in a randomized fashion from their biological parents but will increasingly be subject to explicit design by institutions or individual hu-man engineers through the use of neuroprosthetics and related technologies such as genetic engineering and nanorobotics that support it.[14] Besides the major moral and legal questions arising from such possibilities, there are also operational issues that confront organizations whose workforce includes hu-man agents who have been engineered in such ways. For example, the im-plantation of identical mass-produced neuroprostheses into a large number of workers may create synthetic cognitive or physical characteristics that are shared broadly across that population and which reduce its genotypic diver-sity; such homogenization may render the population more vulnerable to bi-ological or electronic hacking attempts (and may make such attempts more profitable and attractive for would-be adversaries). On the other hand, such standardization of biological structures and cognitive processes may make it easier for effective information security mechanisms (e.g., ones designed to detect and counteract social engineering) to be developed and deployed across the population.[15]

---

[13] Although, for example, factors such as diet, exercise and training, environmental conditions, pharmaceuticals, and medical procedures can extensively modify the form of a human body, the extent to which an existing biological human body can be restructured before ceasing to function is nonetheless relatively limited.

[14] For different perspectives on such possibilities, see, e.g., De Melo-Martín, "Genetically Modi-fied Organisms (GMOs): Human Beings" (2015); Regalado, "Engineering the perfect baby" (2015); Lilley, *Transhumanism and Society: The Social Debate over Human Enhancement* (2013); Nouvel, "A Scale and a Paradigmatic Framework for Human Enhancement" (2015); Section B ("Enhance-ment") in *The Future of Bioethics: International Dialogues*, edited by Akabayashi (2014); Mehlman, *Transhumanist Dreams and Dystopian Nightmares: The Promise and Peril of Genetic Engineering* (2012); and Bostrom, "Human Genetic Enhancements: A Transhumanist Perspective" (2012).

[15] For the relationship between the heterogeneity of information systems and their information security, see Gladden, *The Handbook of Information Security for Advanced Neuroprosthetics* (2015), p. 296, and *NIST Special Publication 800-53, Revision 4: Security and Privacy Controls for Federal Information Systems and Organizations* (2013), p. F-204.

Upgradeability of the Physical Structure of Human Agents

▶ **Historical assumption:** The capacities of human workers can be upgraded through education, training, exercise, and experience.

▶ **Impact of neuroprosthetic augmentation:** The capacities of human workers can potentially be upgraded by swapping electronic components or installing software updates.

Historically, enterprise architects have assumed that 'upgrading' the physical capacities and performance of human workers can only be achieved through age-old means such as education, training, physical exercise, and firsthand experience.[16] This differs from the situation of contemporary computers, which often can easily be upgraded through the addition or replacement of physical components that allow a computer to receive, for example, new sensory mechanisms, new forms of actuators for manipulating the external environment, an increase in processing speed, an increase in RAM, or an increase in the size of a computer's available space for the non-volatile long-term storage of data.[17]

Human workers have not demonstrated the sort of radical physical 'upgradeability' that might involve, for example, the implantation of additional memory capacity into the brain, an alteration of the fundamental rate of electrochemical communication between neurons to increase the brain's 'processing speed,' the addition of new sensory capacities (such as infrared vision), or the addition of new or different limbs or effectors (such as wheels instead of legs).[18]

However, for organizations that deploy posthumanizing neuroprostheses, the growing use of such technologies – perhaps supported, for example, by technologies for somatic cell gene therapy – may increasingly allow the physical components and cognitive capacities of human agents to be upgraded

---

[16] There are many congenital medical conditions that can be treated through conventional surgical procedures, medication, the use of traditional prosthetics, or other therapies. The application of such technologies could be understood as a form of 'augmentation' or 'enhancement' of one's body as it was naturally formed; however, such technologies are more commonly understood as 'restorative' approaches, insofar as they do not grant an individual physical elements or capacities that surpass those possessed by a typical human being. See Gasson (2012).

[17] See, e.g., Mueller, *Upgrading and Repairing PCs, 20th Edition* (2012).

[18] For the possibility of neuroprosthetically facilitated alteration to the physical form and structure of the human body, see, e.g., Warwick, "The Cyborg Revolution" (2014), and Gladden, "Cybershells, Shapeshifting, and Neuroprosthetics: Video Games as Tools for Posthuman 'Body Schema (Re)Engineering'" (2015).

and expanded even after the agents have reached a stage of physical and cognitive maturity.[19] In effect, the use of implantable neuroprostheses that possess swappable or modifiable components and whose behavior can be modified through software updates[20] may allow a human being's physical capacities and cognitive processes to be regularly updated and 'upgraded' in a way similar to that of desktop computers or other electronic hardware/software platforms.

Length of Tenure of Human Agents

▶ **Historical assumption:** A human worker can only fill an organizational role for a limited period of time, due to factors that are largely beyond the control of enterprise architecture.

▶ **Impact of neuroprosthetic augmentation:** A human worker (or functionally equivalent artificial replica of that worker) may be able to fill an organizational role on an extended or even permanent basis.

The period of service (or 'lifespan') of an organization's computerized information systems is conceptualized differently from the period of service (or 'employee tenure') of the organization's human workers. A typical computer does not possess a maximum lifespan beyond which it cannot be made to operate: as a practical matter, individual computers may eventually become obsolete because their functional capacities are inadequate to perform tasks that the computers' operator needs them to perform or because cheaper, faster, and more powerful types of computers have become available to carry out those tasks. Similarly, the failure of an individual component within a

---

[19] See, e.g., Panno, *Gene Therapy: Treating Disease by Repairing Genes* (2005); *Gene Therapy of the Central Nervous System: From Bench to Bedside*, edited by Kaplitt & During (2006); and Bostrom (2012).

[20] Note that it is by no means certain that all or even most posthumanizing neuroprostheses will possess such modifiability; a neuroprostheses which, for example, takes the form of a physical artificial neural network implanted deep within the brain may have no physical components that can be replaced after implantation and no conventional executable 'software' governing its behavior that can be remotely updated. For the possibility of neuroprosthetic devices that involve biological components or store information in biological or biomimetic neural networks, see Merkel et al. (2007); Rutten et al., "Neural Networks on Chemically Patterned Electrode Arrays: Towards a Cultured Probe" (2007); and Stieglitz, "Restoration of Neurological Functions by Neuroprosthetic Technologies: Future Prospects and Trends towards Micro-, Nano-, and Biohybrid Systems" (2007).

Conversely, in some cases, periodic physical upgrades to an implanted neuroprosthesis may be required in order to maintain the device's functional capabilities and ensure the health and information security of its human host. See *Postmarket Management of Cybersecurity in Medical Devices: Draft Guidance for Industry and Food and Drug Administration Staff* (2016) and Gladden, "Information Security Concerns as a Catalyst for the Development of Implantable Cognitive Neuroprostheses" (2016).

computer may render it temporarily nonfunctional; however, the ability to repair, replace, upgrade, or expand a computer's physical components means that a computer's operability can generally be maintained indefinitely, if its owner or operator wishes to do so.[21]

On the other hand, an EA plan needs to account for the fact that the identity of the human agent filling a particular role within an organization is expected to change periodically, due to circumstances beyond an enterprise architect's control – such as the relatively short average period of time during which human workers serve as members of an organization.[22] Some such changes result from workers joining or leaving an organization through processes of hiring, resignation, or termination. Other such changes may unfortunately sometimes occur as a result of the death or disability of an individual worker. While the biological lifespan of a particular human worker can be shortened or extended to some degree as a result of environmental, behavioral, or other factors, the human organism is generally understood to possess a finite biological lifespan that cannot be extended indefinitely through natural biological means.[23] A human being who has exceeded his or her maximum lifespan is no longer alive – that is, he or she will have expired – and cannot be repaired and revived by technological means to make him or her available once again for future organizational use. Instead, he or she will be replaced by another human agent who may possess significantly different characteristics. The possibility of such eventualities is especially significant, for example, for some types of military organizations in which human workers operate in highly dangerous circumstances where the loss of life is not a rare occurrence, especially during periods of intense activity such as combat operations. In these cases, enterprise architects have historically needed to account for periodic non-continuous changes in the characteristics of the human agent filling a particular role.

However, a human agent whose neurons and neural functioning can be maintained, supplemented, or superseded by neuroprosthetic technologies after they have deteriorated or become damaged – or which can be protected from undergoing damage or deterioration in the first place – could potentially

---

[21] For an overview of issues relating to computer reliability, availability, and lifespan, see Siewiorek & Swarz, *Reliable Computer Systems: Design and Evaluation* (1992), and Băjenescu & Bâzu (1999).

[22] For example, in the United States "the median number of years that wage and salary workers had been with their current employer was 4.2 years in January 2016, down from 4.6 years in January 2014." See "Employee Tenure Summary" (2016).

[23] For a discussion and comparison of biologically and nonbiologically based efforts at human life extension, see Koene, "Embracing Competitive Balance: The Case for Substrate-Independent Minds and Whole Brain Emulation" (2012).

experience an extended or even indefinite lifespan, although such engineering might result in detrimental side-effects that render such lifespan extension highly undesirable. Neuroprosthetic technologies may also facilitate practices such as 'mind uploading' or the creation of virtual, artificially intelligent replicas of particular human workers, thereby allowing such quasi-human agents to fill an organizational role permanently and without undergoing processes of development or decline that alter their basic capacities over time.[24]

Developmental and Operational Cycles

▶ **Historical assumption:** A human worker's capacities and performance characteristics are continually evolving in ways that may or may not be desirable and which are largely beyond the control of enterprise architecture.

▶ **Impact of neuroprosthetic augmentation:** The natural developmental cycles of human workers might potentially be slowed, accelerated, or bypassed through the use of neuroprostheses. Human workers might be able to function for extended periods of time without demonstrating undesired changes in their capacities or performance.

A computer's physical form is highly stable: although a computer's components can be physically upgraded or altered by the device's owner or operator, a computer does not physically upgrade or alter itself without its operator's knowledge or permission;[25] a computer does not undergo the sort of developmental cycle of conception, growth, maturity, and senescence demonstrated by biological organisms. In general, the physical alterations made to a computer are reversible: a chip that has been installed to increase the computer's RAM can be removed; a peripheral device that has been added can be disconnected. This allows a computer to be restored to a previous physical and functional state.

Expectations for the operational cycles of human workers have always been of a different sort. Traditionally, enterprise architecture has assumed

---

[24] Transhumanist perspectives on mind uploading are presented, e.g., in Moravec, *Mind Children: The Future of Robot and Human Intelligence* (1990), and Koene (2012). For more critical perspectives on problems inherent in visions of technologically facilitated life extension or the replacement of a human being's entire original biological body via mind uploading, see Proudfoot, "Software Immortals: Science or Faith?" (2012); Pearce, "The Biointelligence Explosion" (2012); and Hanson, "If uploads come first: The crack of a future dawn" (1994).

[25] An exception would be the case of malware that can cause a computer to disable or damage some of its internal components or peripheral devices without the owner or operator's permission. See, e.g., Kerr et al., "The Stuxnet Computer Worm: Harbinger of an Emerging Warfare Capability" (2010).

that a role within an organization cannot be filled permanently by a single human agent possessing a stable and unchanging set of capacities; the unique characteristics of the human agent filling a role are transformed over time. This is due largely to the fact that the physical structure and capacities of a human being do not remain unaltered from the moment of an individual's conception to the moment of his or her death; instead, a human being's physical form and abilities undergo continuous change as the individual develops through a cycle of infancy, adolescence, adulthood, and senescence.[26] From the perspective of enterprise architecture, human beings are only capable of serving as workers during particular phases of this developmental cycle, and the unique strengths and weaknesses displayed by human workers vary as they move through that cycle.

An organization that deploys posthuman neuroprostheses might conceivably exploit such technologies to speed the natural biological processes that contribute to physical growth and cognitive development or to slow or block processes of physical and cognitive decline. Scholars also envision the possibility of neuroprostheses being used to allow human beings to instantly acquire new knowledge or skills through the implantation of memory chips or the downloading of information into one's mind; if indeed feasible, this could allow human cognitive capacities to be maintained or enhanced in a way that bypasses typical human processes of cognitive development and learning.[27]

Spatial Extension and Locality

▷ **Historical assumption:** A human worker occupies a single physical location at any given moment.

▷ **Impact of neuroprosthetic augmentation:** A human worker may virtually inhabit multiple environments simultaneously and interact with them using a human-like or non-human-like body.

The systems defined by an enterprise architecture are not abstract immaterial entities; they are embodied within concrete physical objects. The size, form, and mobility of such objects and the nature of their physical interfaces largely determine the kinds of organizational processes that can be executed by such systems.

---

[26] See Thornton, *Understanding Human Development: Biological, Social and Psychological Processes from Conception to Adult Life* (2008), and the *Handbook of Psychology, Volume 6: Developmental Psychology*, edited by Lerner et al. (2003).

[27] Such possibilities are discussed, e.g., in Spohrer, "NBICS (Nano-Bio-Info-Cogno-Socio) Convergence to Improve Human Performance: Opportunities and Challenges" (2002); McGee (2008); and Warwick (2014), p. 267. Experimental technologies being tested in mice that allow the manipulation of memories are presented in Ramirez et al., "Creating a False Memory in the Hippocampus" (2013).

It is possible for a computer to – like a human being – possess a body that comprises a single unitary, spatially compact physical unit: computerized devices such as a typical desktop computer, smartphone, assembly-line robot, or server may possess a physical form that is clearly distinct from the device's surrounding environment and which is located in only a single place at any given time. However, other computers can – unlike a human being – possess a body comprising disjoint, spatially dispersed elements that exist physically in multiple locations at the same time. The creation of such computerized entities comprising many spatially disjoint and dispersed 'bodies' has been especially facilitated in recent decades by the development of the diverse networking technologies that undergird the Internet and, now, the nascent Internet of Things.[28] The destruction, disabling, or disconnection of one of these bodies that contributes to the form of such an entity may not cause the destruction of – or even a significant degradation in functionality for – the computerized entity as a whole.

On the other hand, enterprise architecture has traditionally presumed that a particular human being occupies or comprises a particular physical biological body. Because that body is unitary – consisting of a single spatially compact unit – a human being is able to inhabit only one space at a given time. While the use of technologies such as telephony, email, instant messaging, and videoconferencing allows a human worker to engage with colleagues around the world and to be 'present' in distant locations in a limited and metaphorical sense, a human being cannot, for example, literally be physically present in multiple cities simultaneously.[29]

When posthumanizing neuroprostheses are deployed, the situation becomes more complex. The use of sensorimotor neuroprosthetic devices and virtual reality technologies may effectively allow a human agent to 'inhabit' a location different from that in which his or her physical body is being housed. The inhabited environment may be either a virtual representation of some distant real-world location or a fabricated virtual environment (such as that presented within a massively multiplayer online roleplaying game, or

---

[28] Regarding the Internet of Things, see Evans, "The Internet of Everything: How More Relevant and Valuable Connections Will Change the World" (2012). For one aspect of the increasingly networked nature of robotics and AI, see Coeckelbergh, "From Killer Machines to Doctrines and Swarms, or Why Ethics of Military Robotics Is Not (Necessarily) About Robots" (2011). Regarding the development of computerized entities with physically disjoint bodies, see Gladden, "The Diffuse Intelligent Other: An Ontology of Nonlocalizable Robots as Moral and Legal Actors" (2016).

[29] The extent to which telepresence in remote locations is cognitively and physically possible for human beings is discussed, e.g., in Salvini et al., "From robotic tele-operation to tele-presence through natural interfaces" (2006).

MMORPG) that does not depict a real-world location. Within such virtualized environments, a human agent might occupy multiple bodies that are potentially of a radically nonhuman nature.[30] He or she might also exist in a way that is extremely 'multilocal' by being present in and interacting with many different environments simultaneously.[31]

Permanence of Physical Substrates

▶ **Historical assumption:** All of the actions of a single human agent are performed by and associated with a single physical substrate – that human being's biological body.

▶ **Impact of neuroprosthetic augmentation:** Some forms of human (or quasi-human) agency may exist and act independent of a fixed biological substrate.

The physical substrates (i.e., biological bodies) within which human agents subsist and perform their work within the world have a much different nature than those electronic and electromechanical physical substrates within which computers perform their computation and other information-processing activities.

Because they are stored in an electronic digital form that can easily be read and written, the data that constitute a particular computer's operating system, applications, configuration settings, activity logs, user files, and other information that has been received, generated, or stored by the device can easily be copied to different storage components or to a different computer altogether.[32] This means that the computational substrate or 'body' of a given computerized system can be replaced with a new body without causing functional changes in the system's memory or behavior.[33] In the case of comput-

---

[30] Regarding the potential for and ramifications of long-term immersion in virtual reality environments, see, e.g., Bainbridge, *The Virtual Future* (2011); Heim, *The Metaphysics of Virtual Reality* (1993); and Koltko-Rivera, "The potential societal impact of virtual reality" (2005). Regarding psychological, social, and political questions relating to repetitive long-term inhabitation of virtual worlds through a digital avatar, see, e.g., Castronova, "Theory of the Avatar" (2003). On implantable systems for augmented or virtual reality, see Sandor et al., "Breaking the Barriers to True Augmented Reality" (2015), pp. 5-6. The potential role of neuroprosthetics in granting human beings virtual bodies of a radically nonhuman nature is analyzed in Gladden, "Cybershells, Shapeshifting, and Neuroprosthetics" (2015).

[31] For a discussion of multilocality, see Gladden, "The Diffuse Intelligent Other" (2016).

[32] An overview of various aspects of information storage in electronic systems is presented in *Information Storage and Management: Storing, Managing, and Protecting Digital Information in Classic, Virtualized, and Cloud Environments*, edited by Gnanasundaram & Shrivastava (2012).

[33] Such abilities are exploited to allow organizations to implement 'failover capacity' by which a

erized systems that are typically accessed remotely (e.g., a cloud-based storage device accessed through the Internet), a system's hardware could potentially be replaced by copying the device's data to a new device without remote users or operators ever realizing that the system's physical computational substrate had been swapped.[34]

On the other hand, enterprise architecture has traditionally assumed that the unique set of skills, knowledge, and memories possessed by a particular human worker are associated with and contained in the physical body of that worker. Although to some limited extent it is possible to modify or replace physical components of a human body, it is not possible for a human being to exchange his or her entire body for another.[35] The body with which a human being was born will – notwithstanding the natural changes that occur as part of its lifelong developmental cycle or any minor intentional modifications – serve as a single permanent substrate within which all of the individual's information processing and cognition will occur and in which all of the individual's sensory and motor activity will take place until the end of his or her life. The dissolution of that body entails the end of that human being's ability to act as an agent within the environment.

The deployment of posthumanizing neuroprostheses challenges these assumptions. Ontologically and ethically controversial practices such as the development of artificial neurons to replace natural biological neurons and various approaches to 'mind uploading' (many of which are facilitated by advanced neuroprosthetics) might someday allow a single human agent's agency to exist and act beyond the physical confines of the agent's original biological physical substrate – but only if terms such as 'agent' and 'agency' are understood in a transhumanist fashion whose coherency and validity are deeply disputed.[36] Similarly, the use of neuroprosthetically mediated cyber-

---

standby duplicate information system can automatically be brought online to take over the work of a primary information system that has suffered a major fault. See, e.g., *NIST SP 800-53* (2013), p. F-231.

[34] The ability to replace or reconfigure remote networked hardware without impacting web-based end users is widely exploited to offer cloud-based services employing the model of infrastructure as a service (IaaS), platform as a service (PaaS), or software as a service (SaaS); for more details, see the *Handbook of Cloud Computing*, edited by Furht & Escalante (2010).

[35] For problems and complications relating to proposed body-replacement techniques such as mind uploading, see Proudfoot (2012); for particular problems that would result from the attempt to adopt a nonhuman body, see Gladden, "Cybershells, Shapeshifting, and Neuroprosthetics" (2015).

[36] Critical posthumanism offers a vigorous critique of such transhumanist and 'techno-idealist' notions of human agency; for a presentation of critical posthumanist positions, see Hayles, *How*

netic networks to create 'hive minds' or other forms of collective agency involving human agents might allow such multi-agent systems or 'super-agents' to survive and function despite the fact that a continual addition and loss of biological substrates may mean that the entity's substrate at one moment in time shares no components in common with its substrate at a later point in time.[37]

Identity of Human Agents

▶ **Historical assumption:** Roles can be assigned to and decisions and actions attributed to individual human beings.

▶ **Impact of neuroprosthetic augmentation:** Roles may be filled by new types of collective human entities; actions may be unattributable to particular human beings.

In a practical sense, enterprise architecture (and the systems and processes that it administers) must often robustly track and account for the 'identity' of desktop computers, smartphones, and other devices operating within an organization's digital-physical ecosystem; various mechanisms are employed for such purposes. However, on a deeper philosophical level it is unclear wherein the unique identity of a conventional computer or computerized entity subsists, or even if such an identity exists.[38] A computer's identity does not appear to be tied to any critical physical component, as such components can be replaced or altered without destroying the computer. Similarly, a computer's identity does not appear to be tied to a particular set of digital data that comprises the computer's operating system, applications, and user data, as that data can be copied with perfect fidelity to other devices, creating computers that are functionally clones of one another.

The identity of human workers has traditionally been easier to track. Enterprise architecture has historically incorporated the assumption that a human being's body creates (or at least, plays a necessary role in creating) a

---

We Became Posthuman: Virtual Bodies in Cybernetics, Literature, and Informatics (1999); Herbrechter (2013), pp. 94, 185-86; and Gladden, Sapient Circuits and Digitalized Flesh (2016).

[37] For classification systems for potential types of hive minds, see Chapter 2, "Hive Mind," in Kelly, Out of Control: The New Biology of Machines, Social Systems and the Economic World (1994); Kelly, "A Taxonomy of Minds" (2007); Kelly, "The Landscape of Possible Intelligences" (2008); Yonck, "Toward a standard metric of machine intelligence" (2012); and Yampolskiy, "The Universe of Minds" (2014). For critical perspectives on the possibility of hive minds, see, e.g., Bendle, "Teleportation, cyborgs and the posthuman ideology" (2002), and Heylighen, "The Global Brain as a New Utopia" (2002).

[38] From the perspective of information security, techniques for device attestation are utilized to identify and authenticate a device on the basis of its configuration and unique operating state; see NIST SP 800-53 (2013), p. F-94. Such techniques are not infallible, however.

single identity for the individual that persists over time. The fact that each human body is unique and is identifiable to other human beings (e.g., that such a body is not invisible, microscopic, or 'flickering' in and out of existence from moment to moment) means that it is possible to associate some human action with the particular human being who performed it and to assign roles and functions within an organization to a particular identifiable human worker.[39]

For organizations that have deployed posthumanizing neuroprosthetic technologies, it may become difficult or impossible to attribute actions to a specific human agent or even to identify which human agent is occupying and utilizing a particular physical body in a given moment. For example, a single networked effector (such as a robotic armature, remote-controlled drone, or LED display screen[40]) might simultaneously belong to the bodies of multiple human agents who jointly or alternately control its behavior, and a single artificial eye might in effect simultaneously belong to the bodies of multiple human users whose minds each receive a live stream of the visual input received by the device. The ability of neuroprosthetically mediated cybernetic networks to create hive minds and other forms of collective consciousness among human and artificial agents may also make it difficult to identify which human agent, if any, is present in a particular physical or virtual environment and is carrying out the behaviors observed there. Actions may instead be determined by and attributable to the system as a whole.[41]

Information Processing and Cognition

Posthumanizing neuroprosthetics has the potential to dramatically reshape the range of information-processing mechanisms and behaviors available to human workers – and thus the forms and degree of cognition that they are capable of displaying. Below we consider a number of intellectual characteristics of human agents and the ways in which the historical assumptions regarding the nature of those characteristics that have been relied upon by

---

[39] For a discussion of philosophical issues relating to the type of personal identity possessed by human beings, see Olson, "Personal Identity" (2015); see also Friedenberg, *Artificial Psychology: The Quest for What It Means to Be Human* (2008), p. 250.

[40] Regarding the existing or potential use of neuroprostheses to control external systems as diverse as desktop computers, vehicles, smart buildings, 3D printers and other manufacturing systems, domestic robots, microphones and speakers, cameras and displays, or game systems, see McGee (2008), pp. 213-15; Warwick (2014), p. 266; Gladden, "Cybershells, Shapeshifting, and Neuroprosthetics" (2015); and Gladden, "Enterprise Architecture for Neurocybernetically Augmented Organizational Systems" (2016).

[41] For the possibility of neuroprosthetically facilitated networks that make collective decisions, see the previously noted sources relating to hive minds as well as Gladden, "Utopias and Dystopias as Cybernetic Information Systems: Envisioning the Posthuman Neuropolity" (2015). A more general discussion of collectively conscious networks and a "post-internet sentient network" is found in Callaghan, "Micro-Futures" (2014).

enterprise architects will become increasingly outdated within organizations utilizing posthumanizing neuroprostheses.

Sapience

▶ **Historical assumption:** Human workers possess a self-awareness that allows them to recognize, analyze, and respond to ambiguous or unexpected circumstances.

▶ **Impact of neuroprosthetic augmentation:** Human workers' self-awareness may be impaired or destroyed in ways that are not necessarily detectable to external observers.

Enterprise architecture has historically presumed that a conventional computer does not possess sapient self-awareness or a subjective conscious experience of reality,[42] while a human worker does. The typical human worker possesses a subjective conscious experience that is not simply sensations of physical reality but a conceptual 'awareness of' and 'awareness that.' These characteristics are not found, for example, in infants or in adult human beings suffering from certain medical conditions such as a coma.[43] In a sense, a typical adult human being can be said to possess sapient self-awareness as a capacity even when the individual is unconscious (e.g., during sleep), although in that moment the capacity is latent and is not being actively utilized or experienced.[44] This sort of self-awareness allows human workers to, for example, recognize (and recognize the need to *respond* to) changing environmental conditions, emergencies, and other exigencies that fall outside of whatever narrow task instructions they might have received from their organizational supervisors.

This situation may be altered through the deployment of posthumanizing neuroprostheses within an organization. By interfering with or altering the biological mechanisms that support consciousness and self-awareness within the brain, neuroprostheses could deprive particular human agents of sapi-

---

[42] For different perspectives on the characteristics that a computer or other artificial system would need to have in order for it to possess sapient self-awareness and a subjective conscious experience of reality, see Friedenberg (2008), pp. 163-78.

[43] Regarding the ways in which consciousness is (perhaps only temporarily) impaired during a coma, see, e.g., *Coma Science: Clinical and Ethical Implications*, edited by Laureys et al. (2009), and *Comas and Disorders of Consciousness*, edited by Schnakers & Laureys (2012).

[44] Such issues are discussed, e.g., in Siewert, "Consciousness and Intentionality" (2011); Fabbro et al., "Evolutionary aspects of self-and world consciousness in vertebrates" (2015); and Boly et al., "Consciousness in humans and non-human animals: recent advances and future directions" (2013).

ence, even if those agents outwardly appear to remain fully functional as human beings; for example, a human agent might retain his or her ability to engage in social interactions with longtime friends – not because the agent's mind is conscious and aware of such interactions, but because a sufficiently sophisticated artificially intelligent neuroprosthetic device is orchestrating the agent's sensorimotor activity.[45]

Autonomy

▷ **Historical assumption:** Human workers display a high degree of intellectual and physical autonomy that allows them to act without direct supervision or support.

▷ **Impact of neuroprosthetic augmentation:** The degree of intellectual and physical autonomy possessed by human workers may be artificially constrained or expanded by means of neuroprostheses.

The forms of autonomy displayed by human workers and computerized systems have historically differed greatly. For computerized devices such as robots, autonomy can be understood as the state of being "capable of operating in the real-world environment without any form of external control for extended periods of time."[46] Such autonomy does not simply involve the ability to perform cognitive tasks like setting goals and making decisions; it also requires an entity to successfully perform physical activities such as securing energy sources and carrying out self-repair without human intervention. Applying this definition, we can say that current computerized devices are typically either nonautonomous (e.g., telepresence robots that are fully controlled by their human operators) or semiautonomous (e.g., robots that require 'continuous assistance' or 'shared control' in order to fulfill their intended purpose).[47] Although some contemporary computerized systems can be understood as being autonomous with regard to fulfilling their intended purpose – in that they can receive sensory input, process information, make decisions, and perform actions without direct human control – they are not autonomous in the full sense of the word, insofar as they are generally not capable of, for example, securing energy sources within the environment or repairing physical damage to themselves.[48]

---

[45] The potential for such misuse of neuroprosthetic technologies is discussed in Gladden, *The Handbook of Information Security for Advanced Neuroprosthetics* (2015), pp. 98, 220.

[46] Bekey, *Autonomous Robots: From Biological Inspiration to Implementation and Control* (2005), p. 1.

[47] See Murphy, *Introduction to AI Robotics* (2000).

[48] The degree of autonomy of different types of robots is analyzed in Gladden, "Managerial Robotics: A Model of Sociality and Autonomy for Robots Managing Human Beings and Machines" (2014).

On the other hand, enterprise architects have historically been able to count on the fact that human workers possess a high degree of autonomy that is reflected at both the intellectual and physical levels. Through the regular action of his or her mind and body, a typical human being is able to secure energy sources and information from the external environment, set goals, make decisions, perform actions, and even (to a limited extent) repair damage that might occur to himself or herself during the course of daily activities, all without direct external guidance or control by other human agents.

Some kinds of posthumanizing neuroprostheses that may someday be deployed within organizations might weaken the desires or strategic planning capacities of human agents or subject them to the control of external agents, thereby reducing their autonomy. New kinds of neuroprosthetically facilitated social network topologies that link the minds of human agents to create hive minds or other forms of merged consciousness can also reduce the autonomy of their individual members. Neuroprosthetic augmentation that renders human agents dependent on their employer or other organizations for ongoing hardware or software upgrades or medical support would similarly reduce the autonomy of those agents at the social and economic levels.[49] On the other hand, technologies that allow human agents to survive and operate in hostile environments or to reduce or repair physical damage to their bodies would enhance such agents' autonomy.

Volitionality

▶ **Historical assumption:** Human workers possess a conscience that allows them to self-correct unethical or otherwise problematic workplace behaviors and autonomously optimize their performance.

▶ **Impact of neuroprosthetic augmentation:** Neuroprostheses might artificially impair, enhance, or otherwise alter users' exercise of conscience with regard to work-related or other activities.

Enterprise architecture has historically incorporated certain assumptions regarding human workers' degree of volitionality, which relates to an entity's

---

[49] For the possibility that neuroprosthetic devices will render their human hosts biologically, psychologically, financially, or socially dependent on the organizations that provide or maintain such technologies, see Koops & Leenes, "Cheating with Implants: Implications of the Hidden Information Advantage of Bionic Ears and Eyes" (2012), p. 125; McGee (2008), p. 213; and Gladden, "Neural Implants as Gateways to Digital-Physical Ecosystems and Posthuman Socioeconomic Interaction" (2016).

ability to self-reflexively shape the intentions that guide its actions.[50] An entity is *nonvolitional* when it possesses no internal goals or 'desires' for achieving particular outcomes nor any expectations or 'beliefs' about how performing certain actions would lead to particular outcomes. An entity is *volitional* if it combines goals with expectations: in other words, it can possess an intention,[51] which is a mental state that comprises both a desire and a belief about how some act that the entity is about to perform can contribute to fulfilling that desire.[52]

Many conventional computerized devices are nonvolitional; however, a growing number of contemporary computerized devices – including a wide variety of robots used in commercial contexts – are volitional. For example, a therapeutic social robot might possess the goal of evoking a positive emotional response in its human user, and its programming and stored information tells it that by following particular strategies for social interaction it is likely to evoke such a response.[53]

Typical adult human beings, meanwhile, can be described as *metavolitional:* they possess what scholars have referred to as 'second-order volitions,' or intentions about intentions.[54] In human beings, this metavolitionality manifests itself in the form of conscience: as a result of possessing a conscience, human agents are able to determine that they do not wish to possess some of the intentions that they are currently experiencing, and they can resolve to change those intentions. Such metavolitionality allows human workers, for example, to avoid unethical behaviors or to decide that they wish to change their workplace behaviors in order to improve their performance or achieve particular goals.

Within organizations deploying posthumanizing neuroprostheses, workers' possession and manifestation of such cognitive dynamics may be altered. Researchers have already observed ways in which certain kinds of neuroprosthetic devices can affect their human host's capacity to possess desires,

---

[50] For a discussion of the volitionality of agents, see Calverley, "Imagining a non-biological machine as a legal person" (2008), pp. 529-35, and Gladden, "The Diffuse Intelligent Other" (2016).

[51] The term 'intentionality' is often employed in a philosophical sense to describe an entity's ability to possess mental states that are directed toward (or 'about') some object; that is a broader phenomenon than the possession of a particular 'intention' as defined here.

[52] See Calverley (2008), p. 529.

[53] The nature and capacities of such social robots are discussed, e.g., in Breazeal, "Toward sociable robots" (2003); Gockley et al., "Designing Robots for Long-Term Social Interaction" (2005); Kanda & Ishiguro, *Human-Robot Interaction in Social Robotics* (2013); *Social Robots and the Future of Social Relations*, edited by Seibt et al. (2014); *Social Robots from a Human Perspective*, edited by Vincent et al. (2015); and *Social Robots: Boundaries, Potential, Challenges*, edited by Nørskov (2016).

[54] See Calverley (2008), pp. 533-35.

knowledge, and belief;[55] insofar as technologies disrupt or control such abilities, they may impair their human host's exercise of his or her conscience, which depends on the possession of these capacities. This may result in the existence of human agents that are no longer fully metavolitional but instead merely volitional or nonvolitional. Conversely, neuroprostheses that enhance such basic cognitive capacities may strengthen their users' manifestations of conscience. An artificially intelligent neuroprosthesis might also serve as an external 'supplemental conscience' that detects its user's plan to perform some unethical or undesirable action, alerts the user to that fact, and attempts to persuade the user to adopt a different course of action.[56] The use of neuroprosthetics, virtual reality, and other technologies to create hive minds and other forms of collective consciousness among human agents may also impair the volitionality of those agents participating in such systems and reduce them to a state that is less than metavolitional; each agent may no longer nurture its own individual conscience but may instead help to form (and be guided by) the conscience of the multi-agent system as a whole.

Knowledge Acquisition

▶ **Historical assumption:** Human workers acquire new knowledge through processes of learning, study, and experience that require an extended period of effort.

▶ **Impact of neuroprosthetic augmentation:** It may be possible to instantaneously 'program' workers with new knowledge or for workers themselves to download new skill-sets or acquire any desired knowledge instantly online through a mere act of will.

Many approaches exist for rapidly expanding the quantity of declarative and procedural knowledge available to a computerized information system. For example, a conventional computer may have software programs and data files copied onto its storage media, thereby instantaneously gaining new capacities and the possession of new information.[57] Alternatively, a traditional

---

[55] Regarding the possibility of developing neuroprostheses that affect emotions and perceptions of personal identity and authenticity, see Soussou & Berger, "Cognitive and Emotional Neuroprostheses" (2008); Hatfield et al., "Brain Processes and Neurofeedback for Performance Enhancement of Precision Motor Behavior" (2009); Kraemer, "Me, Myself and My Brain Implant: Deep Brain Stimulation Raises Questions of Personal Authenticity and Alienation" (2011); Van den Berg, "Pieces of Me: On Identity and Information and Communications Technology Implants" (2012); and McGee (2008), p. 217.

[56] The notion that some types of future neuroprostheses might in effect be capable of serving as advisors or counselors for their human hosts is raised in Gladden, "Neural Implants as Gateways" (2016), and Koops & Leenes (2012), p. 119.

[57] For a discussion of the ways in which the electronic components of traditional computers carry

computer may be directly programmed or configured by a human operator. It does not 'learn' through experience, nor does it undergo a long-term formative process of education in order to acquire new knowledge or information.[58]

On the other hand, enterprise architecture has historically presumed that a very different range of practices is available for instilling new knowledge in the minds of human workers. The cognitive processes and knowledge of a human being are shaped through an initial process of concentrated learning and formal and informal education that lasts for several years and through an ongoing process of learning that lasts throughout the individual's lifetime.[59] Human beings can learn empirically through the firsthand experience of interacting with their environment or by being taught factual information or theoretical knowledge. A human being cannot instantaneously 'download' or 'import' a large body of information into his or her memory in the way that a data file can be copied to a computer's hard drive, nor can he or she be directly 'programmed' by a software developer.

Within organizations that have deployed posthumanizing neuroprostheses, workers may have additional means of acquiring new knowledge. The use of neuroprosthetic devices to monitor, control, or bypass the natural cognitive activity of a human agent may result in agents that do not need to be trained or educated but which can simply be 'programmed' to perform certain tasks or even remotely controlled by external systems to guide them in the performance of those tasks.[60] Moreover, through the use of their neuroprostheses, human workers may be able to either literally download new skill-sets and knowledge into their minds or to effectively gain the same ability by

---

out the work of and are controlled by executable programs – as well as an overview of the ways in which alternative architectures such as that of the neural network can allow computers to learn through experience – see Null & Lobur (2006). A more detailed presentation of the ways in which neural networks can be structured and learn is found in Haykin, *Neural Networks and Learning Machines* (2009). For a review of forms of computer behavior whose outcomes can be hard to predict (e.g., the actions of some forms of evolutionary algorithms or neural networks) as well as other forms of biological or biologically inspired computing, see Lamm & Unger, *Biological Computation* (2011).

[58] This description of conventional computers does not apply, for example, to artificially intelligent systems that process information by means of an artificial neural network that learns over time. For a discussion of such systems, see Friedenberg (2008), pp. 55-72.

[59] See Thornton (2008) and *Handbook of Psychology, Volume 6* (2003).

[60] Regarding the 'programming' of human beings through the intentional, targeted modification of their memories and knowledge, see, e.g., McGee (2008); Pearce (2012); and Spohrer (2002). Regarding the remote control of human bodies by external systems, see Gladden, "Neural Implants as Gateways" (2016), and Gladden, *The Handbook of Information Security for Advanced Neuroprosthetics* (2015).

being able to instantly seek out and access any online information through a mere act of will.[61]

Locus of Information Processing and Data Storage

▶ **Historical assumption:** A human worker stores and processes data within the neural network of his or her biological brain.

▶ **Impact of neuroprosthetic augmentation:** A human worker's storage and processing of data may be performed by his or her brain's neural network in a manner enhanced or controlled by a neuroprosthesis or may be 'outsourced' to external systems by means of a neuroprosthesis.

Historically, enterprise architecture has assumed that the storage of data and processing of information performed by the mind of a human worker take place within structures whose nature differs greatly from that of the organizational desktop computers, smartphones, web servers, and other electronic information systems that EA typically deals with. A conventional contemporary computer is based on a Von Neumann architecture comprising memory, I/O devices, and one or more central processing units connected by a communication bus.[62] Although one can be made to replicate the functioning of the other, the linear method by which such a CPU-based system processes information is fundamentally different from the parallel processing method utilized by a physical neural network such as that constituted by the human brain.[63]

While some information processing takes part in other parts of the body (e.g., the transduction of proximal stimuli into electrochemical signals by neurons in the sensory organs[64]), the core of a human being's information processing is performed by the neural network comprising interneurons in the individual's brain, which also stores memories in the form of engrams.[65]

---

[61] For the potential ability of neuroprostheses to provide their users with hands-free, thought-controlled access to online reference texts, see Gladden, "Information Security Concerns as a Catalyst for the Development of Implantable Cognitive Neuroprostheses" (2016).

[62] See Dumas (2006) and Friedenberg (2008), pp. 27-29.

[63] See Friedenberg (2008), pp. 30-32. The development of memristors represents one effort to design artificial computerized systems that process information using a physical network similar to that of the human brain, rather than a conventional CPU-based computing architecture. See, e.g., *Advances in Neuromorphic Memristor Science and Applications*, edited by Kozma et al. (2012), and Lohn et al., "Memristors as Synapses in Artificial Neural Networks: Biomimicry Beyond Weight Change" (2014).

[64] Such processes of transduction are discussed in Smith, *Biology of Sensory Systems* (2008), and Møller, *Sensory Systems: Anatomy and Physiology* (2014).

[65] For an overview of such processes, see, e.g., *Cognitive Psychology*, edited by Braisby & Gellatly (2012), pp. 229-65; Schwartz, *Memory: Foundations and Applications* (2014); Radvansky, *Human*

The brain constitutes an immensely large and sophisticated neural network, and despite ongoing advances in the field of neuroscience, profound mysteries remain regarding the structure and behavior of this neural network's components and of the network as a whole.[66] The mechanisms by which this biological neural network processes the data provided by sensory input and stored memories to generate motor output and new memories are highly nonlinear and complex; they are not as easy to analyze or control as the dynamics of a CPU-based computer running an executable software program.

An organization's deployment of posthumanizing neuroprostheses may significantly alter the structures within which human workers' data storage and information processing take place. Such activities may occur not within the physical neural network that comprises natural biological neurons in the agent's brain but in other electronic or biological substrates, including neuroprosthetic devices and implantable computers that utilize traditional CPU-based technologies.[67] While some forms of neuroprostheses (such as those comprising physical artificial neurons that are wholly integrated with the brain's own neural network) may potentially be able to participate directly in the storage of data in the form of engrams within their host's own brain,[68] other neuroprostheses may store memories in the form of exograms whose mnemonic contents are 'recalled' by being supplied to their user's brain through sensory organs or pathways; while such neuroprostheses might be physically located within the brain, from a functional perspective they would

---

Memory (2016); and Dudai, "The Neurobiology of Consolidations, Or, How Stable Is the Engram?" (2004).

[66] For example, there is still ongoing debate about the extent to which the brain's structures and processes for storing long-term memories display holographic characteristics, the role of interneuronal communication involving structures other than classical axodendritic synapses, and the role, if any, that quantum-level effects may play in neural dynamics. For a discussion of such issues, see, e.g., Longuet-Higgins, "Holographic Model of Temporal Recall" (1968); Pribram, "Prolegomenon for a Holonomic Brain Theory" (1990); Mulhauser, "On the end of a quantum mechanical romance" (1995); Andrew, "The decade of the brain: further thoughts" (1997); Pribram & Meade, "Conscious Awareness: Processing in the Synaptodendritic Web – The Correlation of Neuron Density with Brain Size" (1999); and Hameroff & Penrose, "Consciousness in the universe: A review of the 'Orch OR' theory" (2014).

[67] See, e.g., Merkel (2007); Warwick & Gasson, "Implantable Computing" (2008); and Soussou & Berger (2008).

[68] Regarding the possibility of neuroprosthetic devices that store information in the form of engrams or which participate in the brain's processes relating to its own natural biological storage of engrams, see Warwick (2014), p. 267; and Gladden, "Neural Implants as Gateways" (2016). For questions about the extent to which technological devices that directly store memories can ever become a part of the human mind, see Clowes, "The Cognitive Integration of E-Memory" (2013).

display many characteristics of external memory-storage technologies like books, films, or online databases.[69]

Emotionality

▶ **Historical assumption:** The emotions of human workers play an important role in the success or failure of an organization's enterprise architecture plans but cannot easily be directly shaped by the organization.

▶ **Impact of neuroprosthetic augmentation:** Organizations may modulate the emotional behaviors of human workers by means of neuroprostheses in order to generate emotions that advance (or impede) the performance of work-related tasks.

Enterprise architecture must take into account the psychological, social, political, and cultural factors at work within an organization that may impede or support the design and implementation of a new enterprise architecture. Emotional factors constitute one piece of that picture.[70]

Although the rise of social robotics has begun to change the situation, it has generally been safely assumed that the kinds of conventional computerized systems administered through enterprise architecture plans do not in themselves manifest or detect emotions: a traditional computer does not possess emotions that are grounded in the current state of the computer's body, are consciously experienced by the computer, and influence the contents of its decisions and behavior.[71] Although a piece of software may run more slowly or have some features disabled when executed on particular computers, the nature of the software's decision-making is not influenced by factors of mood, emotion, or personality determined by a computer's hardware. A software program will typically either run or not run on a given computer; if it runs at all, it will run in a manner that is determined by the internal logic

---

[69] The potential of neuroprostheses to store information in the form of exograms is raised, e.g., in Koops & Leenes (2012), pp. 115, 120, 126; McGee (2008), p. 217; and Gladden, "Neural Implants as Gateways" (2016).

[70] The impact of psychological, social, and cultural factors on the implementation of enterprise architecture plans is discussed, e.g., in Magoulas et al., "Alignment in Enterprise Architecture" (2012); Weiss & Winter, "Development of Measurement Items for the Institutionalization of Enterprise Architecture Management in Organizations" (2012); Stephan Aier, "The Role of Organizational Culture for Grounding, Management, Guidance and Effectiveness of Enterprise Architecture Principles" (2014); and Gladden, "Enterprise Architecture for Neurocybernetically Augmented Organizational Systems" (2016).

[71] For the distinction between the relatively straightforward circumstance of computers possessing some superficial display of 'emotion' simply as a function versus the more doubtful possibility that computers could undergo 'emotion' as a conscious experience, see Friedenberg (2008), pp. 191-200.

and instructions contained within the software code and is not qualitatively determined by the computer's particular physical state.

On the other hand, the successful implementation of an EA plan requires enterprise architects to understand and account for the complex emotional needs and behaviors of human workers. Within the contemporary workplace, it is assumed that the possession and manifestation of emotions is not an extraneous supplement (or obstacle) to the rational decision-making of human beings but is instead an integral component of it.[72] Indeed, some researchers suggest that the possession of emotions is necessary in order for an artificially intelligent embodied entity to demonstrate general intelligence at a human-like level.[73]

By deploying posthumanizing neuroprosthetics within its human workforce, an organization can alter and potentially control the emotional dynamics expressed by its workers. It is already known, for example, that neuroprosthetic technologies such as those for deep brain stimulation (DBS) can significantly affect the emotional behaviors of their users.[74] The use of advanced neuroprosthetic devices that can heighten, suppress, or otherwise modify the emotions of human beings may potentially result in populations of human agents whose programmatically controlled emotional behavior – or lack of emotional behavior – more closely resembles the functioning of computers than that of natural human beings.[75] Many serious ethical, legal, and operational questions would arise from efforts to develop and implement such technologies within a workforce. Just as an organization might employ neuroprostheses in an attempt to modulate the emotions of its own workers in a way that advances the implementation of its EA plan, attention must also be paid to the possibility that a specialized organization such as a military de-

---

[72] See, e.g., the influential discussion of emotional intelligence in Goleman, "What Makes a Leader?" (2004).

[73] See Friedenberg (2008), pp. 32-33, 179-200, and the literature on embodied embedded cognition – e.g., Wilson, "Six views of embodied cognition" (2002); Anderson, "Embodied cognition: A field guide" (2003); Sloman, "Some Requirements for Human-like Robots: Why the recent overemphasis on embodiment has held up progress" (2009); and Garg, "Embodied Cognition, Human Computer Interaction, and Application Areas" (2012).

[74] See Daigle, "Manipulating the Mind: The Ethics of Cognitive Enhancement" (2010), pp. 35-36; Kraemer (2011); Bublitz, "If Man' s True Palace Is His Mind, What Is Its Adequate Protection? On a Right to Mental Self-Determination and Limits of Interventions into Other Minds" (2011), p. 116; and Van den Berg (2012).

[75] For the possibility of developing emotional neuroprostheses of varying types, see Soussou & Berger (2008); McGee (2008), p. 217; Hatfield et al. (2009); Kraemer (2011); Fairclough, "Physiological Computing: Interfacing with the Human Nervous System" (2010); and Gladden, "Enterprise Architecture for Neurocybernetically Augmented Organizational Systems" (2016).

partment (or a corporation's competitive intelligence unit engaged in unlaw-ful commercial espionage or sabotage against a rival) might attempt to ma-nipulate the neuroprosthetic devices deployed within an adversarial organi-zation in order to generate within its workforce emotional behaviors that dis-rupt that organization's normal functioning and undermine the implementa-tion of its own EA plan.[76]

Cognitive Biases

▶ **Historical assumption:** A wide range of ubiquitous cognitive biases cause hu-man workers to behave irrationally and undermine the successful implementa-tion of EA plans.

▶ **Impact of neuroprosthetic augmentation:** Cognitive biases can be monitored and prevented and their effects minimized; workers can be trained to avoid them by means of neuroprostheses.

Enterprise architects have historically had to account for the fact that the human beings responsible for successfully implementing an EA plan regularly demonstrate thought and behaviors that are irrational and counterproduc-tive as a result of cognitive biases. The conventional computing devices ad-dressed by EA plans do not display such phenomena: a conventional com-puter is not inherently subject to human-like cognitive biases, as its decisions and actions are determined by the logic and instructions contained within its operating system and application code and not by the use of evolved heuristic mechanisms that are a core element of human psychology.[77] However, human beings are subject to a common set of cognitive biases that distort individuals' perceptions of reality and cause them to arrive at decisions that are objec-tively illogical and suboptimal.[78] While in earlier epochs such biases may have created an evolutionary advantage that aided the survival of those human be-ings who possessed them (e.g., by providing them with heuristics that al-lowed them to quickly identify and avoid potential sources of danger), these

[76] For concerns that neuroprostheses and related neurotechnologies might be used for attempted mind control (potentially by unauthorized adversaries who have gained access to such devices), see, e.g., Kohno et al., "Security and Privacy for Neural Devices" (2009); Bublitz (2011), pp. 97-98; Talan, "DARPA: On the Hunt for Neuroprosthetics to Enhance Memory" (2014), pp. 9-10; Krish-nan, "From Psyops to Neurowar: What Are the Dangers?" (2014), p. 10; and Gladden, "Neuromar-keting Applications of Neuroprosthetic Devices: An Assessment of Neural Implants' Capacities for Gathering Data and Influencing Behavior" (2016).

[77] It is possible, however, for a computer to indirectly demonstrate human-like cognitive biases if the human programmers who designed the computer's software were not attentive to such considerations and inadvertently programmed the software to behave in a manner that manifests such biases. For a discussion of such issues, see, e.g., Friedman & Nissenbaum, "Bias in Computer Systems" (1997).

[78] For an overview of human cognitive biases in relation to organizational management, see Ki-nicki & Williams, *Management: A Practical Introduction* (2010), pp. 217-19.

biases cause contemporary human workers to frequently err when evaluating factual claims or attempting to anticipate future events or manage risk. To some extent, such biases can be counteracted through conscious awareness, training, and effort; however, they cannot be wholly eradicated.

A future organization could potentially deploy posthumanizing neuro-prostheses that are designed to detect and directly modify or interrupt patterns of thought that reflect cognitive biases. Alternatively, a neuroprosthesis could be used to monitor the cognitive processes of a human being and alert that individual whenever the device detects that he or she is about to undertake a decision or action that is flawed or misguided because the person's cognitive processes have been influenced by a cognitive bias; beyond directly intervening to prevent the effects of cognitive biases in this manner, such a device could potentially also train the mind over time to recognize and avoid cognitive biases on its own.[79]

Fidelity of Data Storage Provided by Memory Systems

▶ **Historical assumption:** Human workers store data within their minds in the form of memories that are impressionistic and compressed and which degrade significantly over time.

▶ **Impact of neuroprosthetic augmentation:** Neuroprostheses may provide human workers with the ability to store data in the form of memories that are of perfect fidelity, do not degrade over time, and can be copied to or from external systems.

Enterprise architecture has historically relied on the fact that conventional contemporary computers are able to store data in a stable electronic digital form that is practically lossless, does not degrade rapidly over time, can be copied to other devices or media and backed up with full fidelity, and does not require a continuous power supply in order to preserve the data.[80] However, data stored within the minds of human workers is historically of a much

---

[79] A neuroprosthetic device may potentially be able to serve as a business advisor or personal coach to its human host – for example, by researching available options and providing information through an augmented reality display to support decision-making or by warning its host when he or she is about to make a flawed decision that is influenced by some human cognitive bias. See, e.g., Gladden, "Neural Implants as Gateways" (2016); Gladden, "Enterprise Architecture for Neurocybernetically Augmented Organizational Systems" (2016); Koops & Leenes (2012), p. 119; and Calverley (2008).

[80] Regarding the creation, storage, and transfer of digital data files by computers and other electronic devices, see, e.g., Austerberry, *Digital Asset Management* (2013); Coughlin, *Digital Storage in Consumer Electronics: The Essential Guide* (2008); and *Information Storage and Management* (2012).

different qualitative sort. The human mind does not store a perfect audiovisual record of all the sensory input, thoughts, and imaginings that it experiences during a human being's lifetime; the brain's capacities for both the retention and recall of information are limited.[81] Not only are memories stored in a manner which from the beginning is compressed, impressionistic, and imperfect, but memories also degrade over time.[82] Historically, the only way to transfer memories stored within one human mind to another human mind has been for the memories to be described and expressed through some social mechanism such as oral speech or written text.

By deploying posthumanizing neuroprostheses among its personnel, an organization may be able to significantly alter the quality of information stored within its workers' minds. The use of neuroprosthetic devices to control, supplement, or replace the brain's natural memory mechanisms could result in human agents that possess memory that is effectively lossless, does not degrade over time, and can potentially be copied to or from external systems.[83] The processes for copying large and complex arrays of stored data from a human brain to an external system may differ radically in their nature and complexity from the processes of copying such data from an external system to a human brain, if either type of process is theoretically possible at all.[84]

---

[81] The concept that some people possess the power of perfect 'photographic memory' exists as a notion within popular culture; however, the existence of such a phenomenon in natural human beings has yet to be scientifically demonstrated. Cases of eidetic memory (understood in the technical definition of the term) and hyperthymesia (or 'superior autobiographical memory') have indeed been documented; however, these phenomena involve a shorter duration of retention or a more constrained subject matter than supposed cases of 'photographic memory.' Regarding hyperthymesia, see Taylor, "Hyperthymesia" (2013); regarding cases of supposed eidetic memory, see Moxon, *Memory* (2000), p. 15, and Schwartz (2014), p. 172.

[82] See Dudai (2004).

[83] Regarding genetic and neuroprosthetic technologies for memory alteration in biological organisms, see Han et al., "Selective Erasure of a Fear Memory" (2009); Josselyn, "Continuing the Search for the Engram: Examining the Mechanism of Fear Memories" (2010); and Ramirez et al. (2013). Regarding the use of neuroprosthetic systems to store memories as effectively lossless digital exograms, see Merkel et al. (2007); Robinett, "The consequences of fully understanding the brain" (2002); McGee (2008), p. 217; and Gladden, "Neural Implants as Gateways" (2016).

[84] Regarding the difficulty of analyzing, interpreting, or purposefully editing the contents of information stored in complex biological or biomimetic neural networks, see, e.g., Friedenberg (2008), pp. 31-32, and Gladden, "Information Security Concerns as a Catalyst for the Development of Implantable Cognitive Neuroprostheses" (2016). If holographic models of memory storage in the brain are correct, the challenge of manipulating memories in such ways would be even greater; for discussion of such models, see, e.g., Longuet-Higgins (1968) and Pribram (1990).

Predictability of Behavior

▶ **Historical assumption:** The future behavior of an individual human being within a workplace setting is difficult to predict with much certainty.

▶ **Impact of neuroprosthetic augmentation:** Neuroprostheses may make it easier to predict the future behavior of human workers – either by gathering more detailed data about a worker's cognitive functioning or by making it possible to influence or control behavior by means of such devices.

Enterprise architects have traditionally presumed that the degree of predictability demonstrated by human workers differs significantly from that manifested by electronic information systems. Computerized devices can be affected by a wide range of component failures and bugs resulting from hardware or software defects or incompatibilities; however, because a typical computer is controlled by discrete linear executable code that can be easily accessed – and because there exist diagnostic software, software debugging techniques, established troubleshooting practices, and methods for simulating a computer's real-world behaviors in development and testing environments – it is generally easier to analyze and reliably predict the behavior of a computer than that of a human being.[85] While all human beings demonstrate basic similarities in their behavior – and individual human beings possess unique personalities, habits, and psychological and medical conditions that allow their reactions to particular stimuli or future behavior to be predicted with some degree of likelihood – it is not possible to predict with full accuracy and certainty the future actions of a particular human being.

An organization may be able to modify such traditional characteristics by deploying neuroprosthetic devices among its human workers. Such devices may allow the organization to gather detailed information about the cognitive structures and processes of individual human workers that was previously inaccessible.[86] Moreover, human agents whose actions are influenced

---

[85] Even the behavior of sophisticated artificially intelligent computerized systems can be easy to predict and debug, if it is controlled by a conventional executable program rather than, e.g., the actions of a physical artificial neural network. For a discussion of different models for generating artificial intelligence through hardware and software platforms, see Friedenberg (2008), pp. 27-36.

[86] For the possibility that neuroprostheses might be used for (potentially illegal and unethical) surveillance of their hosts' cognitive activity and behaviors, see, e.g., Bonaci et al., "App Stores for the Brain" (2015), p. 35; Brunner et al., "Current Trends in Hardware and Software for Brain-Computer Interfaces (BCIs)" (2011); Gladden, *The Handbook of Information Security for Advanced Neuroprosthetics* (2015); and Gladden, "Neuromarketing Applications of Neuroprosthetic Devices: An Assessment of Neural Implants' Capacities for Gathering Data and Influencing Behavior" (2016).

or controlled by neuroprostheses may produce behavior that is more predictable and is easily 'debugged' in a straightforward and precise manner that has traditionally been possible only when dealing with computers.[87] Serious ethical and legal questions are raised by the potential use of such technologies.

Information Security Vulnerabilities

▶ **Historical assumption:** Human workers are not subject to the forms of electronic hacking and malware to which computerized information systems are vulnerable.

▶ **Impact of neuroprosthetic augmentation:** Possession of a neuroprosthesis may render a human worker vulnerable to electronic hacking and computer viruses and worms; at the same time, neuroprostheses may provide organizational operatives with new tools for detecting and countering InfoSec threats.

Maintaining information security is a major concern of enterprise architecture. For example, Hoogervorst identifies information security as a component of the element of 'quality' within the information architecture domain.[88] Other EA frameworks, such as the Siemens Framework described by Rohloff, do not explicitly identify information security as a domain or building block – not because it is unimportant, but because it underlies *all* of the architectural building blocks.[89]

An organization's computerized systems are vulnerable to a wide variety of electronic hacking techniques and other attacks that can compromise the confidentiality, integrity, and availability of information that is received, generated, stored, or transmitted by a system and can even result in unauthorized parties gaining complete control over the system.[90] Human workers have traditionally been vulnerable to efforts at compromising information security

---

[87] Regarding the testing and debugging of neuroprosthetic devices see Gladden, *The Handbook of Information Security for Advanced Neuroprosthetics* (2015), pp. 176-77, 181-84, 213-14, 242-43, 262. Regarding potential efforts to employ neuroprostheses to 'debug' the behaviors of their human hosts, see Gladden, *Sapient Circuits and Digitalized Flesh* (2016), p. 181.

[88] See Hoogervorst (2004), p. 15.

[89] Rohloff (2008), p. 5.

[90] For an overview of such possibilities (as well as related preventative practices and responses), see Rao & Nayak, *The InfoSec Handbook* (2014). Information security's fundamental 'CIA Triad' of ensuring the confidentiality, integrity, and availability of information is discussed in Parker, "Toward a New Framework for Information Security" (2002); "Security Risk Assessment Framework for Medical Devices" (2014); *NIST Special Publication 1800-1b: Securing Electronic Health Records on Mobile Devices: Approach, Architecture, and Security Characteristics* (2016), p. 9; and Gladden, "Information Security Concerns as a Catalyst for the Development of Implantable Cognitive Neuroprostheses" (2016).

involving social engineering, blackmail, or other psychologically based attacks.[91] However, because human workers possess biological rather than electronic components and their minds conduct information processing through
the use of an internal biological neural network rather than executable software programs stored in binary digital form, it has historically not been possible for adversaries to 'hack into' a human worker's neural information-processing system in order to access, steal, or manipulate the individual's
thoughts or memories using the same electronic hacking techniques that are
applied to electronic computers and computer-based systems.

The deployment of posthumanizing neuroprostheses within an organization may significantly alter the organization's ability to maintain information
security. Human beings who possess electronic neuroprosthetic devices will
be vulnerable to computer viruses, worms, and electronic hacking attacks
similar to those that target conventional computers and may become reliant
on some of the same security measures used to protect typical electronic computers (such as antivirus software and electronic firewalls).[92] Moreover, advanced technologies for genetic engineering and the production of customized biopharmaceuticals and biologics may allow the 'biohacking' even of human agents that do not possess electronic neuroprosthetic components. At
the same time, neuroprostheses may provide organizational operatives with
powerful new tools for identifying InfoSec vulnerabilities within an organization and detecting and counteracting information security threats.[93]

---

[91] Social engineering attacks are discussed in Sasse et al., "Transforming the 'weakest link' – hu-
man/computer interaction approach to usable and effective security" (2001), and Rao & Nayak
(2014), pp. 307-23.

[92] Such issues are raised, e.g., in ISO 27799:2016, Health informatics – Information security management in health using ISO/IEC 27002 (2016); Luber et al., "Non-invasive brain stimulation in
the detection of deception: Scientific challenges and ethical consequences" (2009); Denning et
al. (2009); "Cybersecurity for Medical Devices and Hospital Networks: FDA Safety Communication" (2013), p. 1; Bonaci et al. (2015), p. 35; Gladden, The Handbook of Information Security for
Advanced Neuroprosthetics (2015); and Gladden, "Neuromarketing Applications of Neuroprosthetic Devices" (2016). Regarding the possibility of hybrid biological-electronic computer viruses
and other attacks, see Gladden, The Handbook of Information Security for Advanced Neuroprosthetics (2015), p. 53, and Gladden, Sapient Circuits and Digitalized Flesh (2016), p. 182.

[93] The potential use of neuroprostheses by specialized human agents such as military personnel
or cybersecurity operatives in order to advance the information security of their employers (and
human society more broadly) is discussed in Falconer (2003); Moreno (2004); Clancy (2006);
Schermer (2009); Brunner & Schalk (2009); "Bridging the Bio-Electronic Divide" (2016); Szoldra,
"The government's top scientists have a plan to make military cyborgs" (2016); and Gladden, "Information Security Concerns as a Catalyst for the Development of Implantable Cognitive Neuroprostheses" (2016). It should be noted that the use of neuroprostheses to treat or reverse the
effects of conditions such as Alzheimer's disease that impair access to human beings' long-term

Social Engagement

The capacity for social behaviors, interactions, and relations possessed by human workers may be altered significantly by an organization's deployment of neuroprosthetic devices among those workers. Below we consider a number of social characteristics of human agents and the ways in which the historical assumptions regarding the nature of those characteristics that enterprise architects have relied upon will become increasingly outdated within organizations utilizing posthumanizing neuroprostheses.

Degree of Sociality

▷ **Historical assumption:** While it suffers from being slow, inefficient, and imprecise, social interaction is the critical means by which human workers communicate and collaborate to perform work-related tasks.

▷ **Impact of neuroprosthetic augmentation:** Neuroprostheses may allow human workers to receive and exchange information in a direct and instantaneous manner that bypasses the need for social interaction; such devices may also either impair or enhance workers' capacity for engaging socially with others.

Enterprise architecture has traditionally drawn sharp distinctions between the ways in which electronic computerized systems interact with one another and the ways in which human workers interact. Conventional computers may display social behaviors and engage in short-term, isolated social interactions with human beings or other computers, but they do not participate in long-term social relations that deepen over time as a result of their experience of such engagement and which are shaped by society's expectations for social roles to be filled by the computers.[94] On the other hand, human beings display social behaviors, engage in isolated and short-term social interactions, and participate in long-term social relations that evolve over time and are shaped by society's expectations for the social roles to be filled by a particular individual.[95] Although the social content and nature of complex communicative human actions such as speaking and writing are obvious,

---

memories would, in effect, increase such individuals' information security by enhancing the integrity and availability of the information contained within such memories. See Gladden, "Information Security Concerns as a Catalyst for the Development of Implantable Cognitive Neuroprostheses" (2016).

[94] Although there already exist telepresence robots (e.g., Ishiguro's Geminoids) that manifest highly sophisticated, human-like levels of sociality, such sociality is technically possessed not by the robot itself but by the hybrid human-robotic system that it forms with its human operator. Regarding such issues, see Vinciarelli et al., "Bridging the Gap between Social Animal and Unsocial Machine: A survey of Social Signal Processing" (2012) and Gladden, "Managerial Robotics" (2014).

[95] Regarding the distinction between social behaviors, interactions, and relations, see Vinciarelli

even such basic activities as standing, walking, and breathing have social aspects, insofar as they can convey intentions, emotions, and attitudes toward other human beings.[96] The skill and efficiency with which human workers communicate and collaborate by means of social activities vary greatly between individuals.

When deployed among an organization's workforce, neuroprosthetic devices that negatively affect long-term memory processes could make it difficult or impossible for human agents to engage in friendships and other long-term social relationships with other intelligent agents. Such human agents would no longer be fully social but instead semisocial or even nonsocial.[97] On the other hand, posthumanizing neuroprostheses that reduce the impact of cognitive or emotional disorders, enhance memory, and train users to engage socially with others more effectively may optimize workers' patterns of communication and collaboration and improve overall performance. Neuroprostheses may also be used to create links between the brains of human workers that allow information to be transferred into and between workers' minds in a direct and instantaneous manner that does not require mediation through relatively slow and imprecise social behaviors such as speech or the typing of text.[98] Ongoing immersion in virtual worlds or neuroprosthetically enabled cybernetic networks with other human minds or other kinds of intelligent agents could potentially also lead to the atrophying or enhancement of human agents' social capacities.[99]

Relationship to Organizational Culture

▶ **Historical assumption:** Workplace culture is a critical element in the successful implementation of an EA plan, but enterprise architecture possesses only limited tools for diagnosing and shaping such culture.

---

et al. (2012) and Gladden, "Managerial Robotics" (2014).

[96] Forms of nonverbal communication such as oculesics, haptics, kinesics, and proxemics are discussed in Andersen & Andersen, "Measures of Perceived Nonverbal Immediacy" (2005). The informational content of paralanguage is discussed in Johar, *Emotion, Affect and Personality in Speech: The Bias of Language and Paralanguage* (2016).

[97] For ways of describing and classifying degrees of sociality, see Vinciarelli et al. (2012) and Gladden, "Managerial Robotics" (2014).

[98] Regarding such possibilities, see Rao et al., "A direct brain-to-brain interface in humans" (2014), and Gladden, "Utopias and Dystopias as Cybernetic Information Systems" (2015), as well as the literature relating to hive minds cited earlier in this text.

[99] For the possibility that neuroprostheses might provide their hosts with expanded knowledge in some spheres while simultaneously diminishing their social or intellectual capacities in other areas, see Bostrom & Sandberg, "Cognitive Enhancement: Methods, Ethics, Regulatory Challenges" (2009), pp. 322-23, and Gladden, "Managing the Ethical Dimensions of Brain-Computer Interfaces in eHealth: An SDLC-based Approach" (2016).

> ▶ **Impact of neuroprosthetic augmentation:** Neuroprostheses may provide an organization with new tools for analyzing its existing workplace culture, modifying that culture, or even fabricating an entirely new workplace culture.

Understanding and optimizing workplace culture is critical for successfully designing and implementing an enterprise architecture plan within an organization.[100] Historically, questions of culture have been associated almost exclusively with the behavior of human agents working within an organization and not directly with the behaviors of electronic information systems. While a large number of computers can be linked to form networks that may constitute a form of computerized multi-agent society, such aggregations of conventional computers do not create or experience their own 'cultures.'[101] On the other hand, human beings create and exist within unique cultures that include particular forms of art, literature, music, architecture, history, sports and recreation, technology, ethics, philosophy, and theology. Such cultures also develop and enforce norms regarding the ways in which organizations such as businesses should or should not operate.

As sensors implanted within a workforce that gather diverse types of real-time data about the activities and interactions of human workers, neuroprostheses may provide an organization with a direct and powerful new means of analyzing the realities of its workplace culture. The deployment of posthumanizing neuroprostheses within an organization may also transform the nature of the workplace culture and the means by which it is created and experienced. Human agents whose thoughts, dreams, and aspirations have been attenuated or even eliminated or whose physical sensorimotor systems are controlled through the use of neuroprosthetic devices may no longer possess a desire or ability to perceive or generate certain kinds of cultural artifacts. Moreover, if a single centralized system (e.g., a server providing a shared virtual reality experience to large numbers of individuals) maintains and controls all of the sensorimotor channels through which human agents are able to create and experience culture, then that automated system may generate substantially all of the aspects of culture within that virtual world, without

---

[100] Regarding the critical role that organizational culture and cultural knowledge play, e.g., in the management of enterprise architecture, see Aier (2014); Hoogervorst (2004); Liu et al., "A Design of Business-Technology Alignment Consulting Framework" (2011); Magoulas et al. (2012), pp. 93-95, 98; and Gladden, "Enterprise Architecture for Neurocybernetically Augmented Organizational Systems" (2016).

[101] Regarding prerequisites for artificial entities or systems to produce their own culture (or collaborate with human beings in the production of a shared human-artificial culture), see, e.g., Payr & Trappl, "Agents across Cultures" (2003).

the human agents who dwell in that world being able to contribute meaningfully to the process.[102]

Economic and Financial Relationship with Employer

▶ **Historical assumption:** An organization commonly uses financial compensation and rewards to acquire workers and incentivize particular workplace behaviors.

▶ **Impact of neuroprosthetic augmentation:** Implantation of a neuroprosthesis into a worker by his or her organization may make the worker permanently financially dependent on the organization and create complications regarding the ownership of intellectual property and other goods or services produced by means of the neuroprosthesis.

Enterprise architecture typically assumes that the human agents working within an organization have an economic and financial relationship to the organization (e.g., the relationship of employee to employer) that enables their participation in the organization's work and may, to some extent, be exploited in order to generate motivation and incentives for certain types of workplace behaviors and disincentives for other types of behaviors.[103]

The deployment of posthumanizing neuroprostheses within an organization may significantly change the economic and financial relationship of the organization to its affected workers. Depending on the precise contractual terms and conditions under which such components were acquired, a human agent whose body has been subject to neuroprosthetic augmentation and is partially composed of electronic components may not even fully 'own' his or her body or the products generated by it, including intellectual property in the form of thoughts and memories.[104] In order to preserve his or her psychological and physical health, a neuroprosthetically augmented individual may

---

[102] Regarding the possibilities of a centralized computerized system shaping culture by mediating and influencing or controlling the communications among neuroprosthetically enabled human minds, see Gladden, "Utopias and Dystopias as Cybernetic Information Systems" (2015), and Gladden, "From Stand Alone Complexes to Memetic Warfare: Cultural Cybernetics and the Engineering of Posthuman Popular Culture" (2016).

[103] The extent to which financial compensation actually provides motivation for workers is a subject of debate. See, for example, Herzberg, "One more time: How do you motivate employees" (1986), and Rynes et al., "The importance of pay in employee motivation: Discrepancies between what people say and what they do" (2004).

[104] Complex legal and regulatory regimes govern the ownership of information that is received, created, stored, or transmitted by neuroprosthetic devices. See Kosta & Bowman, "Implanting Implications: Data Protection Challenges Arising from the Use of Human ICT Implants" (2012); McGee (2008); Mak, "Ethical Values for E-Society: Information, Security and Privacy" (2010); McGrath & Scanaill, "Regulations and Standards: Considerations for Sensor Technologies" (2013); Shoniregun et al., "Introduction to E-Healthcare Information Security" (2010); and Gladden, *The Handbook of Information Security for Advanced Neuroprosthetics* (2015).

also require continual maintenance services, medication, software updates, or replacement parts. If – due to legal, financial, or operational reasons – the individual is only able to obtain such neuroprosthetic support services from his or her employer (or from another particular corporation, government agency, or other institution) and is barred from purchasing goods or services from competing enterprises, the individual may be rendered biologically and psychologically dependent on and financially indebted to that institution throughout the rest of his or her life.[105] The use of neuroprosthetic devices or other technologies that directly affect a human agent's cognitive processes may also impair that agent's ability to make free choices as an autonomous economic actor.

Rights, Responsibilities, and Legal Status of Human Agents

▶ **Historical assumption:** Unlike computers and other property owned by an organization, human workers possess fundamental rights that limit the ability of an EA plan to unilaterally dictate such workers' location, physical structure, and behavior.

▶ **Impact of neuroprosthetic augmentation:** The 'ownership' of workers' neuroprosthetically augmented physical bodies and cognitive processes and workers' legal right to control the nature and disposition of their own bodies may become more ambiguous.

Different types of resources within an organization historically possess very different legal status. An organization's human resources possess many fundamental personal rights as human beings. On the other hand, a desktop computer or web server is a piece of property that can be owned by the organization; it is not a legal person that possesses a recognized set of rights and responsibilities.[106] This distinction means that enterprise architects must

---

[105] Regarding the possibility that use of a neuroprosthesis may create dependencies that would result in psychological, physical, economic, or social harm to its human host if use of the device were to be discontinued, see Bostrom & Sandberg (2009), p. 323; McGee (2008), p. 213; Koops & Leenes (2012), p. 125; Gladden, "Neural Implants as Gateways" (2016); and Gladden, "Managing the Ethical Dimensions of Brain-Computer Interfaces in eHealth" (2016). For the fact that corporations or other organizations that manufacture or provide neuroprostheses may purposefully attempt to create high 'switching costs' that discourage or prevent device hosts from using the support services of competing organizations, see Gladden, *The Handbook of Information Security for Advanced Neuroprosthetics* (2015).

[106] Stahl suggests that a kind of limited 'quasi-responsibility' can be attributed to conventional computers and computerized systems. In this model, it is a computer's human designers, programmers, or operators who are ultimately responsible for the computer's actions; declaring a particular computer to be 'quasi-responsible' for some action that it has performed serves as a sort of temporary moral and legal placeholder until the computer's human designers, programmers, and operators can be identified and final responsibility for the computer's actions assigned

account for differences in the extent to which an EA plan can summarily dictate the location, physical components, and behaviors of different sorts of resources within an organization.

Just as the creation of artificially intelligent entities with the appearance of human-like self-awareness and volitionality creates ethical and legal questions regarding whether such entities possess any form of rights,[107] the creation of joint human-robotic entities (such as that of a human brain maintained within a largely cyborg body) poses new ethical and legal questions – such as whether an individual who has been heavily neuroprosthetically augmented still possesses full legal rights of self-determination and the ability to decline consent for mechanical procedures to be performed upon neuroprosthetically linked electromechanical components that are not owned by the individual in the same way that he or she would be able to decline consent for medical procedures to be performed upon natural biological components of his or her body. (Such a situation grows even more complicated, for example, when it involves neuroprosthetically linked external systems such as video cameras, web servers, vehicles, and other devices that functionally serve as a portion of the individual's cybernetically expanded 'body' but which may also be shared with other users.)

Human agents that have been intentionally neurocybernetically engineered by other human beings or organizations to grant them a radically different physical form or cognitive capacities may be subject to claims (whether ethically and legally valid or not) that they are not full-fledged legal persons but rather wards or even property of those who have created them – especially if the agents have been engineered to possess nonhuman characteristics that clearly distinguish them from 'normal' human beings and whose design is

---

to the appropriate human parties. See Stahl, "Responsible Computers? A Case for Ascribing Quasi-Responsibility to Computers Independent of Personhood or Agency" (2006).

[107] An adult human being is typically recognized by the law as being a legal person who bears responsibility for his or her decisions and actions. In some cases, relevant distinctions exist between legal persons, moral subjects, and moral patients. For example, an adult human being who is conscious and not suffering from psychological or biological impairments would typically be considered both a legal person who is legally responsible for his or her actions as well as a moral subject who bears moral responsibility for those actions. An infant or an adult human being who is in a coma might be considered a legal person who possesses certain legal rights, even though a legal guardian may be appointed to make decisions on the person's behalf; such a person is not (at the moment) a moral agent who undertakes actions for which he or she bears moral responsibility but is still a 'moral patient' whom other human beings have an obligation to care for and to not actively harm. Regarding distinctions between legal persons, moral subjects, and moral patients – especially in the context of comparing human and artificial agents – see, e.g., Wallach & Allen, *Moral machines: Teaching robots right from wrong* (2008); Gunkel, *The Machine Question: Critical Perspectives on AI, Robots, and Ethics* (2012); Sandberg, "Ethics of brain emulations" (2014); and Rowlands, *Can Animals Be Moral?* (2012).

considered to be intellectual property owned by the individual or organization responsible for that engineering. Developing a legal framework that resolves such questions to the satisfaction of all stakeholders is expected to be challenging, given the number and types of institutions and social groups that possess potentially competing financial, political, cultural, or personal stakes in the outcome.[108]

Conclusion

As part of its work of designing and implementing target architectures for organizations, the discipline of enterprise architecture has historically relied on a set of assumptions regarding the physical, cognitive, and social capacities of the human beings serving as organizational members. As we have explored in this text, those traditional assumptions will become increasingly obsolete for those organizations that choose to deploy posthumanizing neuroprosthetic technologies among their personnel. The use of advanced neuroprostheses intensifies the ongoing structural, systemic, and procedural fusion of human personnel with electronic information systems in a way that creates for human workers new capacities and limitations and transforms the roles available to them.

Neuroprostheses have the potential to affect workers especially in the three areas of their physical form, intellect, and social engagement. First, neuroprostheses may affect workers' physical form, as seen in their bodies' physical components, the extent to which their physical form is the subject of organizational design, their length of tenure as workers, the developmental and operational cycles that they experience, their spatial extension and locality, the permanence of their physical substrates, and the nature of their personal identity. Second, neuroprostheses can affect the information processing and cognition of neuroprosthetically augmented workers, as reflected in their degree of sapience, autonomy, and volitionality; their means of knowledge acquisition; their locus of information processing and data storage; their emotionality and cognitive biases; and their fidelity of data storage, predictability of behavior, and information security vulnerabilities. Finally, the deployment of posthumanizing neuroprostheses may affect workers' social engagement, as manifested in their degree of sociality; relationship to organizational culture; economic relationship with their employers; and rights, responsibilities, and legal status.

It is not claimed or expected that the types of neuroprosthetic augmentation discussed here will soon be exploited purposefully by a broad range of organizations: indeed, a complex array of ethical, legal, political, economic,

---

[108] Similar political and legal debates have arisen, for example, surrounding the technologies for genetic engineering that enable the widespread production and commercial use of GMOs.

and functional factors will prevent most organizations from deploying advanced neuroprostheses among their personnel for the foreseeable future. However, selected organizations such as military agencies and departments are already working to develop such technologies and implement them among their personnel for highly specialized purposes. Such organizations will be forced to transform their enterprise architecture practices and plans in order to address the deepening human-computer hybridization brought about by posthumanizing neuroprosthetic technologies.

In closing, it should again be noted that we have not attempted to explore in detail the many serious ethical and legal problems that are created by any effort on the part of organizations to intentionally deploy posthumanizing neuroprostheses among their human personnel. However, this does not mean that such concerns are insignificant. Indeed, it is critical that thoughtful and comprehensive analyses of such organizational practices be undertaken from ethical and legal perspectives, and it is our hope that the investigation of neuroprosthetically facilitated human-computer fusion presented in this text may provide a useful foundation for such analyses.

Chapter Seven

From Virtual Teams to Hive Minds:
Developing Effective Network Topologies for Neuroprosthetically Augmented Organizations

Abstract. This text develops a model based on network topology that can be used to analyze or engineer the structures and dynamics of an organization in which neuroprosthetic technologies are employed to enhance the abilities of human personnel. We begin by defining *neuroprosthetic supersystems* as organizations whose members include multiple neuroprosthetically augmented human beings. It is argued that the expanded sensory, cognitive, and motor capacities provided by 'posthumanizing' neuroprostheses may enable human beings possessing such technologies to collaborate using novel types of organizational structures that differ from the traditional structures that are possible for unaugmented human beings. The concept of network topology is then presented as a concrete approach to analyzing or engineering such neuroprosthetic supersystems. A number of common network topologies such as chain, linear bus, tree, ring, hub-and-spoke, partial mesh, and fully connected mesh topologies are discussed and their relative advantages and disadvantages noted.

Drawing on the notion of different architectural 'views' employed in enterprise architecture, we formulate a topological model that incorporates five views that are relevant for neuroprosthetic supersystems: the (1) physical and (2) logical topologies of the neuroprosthetic devices themselves; (3) the natural topology of social relations of the devices' human hosts; (4) the topology of the virtual environments, if any, created and accessed by means of the neuroprostheses; and (5) the topology of the brain-to-brain communication, if any, facilitated by the devices. Potential uses of the model are illustrated by applying it to four hypothetical types of neuroprosthetic supersystems: (1) an emergency medical alert system incorporating body sensor networks (BSNs); (2) an array of centrally hosted virtual worlds; (3) a 'hive mind' administered by a central hub; and (4) a distributed hive mind lacking a central hub.

It is our hope that models such as the one formulated here will prove useful not only for engineering neuroprosthetic supersystems to meet functional requirements but also for analyzing the legal, ethical, and social aspects of potential or existing supersystems, to ensure that the organizational deployment of neuroprosthetic technologies does not undermine the wellbeing of such devices' human users or of societies as a whole.

Introduction

The power and sophistication of neuroprosthetic technologies are rapidly increasing. While a growing number of organizations are incorporating advanced technologies relating to social robotics, artificial intelligence, and virtual reality into their business operations, the notion that organizations would implant neuroprosthetic devices in the brains of their workers in order to enhance productivity might appear to be one which – for the moment – exists primarily within the domain of science fiction and futurology. And yet a small number of specialized organizations are already actively seeking to develop advanced neuroprosthetic technologies and deploy them among their personnel, not simply for therapeutic medical purposes but in order to support and enhance their workers' job performance. Chief among these organizations are military agencies and departments who see in advanced neuroprostheses a means of enhancing the health, safety, and effectiveness of soldiers who must operate in highly dangerous circumstances and upon whose work depends the security of entire nations.[1] Moreover, as new noninvasive neuroprosthetic technologies are developed that do not require dangerous and expensive implant surgery, it is expected that the organizational deployment of 'posthumanizing' neuroprostheses will eventually become feasible and desirable for a wider range of organizations such as businesses.[2]

By their very nature, neuroprosthetic devices create new relationships of communication and control involving their human hosts, the designers and producers of such devices, the medical and IT personnel who maintain them, and the devices themselves. When such devices are purposefully introduced into an organization and deployed among its personnel, they alter in subtle or dramatic ways the manner in which employees interact with one another

---

[1] Regarding efforts by DARPA (the US military research agency) and other institutions to develop neurotechnologies for increasing soldiers' alertness and reducing their need for sleep, see, e.g., Falconer, "Defense Research Agency Seeks to Create Supersoldiers" (2003); Moreno, "DARPA On Your Mind" (2004); Clancy, "At Military's Behest, Darpa Uses Neuroscience to Harness Brain Power" (2006); and Wolf-Meyer, "Fantasies of extremes: Sports, war and the science of sleep" (2009). Potential military applications of neurotechnologies for human enhancement are also discussed in Schermer, "The Mind and the Machine. On the Conceptual and Moral Implications of Brain-Machine Interaction" (2009); Brunner & Schalk, "Brain-Computer Interaction" (2009); Coker, "Biotechnology and War: The New Challenge" (2004); Graham, "Imagining Urban Warfare: Urbanization and U.S. Military Technoscience" (2008), p. 36; Krishnan, "Enhanced Warfighters as Private Military Contractors" (2015); and Kourany, "Human Enhancement: Making the Debate More Productive" (2013), pp. 992-93.

[2] At present, for example, cochlear implant surgery can cost between $40,000-$100,000 per person, while implantation surgery for deep brain stimulation (DBS) devices can cost between $35,000-$100,000 per recipient. See "Cochlear Implant Quick Facts" and Okun, "Parkinson's Disease: Guide to Deep Brain Stimulation Therapy" (2014).

and carry out their roles. This raises the question of how such an organization's formal array of personnel structures, processes, and systems – that is, its organizational architecture[3] – will need to be altered in order to adapt to the presence and activity of neuroprosthetically augmented personnel. That is the question to be investigated in this text, using the lens of *network topology*.

We begin by defining neuroprosthetic supersystems as organizations whose members include multiple neuroprosthetically augmented human beings. The concept of network topology is then presented as a valuable approach to analyzing or engineering such supersystems. A number of common network topologies are discussed, and their relative advantages and disadvantages noted. Building on enterprise architecture's notion of architectural 'views,' we next formulate a topological model that incorporates five views that are relevant for neuroprosthetic supersystems. Finally, potential uses of the model are illustrated by applying it to four types of neuroprosthetic supersystems that could hypothetically be established through an organization's deployment of neuroprosthetic technologies. While the model formulated in this text can facilitate the design of neuroprosthetic supersystems that meet an organization's operational requirements, it also provides a tool for analyzing the legal, ethical, and social aspects of such supersystems – which can help ensure that any organizational deployment of neuroprosthetic technologies is performed in an appropriate manner that not only safeguards the rights and wellbeing of such devices' human hosts but also advances the organization's larger mission and the welfare of society as a whole.

Defining Neuroprosthetic Supersystems

A neuroprosthesis can be defined as *an artificial device that is integrated into the neural circuitry of a human being*.[4] Such devices typically support or participate in the sensory, cognitive, or motor processes of their human host;[5] however, they can also be employed for such ends as gathering real-time data about the host's biological processes and transmitting it to an external computer for medical, archival, or potentially even surveillance purposes.[6] In

---

[3] Within the 'congruence model' of organizational architecture conceptualized by Nadler and Tushman, structures, processes, and systems constitute the three main elements of an organization that are taken into account. See Nadler & Tushman, *Competing by Design: The Power of Organizational Architecture* (1997), p. 47.

[4] See Lebedev, "Brain-Machine Interfaces: An Overview" (2014), and Gladden, "Enterprise Architecture for Neurocybernetically Augmented Organizational Systems" (2016).

[5] See Lebedev (2014).

[6] See, e.g., Lorence et al., "Transaction-Neutral Implanted Data Collection Interface as EMR Driver: A Model for Emerging Distributed Medical Technologies" (2009); Bonaci et al., "App

principle, neuroprostheses may be either 'invasive' (i.e., surgically implanted in the brain of a human host) or 'non-invasive' (e.g., consisting of an external device worn by a human host); however, significant challenges exist with developing non-invasive technologies that become truly integrated into the neural circuitry of a human being.[7] Thus, according to the definition used in this text, contemporary neuroprostheses can typically be identified with invasive 'neural implants.'

At present, neuroprosthetic devices are used primarily for therapeutic purposes, as a means of restoring some capacity that is absent as a result of injury or illness: for example, cochlear implants, auditory brainstem implants, and retinal prostheses are used to restore sensory functionality to those who have lost the ability to hear or see; robotic prosthetic limbs are used to replace natural biological limbs that have been amputated; robotic exoskeletons and thought-controlled wheelchairs grant a degree of mobility to the paralyzed; and experimental neural bridges are being developed to restore memory function in individuals who are unable to access their long-term memories due to hippocampal damage.[8] However, efforts are also underway to develop and deploy 'posthumanizing' neuroprostheses[9] whose aim is not to restore some capacity that is found in typical human beings but to grant their human

---

Stores for the Brain" (2015), p. 35; Luber et al., "Non-invasive brain stimulation in the detection of deception: Scientific challenges and ethical consequences" (2009); and Gladden, *The Handbook of Information Security for Advanced Neuroprosthetics* (2015).

[7] See Gasson, "Human ICT Implants: From Restorative Application to Human Enhancement" (2012), p. 14, and Panoulas et al., "Brain-Computer Interface (BCI): Types, Processing Perspectives and Applications" (2010).

[8] See, e.g., Ochsner et al., "Human, non-human, and beyond: cochlear implants in socio-technological environments" (2015); Cervera-Paz et al., "Auditory Brainstem Implants: Past, Present and Future Prospects" (2007); Weiland et al., "Retinal Prosthesis" (2005); Viola & Patrinos, "A Neuroprosthesis for Restoring Sight" (2007); Gasson et al., "Human ICT Implants: From Invasive to Pervasive" (2012); Soussou & Berger, "Cognitive and Emotional Neuroprostheses" (2008); and Gladden, "Neural Implants as Gateways to Digital-Physical Ecosystems and Posthuman Socioeconomic Interaction" (2016).

[9] 'Posthumanizing' technologies can be understood as those that bring about an ecosystem in which entities other than natural biological human beings exist as intelligent agents and social actors that create meaning in the world. Technologies relating to artificial intelligence, artificial life, virtual reality, genetic engineering, and neuroprosthetic augmentation are a catalyst for processes of posthumanization; however, non-technological forces of posthumanization also exist. For more details, see Ferrando, "Posthumanism, Transhumanism, Antihumanism, Metahumanism, and New Materialisms: Differences and Relations" (2013); Herbrechter, *Posthumanism: A Critical Analysis* (2013); Miah, "A Critical History of Posthumanism" (2008); Birnbacher, "Posthumanity, Transhumanism and Human Nature" (2008); and Gladden, *Sapient Circuits and Digitalized Flesh: The Organization as Locus of Technological Posthumanization* (2016).

hosts sensory, cognitive, and motor capacities that far exceed those that are possible for natural biological human beings.[10]

A neuroprosthetic device can, in itself, be understood as a type of 'system.' The human host of a neural implant is also a type of system, as is the hybrid biological-electronic entity that is created when a neuroprosthetic device becomes integrated into the neural circuitry of its human host. And finally, a group of human beings who possess neuroprosthetic devices can collectively form a system. Use of the word 'system' when discussing such technologies and their implementation can thus be ambiguous. For purposes of clarity, within this text we will use the term 'neuroprosthetic device' to refer to a neuroprosthesis, 'host-device system' to refer to a neuroprosthetically augmented human being, and 'neuroprosthetic supersystem' to refer to a collection of neuroprosthetically augmented human beings. The creation and maintenance of effective neuroprosthetic devices and host-device systems involves fields such as computer science, biology, biomedical engineering, neurosurgery, and bioethics. The creation and maintenance of effective neuroprosthetic supersystems, on the other hand, relies just as strongly on such disciplines as network design, enterprise architecture, organizational design, and management cybernetics. This text explores one aspect of the design and management of neuroprosthetic supersystems by formulating a topological model that can be used to analyze the structure and behavior of neuroprosthetic supersystems and to design architectures for such supersystems that optimize desired characteristics from functional, financial, legal, ethical, and other perspectives.

The Organization of Neuroprosthetic Supersystems as Multi-Agent Systems

An organization such as a business or government agency can be understood as a specialized type of multi-agent system composed of autonomous intelligent agents. Historically, such agents have primarily been human beings, although some organizations (such as ranches or military units) have long incorporated non-human agents such as trained dogs or horses in certain roles. Today, organizations increasingly incorporate social robots, chatbots, and other artificial agents that perform specialized functions: while such artificial agents may not appear in an organization's formal (human) personnel structure, they must be taken into account by any cybernetic analysis of the organization's internal processes of communication and control.

---

[10] See, e.g., McGee, "Bioelectronics and Implanted Devices" (2008); Warwick & Gasson, "Implantable Computing" (2008); Gasson (2012); Gladden, "Neural Implants as Gateways" (2016); and Gladden, "Enterprise Architecture for Neurocybernetically Augmented Organizational Systems" (2016).

There are many different ways in which an organization's agents might be grouped and their interactions arranged in an effort to maximize the organization's efficiency and productivity. Horling and Lesser note that the range of potential organizational forms available to contemporary organizations includes hierarchies (which can be either simple, uniform, or multi-divisional), holarchies (or 'holonic organizations'), coalitions, teams, congregations, societies, federations (or 'federated systems'), matrix organizations, compound organizations, and sparsely connected graph structures (which may either possess statically defined elements or be an 'adhocracy').[11] Such structures have been developed over time to suit the unique characteristics of the key members that constitute contemporary organizations – i.e., natural biological human beings. However, for organizations that deploy posthumanizing neuroprosthetic technologies among their personnel, new types of organizational structures are expected to become feasible – or perhaps even necessary. For example, an organization that is composed of neuroprosthetically augmented human members may be able to link them by means of a decentralized network that enables the direct sharing of thoughts and emotions between members' minds, allowing information to be disseminated instantaneously and decisions to be made in a collective manner that is not possible for conventional human organizations.[12] However, the fact that such new types of personnel structures may become technologically possible does not necessarily mean that they are desirable or appropriate from an operational, financial, legal, or ethical perspective. The planning and evaluation of organizational structures for neuroprosthetic supersystems can be supported by the use of models that highlight critical aspects of such structures.

---

[11] See Horling & Lesser, "A Survey of Multi-Agent Organizational Paradigms" (2004), and Gladden, *Posthuman Management: Creating Effective Organizations in an Age of Social Robotics, Ubiquitous AI, Human Augmentation, and Virtual Worlds* (2016), p. 122.

[12] Regarding the potential creation of hive minds and neuroprosthetically facilitated collective intelligences, see, e.g., McIntosh, "The Transhuman Security Dilemma" (2010); Roden, *Posthuman Life: Philosophy at the Edge of the Human* (2014), p. 39; and Gladden, "Utopias and Dystopias as Cybernetic Information Systems: Envisioning the Posthuman Neuropolity" (2015). For a classification of different kinds of potential hive minds, see Chapter 2, "Hive Mind," in Kelly, *Out of Control: The New Biology of Machines, Social Systems and the Economic World* (1994); Kelly, "A Taxonomy of Minds" (2007); Kelly, "The Landscape of Possible Intelligences" (2008); Yonck, "Toward a standard metric of machine intelligence" (2012); and Yampolskiy, "The Universe of Minds" (2014). For critical perspectives on hive minds, see, e.g., Maguire & McGee, "Implantable brain chips? Time for debate" (1999); Bendle, "Teleportation, cyborgs and the posthuman ideology" (2002); and Heylighen, "The Global Brain as a New Utopia" (2002).

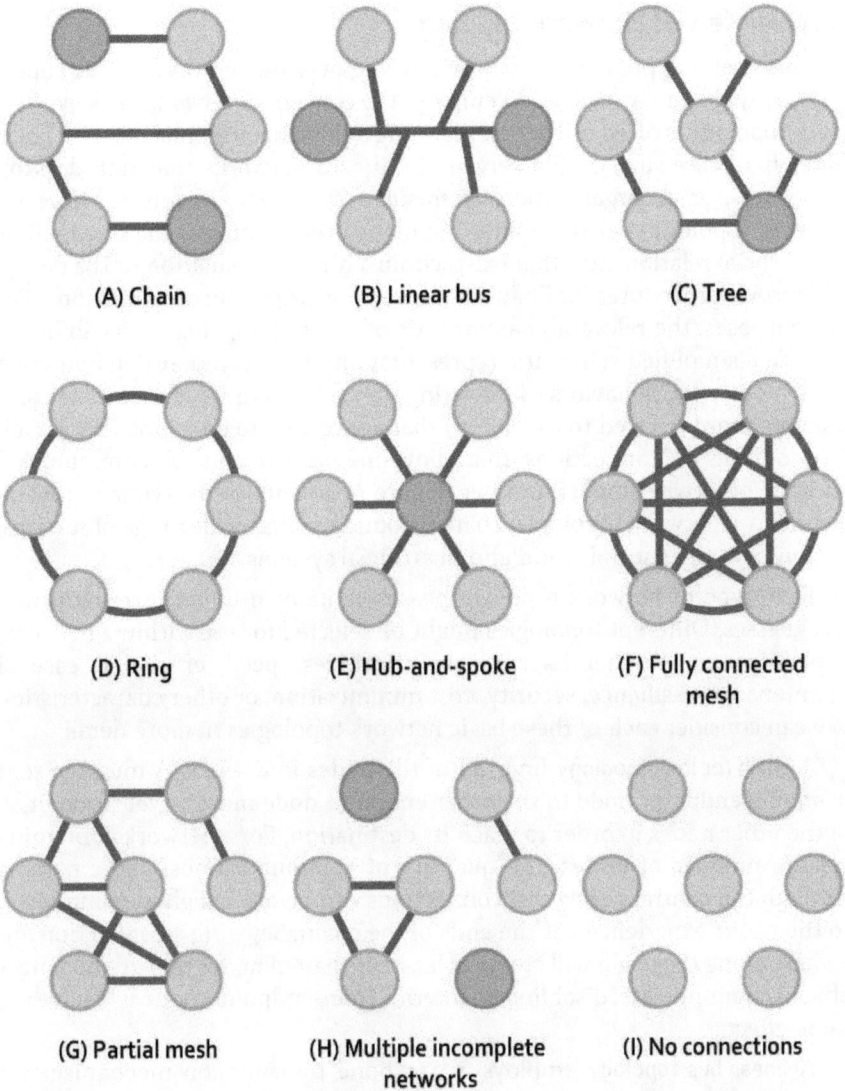

Fig 1: An overview of basic types of network topologies, including the (A) chain (or line), (B) linear bus, (C) tree, (D) ring, (E) hub-and-spoke (or star), (F) fully connected mesh, and (G) partial mesh topologies, along with depictions of (H) multiple incomplete networks that fail to connect all of the nodes and (I) a complete lack of connections between nodes. In some topologies, one or more nodes are highlighted to indicate their special roles, including the endpoints in the (A) chain and (B) linear bus topologies, the root node in the (C) tree topology, the hub in the (E) hub-and-spoke topology, and those nodes that are excluded from any networks in the case of (H) multiple incomplete networks.

An Overview of Network Topology

One useful approach to analyzing or engineering neuroprosthetic super-systems within an organization employs the concept of **network topology**. Network topology is often utilized in planning and managing collections of computer hardware such as file servers, local area networks that link desktop computers, or an organization's infrastructure of Wi-Fi routers. However, network topology is also employed in many other contexts, such as analysis of the social relationships that exist within a human population or the design of approval procedures for financial transactions within an organization.[33] For our purposes, the relevant characteristic of network topology is its ability to provide a simplified schematic representation of the complex dynamics of a system's internal behaviors by depicting the system as a set of objects or components (here referred to as 'nodes') that are related to one another through a set of links or 'connections' that allow one node to control, communicate with, or otherwise impact another. Figure 1 presents an overview of several elementary network topologies that are found within a wide range of settings, both within the natural world and in artificial systems.

Each type of network topology possesses its own unique strengths and weaknesses. Different topologies might be selected for use within a network, depending on whether its engineer prioritizes speed, efficiency, ease of maintenance, resilience, security, cost minimization, or other characteristics. We can consider each of these basic network topologies in more detail.

A **chain (or line) topology** links all of the nodes in a series. A message sent from one endpoint node to the other endpoint node must travel through all of the other nodes in order to reach its destination. For a network containing a large number of nodes, the quantity of communications traffic passing through the central nodes and connections can be quite high in comparison to the traffic experienced at the ends of the chain. Severing a connection anywhere along the chain will create at least one pair of nodes that are no longer able to communicate; disabling an interior (non-endpoint) node will have the same effect.

A **linear bus topology** employs a backbone transmission mechanism (or 'bus') that connects two endpoint nodes, with all of the other nodes also connecting directly to the bus. A message that is transmitted into the bus by a given node will be simultaneously received by all other nodes.

A **tree topology** is a hierarchical structure including a single root node with connections that branch like the boughs of a tree. Travelling away from the

---

[33] For an overview of network topology in the context of designing and maintaining computer networks, see, e.g., Robertazzi, *Networks and Grids: Technology and Theory* (2007), and Sosinsky, *Networking Bible* (2009). For a discussion of network topology in the context of social networks, see McCulloh et al., *Social Network Analysis with Applications* (2013).

root node, a given 'parent' node may have any number of 'child' nodes, but a given child node will have only a single parent. For any two nodes, there will be a single shortest path connecting them; messages sent between the two nodes may need to travel a short or long distance, depending on (for example) whether the nodes lie along the same branch or are the endpoints of different branches that split all the way back at the root node.

A **ring topology** creates a closed circuit linking all of the nodes. In typical implementations of ring networks for computers, the messages sent between nodes only travel in a single direction; in that sort of arrangement, severing a connection anywhere will render at least some of the nodes unable to communicate with one another. In a ring that utilizes bidirectional communication, severing a single connection will increase the distance between some of the nodes and the time needed for messages to travel between them, but it will not leave any of the nodes unable to communicate with others.

A **hub-and-spoke (or star) topology** utilizes as a single node as a hub through which all messages must pass. This model allows for effective centralized control of communications: the hub node may decide which messages are allowed to pass between which nodes. However, the hub also represents a single point of failure: if the hub is disabled, all communication between the remaining hubs is completely severed. A major issue in the design of hub-and-spoke networks is determining exactly where to locate the hub in order to maximize efficiency, security, or other characteristics of the network; Figure 2 depicts such a scenario. Numerous approaches have been developed for using artificial intelligence to optimize the position of hubs within large hub-and-spoke networks.[14]

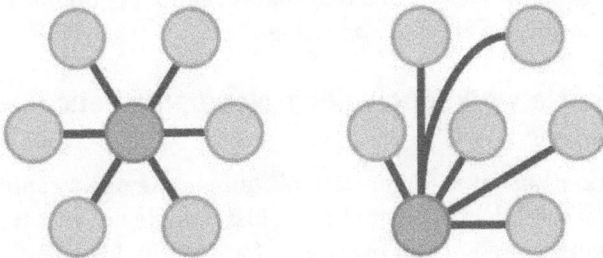

Fig 2: Two different ways in which a hub-and-spoke (or star) network could be constructed to connect the same set of nodes by utilizing different hub placements.

A **fully connected mesh topology** enables direct communication between every pair of nodes in the network. Such a network is highly efficient, insofar

[14] See, e.g., Alumur & Kara, "Network hub location problems: The state of the art" (2008).

as a message between any two nodes always travels along the shortest possible path. Such a network is also maximally robust and resilient: even if many of a node's connections were to be severed, it would retain the ability to exchange messages with every other node in the network, as long as at least one of its connections remained intact. However such networks require the creation and maintenance of a large and complex infrastructure of connections.

A **partial mesh topology** incorporates a variety of topologies to create a network that includes some subregions that are fully connected meshes but which – as a whole – is not fully connected. Some areas of the network may be rich in connections, while other nodes (such as those that are less important or which generate or receive less communication traffic) may only have direct connections to a limited number of other nodes. Use of such a topology allows a network to enjoy some of the benefits of efficiency and resilience found in a fully connected mesh topology but without the creation of such a complex infrastructure of connections.

It is also possible for nodes to comprise **multiple incomplete networks**; in such an arrangement, at least some of the nodes within a system possess connections with other nodes, but there are also some pairs of nodes that are not able to communicate with one another. Such an arrangement might be purposefully implemented (e.g., in order to maximize security by isolating some nodes from the network), or it might arise unintentionally (e.g., as a result of damage to nodes or connections that severs some of the network's internal communication paths).

It is also possible for there to be a **complete lack of connections** between any nodes in the system. This means that nodes cannot communicate with one another or control one another's activities.

The Roles of Network Topologies in Neuroprosthetic Supersystems

The relationships of an organization's human members cannot typically be fully and accurately represented using just a single network topology. For example, Figure 3 presents an overview of a hypothetical organization that includes seven employees: it illustrates the fact that the network topology of the relationships between the organization's members takes on a very different form depending on whether one analyzes the relationships through the lens of the organization's formal personnel structure, the communication enabled by its email system, or the face-to-face interaction that the employees experience in the workplace.

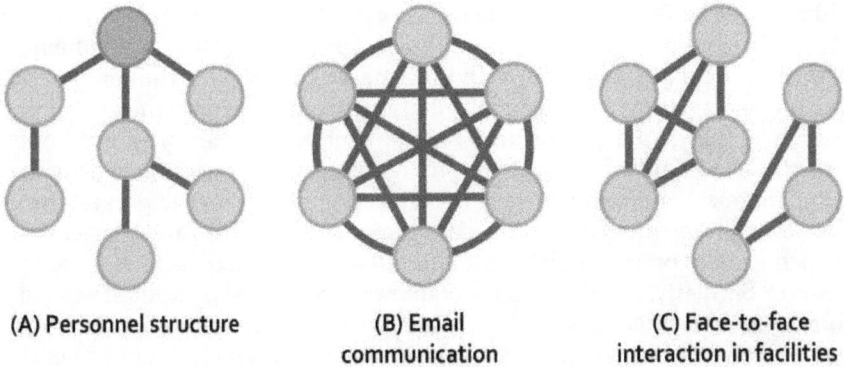

| (A) Personnel structure | (B) Email communication | (C) Face-to-face interaction in facilities |

Fig 3: Three ways in which the network topology of the relationships of an organization's employees may be represented by using different perspectives. The employees can be viewed according to (A) the official personnel structure governing their decision-making authority and reporting relationships, (B) the fact that any employee may communicate with any other by means of the organization's email system, or (C) the fact that the employees work in two different buildings in different countries, which divides the employees into two groups of persons that can interact face-to-face among themselves but not with workers in the other facility.

Enterprise Architecture and Organizational Views

The discipline of enterprise architecture (EA) emerged in the 1980s and 1990s as a response to the fact that organizations deploying new computerized information systems were failing to realize expected gains in productivity and efficiency – in large part, because the acquisition of such systems was not guided by the organizations' larger business strategies and the systems were not being effectively integrated into the organizations' existing structure and dynamics.[15] Enterprise architecture seeks to facilitate the successful integration of such new information technologies into an organization by generating 'alignment' between the organization's electronic information systems, human resources, business processes, workplace culture, mission and strategy, and external ecosystem, which increases the organization's ability to manage complexity, resolve internal conflicts, and adapt proactively to environmental change.[16]

Enterprise architecture formalizes different ways of viewing an organization's network topologies through the concept of 'viewpoints' and the preparation of documents known as 'views.'[17] A number of models exist that define

---

[15] See Magoulas et al., "Alignment in Enterprise Architecture" (2012), p. 89, and Hoogervorst, "Enterprise Architecture: Enabling Integration, Agility and Change" (2004), p. 16.

[16] Regarding different types of alignment generated by EA, see Chan & Reich, "IT alignment: what have we learned?" (2007); Magoulas et al. (2012); and Gladden, "Enterprise Architecture for Neurocybernetically Augmented Organizational Systems" (2016).

[17] For a more detailed exposition of the fundamentals of enterprise architecture, see Chapter 5 of this work.

different views. For example, Kruchten's '4+1' model was developed primarily to support the practice of software design, although it has been influential within the field of EA more broadly: as a means of analyzing the components and internal relationships of an information system, it defines the *logical view* (which employs the perspective of end-user functionality), *process view* (which focuses on communication, integration, and other system behaviors and dynamics), *development view* (which is prepared from the perspective of the design, implementation, and management of software), and *physical view* (which focuses on the deployment and interconnection of hardware components).[18] Similarly, the Siemens EA Framework described by Rohloff explicitly formulates the three perspectives of *component, communication,* and *distribution* views.[19] The generic EA framework that we present in Chapter 5 of this text (and which builds on the approaches of Kruchten and Siemens) incorporates three primary perspectives on architecture domains: the *component, interaction,* and *membership* views.[20]

Defining a Model of Relevant Views for Neuroprosthetic Supersystems

It is possible to define at least five topological views that are relevant for analyzing, designing, and maintaining neuroprosthetic supersystems. Two of these relate primarily to the neuroprosthetic devices themselves:

- The **physical neuroprosthetic device topology** reflects the physical connections that exist between the neuroprostheses possessed by different human hosts within the organization. Such physical connections may be hardwired (e.g., utilizing generic Cat 5 Ethernet or fiber optic cables), may be wireless (e.g., utilizing radio-based Wi-Fi or Bluetooth, infrared, or laser transmissions), or may combine wired and wireless components.

- The **logical neuroprosthetic device topology** reflects the ways in which the software or other control processes within the supersystem allow its neuroprostheses to communicate with one another. This may dif-

---

[18] See Kruchten, "The 4+1 view model of architecture" (1995), and Part 8.1 of *ArchiMate® 2.1 Specification* (2015).

[19] See Rohloff, "Framework and Reference for Architecture Design" (2008), pp. 5-6.

[20] In our generic EA framework, the component view highlights all of the entities that together constitute the enterprise, including employees, physical facilities, computing devices, vehicles, products, and financial, material, and informational resources, as well as the capacities and internal processes that these entities possess. The interaction view highlights the network topology of the ways in which these entities are connected to one another and the processes (such as those of communication and control) by which they interact. The membership view highlights the boundaries and occupants of those spatiotemporal regions (such as physical buildings, countries, time zones, or virtual environments) within which organizational elements are located or operate and of those functional or conceptual groupings (such as corporate departments or project teams) to which elements belong or to whose authority they are subject. See Chapter 5 of this text for more details.

fer from the devices' physical topology. For example, consider a situation in which each of a supersystem's neuroprostheses has the physical capacity to transmit radio signals to and receive radio signals from any other neuroprosthesis in the supersystem, however each device's built-in security software only processes an incoming message and delivers it to the device's host if the message was sent by a neuroprosthesis belonging to that host's direct supervisor or to a direct subordinate. In this case, the neuroprostheses' physical topology would constitute a fully connected mesh, but the logical topology would take the form of a hierarchical tree topology.

Three other relevant views relate primarily to the devices' human hosts, in their role as embodied and embedded intelligences and social actors:

- The **human hosts' natural social topology** reflects social relations between individual human beings within the organization that are not dependent on neuroprosthetic devices. Such relations may already have existed prior to individuals' neuroprosthetic augmentation or may have been developed after their augmentation but without the facilitation of neuroprostheses; they might continue to exist even if the supersystem's neuroprostheses were disabled.

- The topology of the **neuroprosthetically facilitated virtual environment** reflects the 'inhabitants' of the virtual environments, if any, that are created by means of the neuroprostheses and which provide an opportunity for social interaction among the human hosts who spend time in such an environment – even if they do not, in practice, directly interact with one another within that venue.[21] Such immersive virtual environments might include shared virtual spaces that are created temporarily to facilitate the completion of a team's particular tasks, or they might be persistent and massively multiuser virtual worlds that fill a general organizational role.

- The topology of **neuroprosthetically facilitated brain-to-brain communication** reflects the lines of communication, if any, that are established within the supersystem directly between the brains of human hosts by means of their neuroprostheses.[22] Such communication might

[21] For a practical overview of virtual teams, see Zofi, *A Manager's Guide to Virtual Teams* (2012), and Settle-Murphy, *Leading Effective Virtual Teams: Overcoming Time and Distance to Achieve Exceptional Results* (2012). Various aspects of virtual organizations are discussed in Fairchild, *Technological Aspects of Virtual Organizations: Enabling the Intelligent Enterprise* (2004); *Virtual Organizations: Systems and Practices*, edited by Camarinha-Matos et al. (2005); and Shekhar, *Managing the Reality of Virtual Organizations* (2016). The broader implications of long-term immersion in virtual reality environments are discussed in Bainbridge, *The Virtual Future* (2011); Heim, *The Metaphysics of Virtual Reality* (1993); and Koltko-Rivera, "The potential societal impact of virtual reality" (2005).

[22] Regarding such possibilities, see, e.g., Rao et al., "A direct brain-to-brain interface in humans" (2014), and Gladden, "Utopias and Dystopias as Cybernetic Information Systems" (2015).

conceivably take a number of forms, from the simple hands-free composition and sending of text messages by means of participants'
thoughts to the full integration of hosts' sensory experiences, memories, and volitions in order to fashion a 'hive mind' possessing a
shared will.

Additional types of views are certainly possible and may be critical for particular types of systems and organizational contexts. However, taken together, the five basic views described above provide a robust representation
of the network topologies of a neuroprosthetic supersystem that can be useful
for such diverse purposes as system design and engineering, system management, information security planning, and ethical analysis.

The Dynamic Nature of Neuroprosthetic Topologies

When analyzing or engineering network topologies for neuroprosthetic
supersystems, it should be noted that a network may change its topology periodically or even continually. For example, hosts' access to virtual environments may be added or removed as needed, in order to enforce information
security policies[23] or minimize resource consumption. Some networks utilize
ad hoc and dynamically reconfigurable topologies that can be updated as
needed to reflect a network's change in membership, purpose, or operational
context. Networks may incorporate artificially intelligent control mechanisms that autonomously detect changing internal or environmental conditions and adapt the network's design (e.g., by relocating the position of a logical hub) according to fixed rules; other networks may possess configurations
that were not intentionally designed but which were developed through the
use of evolutionary algorithms that optimize desirable performance characteristics.[24]

Applying the Topological Model to Hypothetical Neuroprosthetic Supersystems

In order to illustrate the role of the five basic topological views in describing a neuroprosthetic supersystem, below we use the views to depict the neuroprosthetic supersystems of four very simple hypothetical organizations.

---

[23] For example, the InfoSec principle of purposefully structuring access for least functionality and
least privilege is discussed in *NIST Special Publication 800-53, Revision 4: Security and Privacy
Controls for Federal Information Systems and Organizations* (2013), pp. F-71-F-73, F-179.

[24] Regarding such approaches to networks in various contexts, see, e.g., Gen et al., *Network Models and Optimization: Multiobjective Genetic Algorithm Approach* (2008); *Adaptive Networks:
Theory, Models and Applications*, edited by Gross & Sayama (2009); LoBello & Toscano, "An adaptive approach to topology management in large and dense real-time wireless sensor networks"
(2009); and *Self-Organizing Networks (SON): Self-Planning, Self-Optimization and Self-Healing
for GSM, UMTS, and LTE*, edited by Ramiro & Hamied (2012).

Scenario 1: An Emergency Medical Alert System

Figure 4 depicts the five basic topological views as applied to a neuroprosthetically enabled medical alert system. Each member of the organization possesses a neuroprosthetic device that monitors its host's biological processes and medical condition by means of a body sensor network (BSN) comprising numerous implanted sensors. If a host should experience a debilitating injury or sudden illness that renders him or her unconscious, that host's neuroprosthesis detects that occurrence and transmits a medical alert and call for help that is communicated simultaneously to all of the other hosts' devices by means of a shared bus. The human hosts all interact periodically with one another through non-neuroprosthetic social channels, and the neuroprosthetic network does not create a shared virtual environment or enable direct brain-to-brain communication.

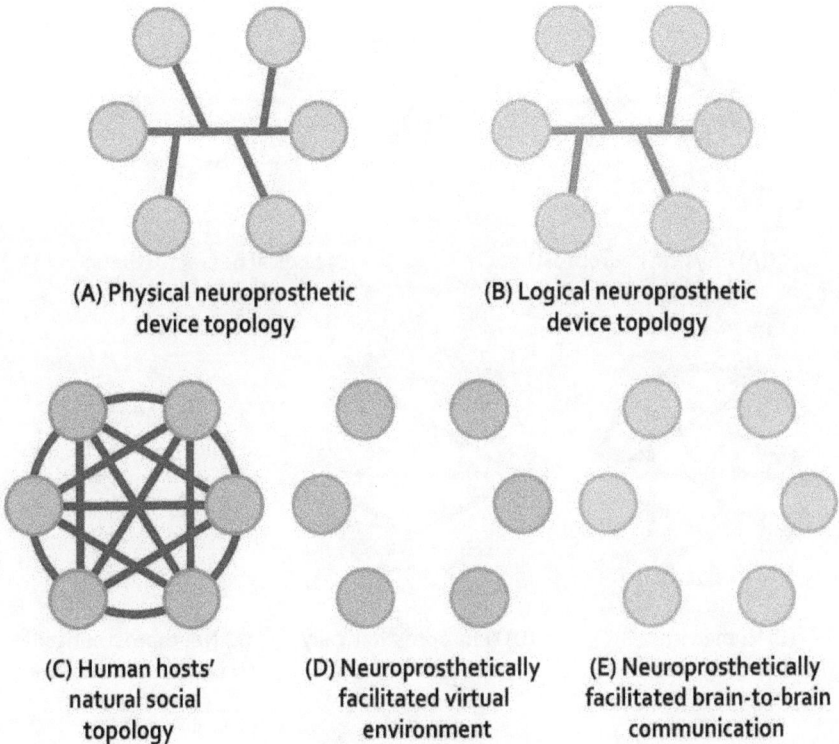

(A) Physical neuroprosthetic device topology

(B) Logical neuroprosthetic device topology

(C) Human hosts' natural social topology

(D) Neuroprosthetically facilitated virtual environment

(E) Neuroprosthetically facilitated brain-to-brain communication

Fig 4: A neuroprosthetic network that provides an emergency medical alert system. The devices themselves are linked through a physical (A) and logical (B) bus; their human hosts all interact periodically with one another through non-neuroprosthetic social channels (C). The neuroprostheses do not create a shared virtual world (D) or provide direct brain-to-brain communication (E).

Scenario 2: Centralized Hosting of Multiple Virtual Worlds

Figure 5 depicts the use of neuroprosthetic devices to create a set of virtual worlds that are administered by a single host's neuroprosthesis. Each of the organization's human members has access to one of two virtual worlds, depending on his or her organizational role – except for the host of the neuroprosthesis serving as the system's hub, who has access to both virtual worlds. All of the hosts interact with one another in their shared physical workplace. The neuroprosthetic network also allows each human host to mentally compose and send text messages to any other host, thus enabling a basic form of brain-to-brain communication. Disabling of the hub neuroprosthesis would shut down the virtual environments and brain-to-brain communication.

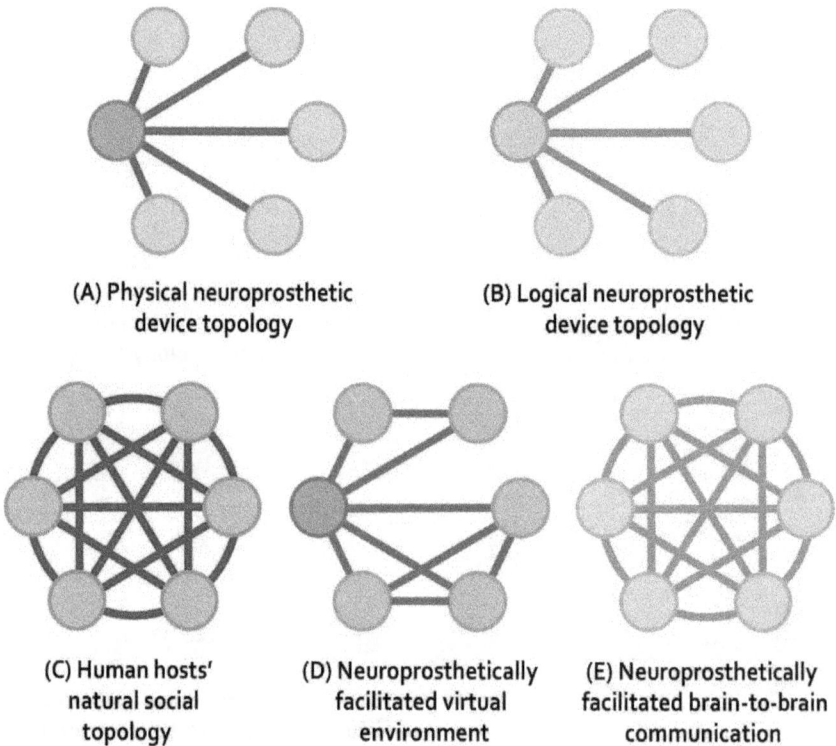

(A) Physical neuroprosthetic
device topology

(B) Logical neuroprosthetic
device topology

(C) Human hosts'
natural social
topology

(D) Neuroprosthetically
facilitated virtual
environment

(E) Neuroprosthetically
facilitated brain-to-brain
communication

Fig 5: A neuroprosthetic network in which a single device serves as the hub of physical (A) and logical (B) hub-and-spoke networks in order to maintain two distinct virtual environments (D) and enable brain-to-brain communication (E) among all hosts – who also interact socially in their shared physical workplace (C).

Scenario 3: A Hive Mind with a Central Hub

Figure 6 depicts the use of neuroprosthetic devices to fashion a 'hive mind' whose human members share memories, emotions, and desires to create a collective entity with a common will. In this case, one neuroprosthetic device serves as the physical hub for the network, allowing all of the other devices to communicate with one another through it. Some of the network's human hosts interact with one another in their physical workplace, but others are physically isolated and interact only by means of the neuroprosthetic network. The neuroprostheses allow direct brain-to-brain communication between hosts, but do not override or interfere with the hosts' natural biological sensory systems in order to create a shared virtual world.

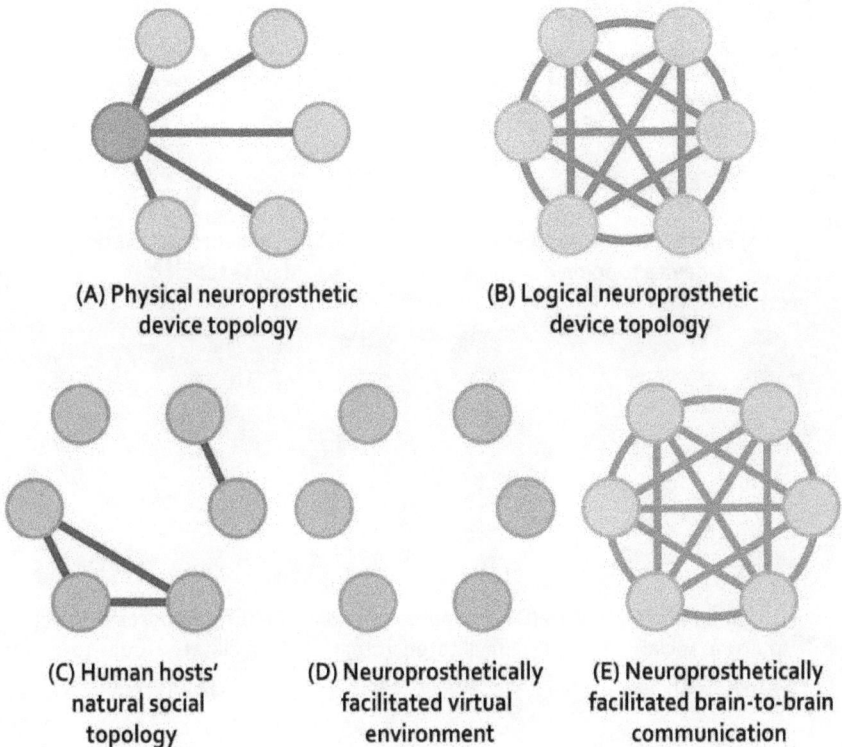

(A) Physical neuroprosthetic device topology

(B) Logical neuroprosthetic device topology

(C) Human hosts' natural social topology

(D) Neuroprosthetically facilitated virtual environment

(E) Neuroprosthetically facilitated brain-to-brain communication

Fig 6: A neuroprosthetic supersystem that links its human members to create a centralized hive mind. One neuroprosthesis serves as the physical hub in a hub-and-spoke topology (A) to create a fully connected logical mesh including all devices (B). The network creates direct brain-to-brain communication (E) but not a shared virtual environment (D) for its human members, only some of whom enjoy social relations in the physical workplace (C).

Scenario 4: A Distributed Hive Mind Lacking a Central Hub

Figure 7 depicts the use of neuroprosthetic devices to create a decentralized hive mind that lacks a central physical or logical hub: any new device joining the network establishes direct physical and logical connections with all other devices, and the disabling of a node or severing of individual connections does not imperil the functioning of the network as a whole. The network's human members do not interact socially in the primary physical world and do not have access to a shared virtual environment; however, by means of the neuroprosthetic network they are able to share their internal monologues,[25] imaginings, and volitions to forge collective decisions.

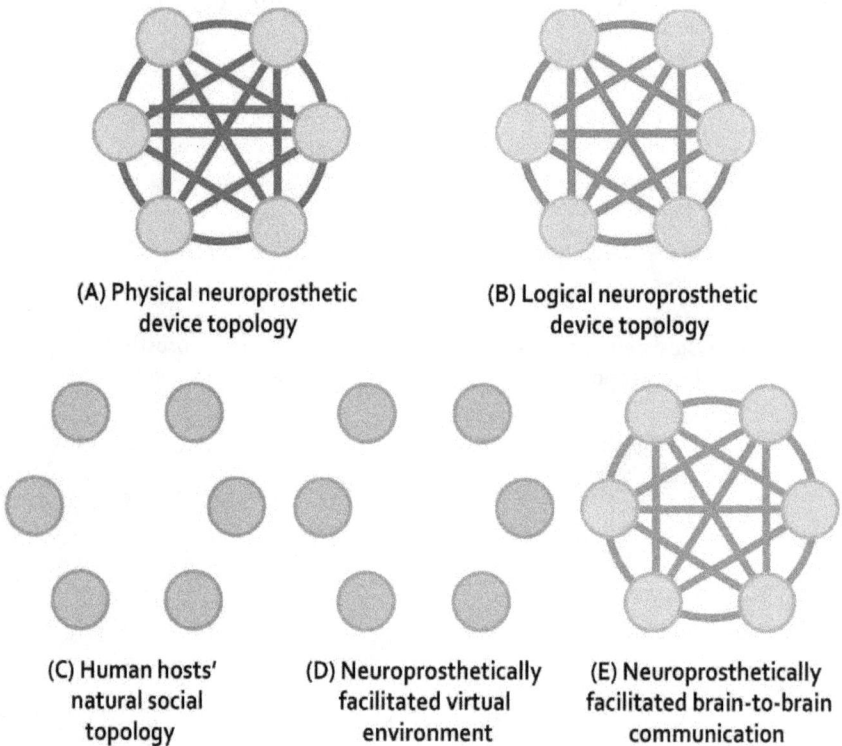

(A) Physical neuroprosthetic device topology

(B) Logical neuroprosthetic device topology

(C) Human hosts' natural social topology

(D) Neuroprosthetically facilitated virtual environment

(E) Neuroprosthetically facilitated brain-to-brain communication

Fig 7: A neuroprosthetically facilitated hive mind whose neuroprostheses all enjoy direct physical (A) and logical (B) connections with one another. The network's human hosts do not interact socially in the primary physical world (C) and do not inhabit a shared virtual environment (D); however, their brains communicate directly in a fully connected mesh (E).

[25] For an overview of such cognitive processes, see "The Internal Monologue" in Butler, *Rethinking Introspection* (2013), pp. 119-47.

Conclusion

For most contemporary organizations, the notion of intentionally integrating posthumanizing neuroprosthetic technologies into the workplace has no relevance as a strategic, operational, or tactical possibility. However, for the select group of specialized organizations (such as military departments) that are actively seeking to develop and deploy such technologies among their personnel, complex questions are emerging that relate not only to such devices' biological and technological elements but also to their *organizational* aspects. For example, how can such technologies be most effectively incorporated into an organization's personnel structures, business processes, and information systems? In this text, we have investigated one facet of that question by developing a model based on the concepts of network topology and enterprise architecture's technique of analyzing an organization through the lens of different formalized 'views.' Our model incorporates five views whose uses we have explored by applying the model to four types of neuroprosthetic supersystems that might be established through the deployment of neuroprosthetic technologies. While such a model can be of immediate use for those organizations that are considering the possibility of implementing posthumanizing neuroprosthetic technologies – or that are already actively working to deploy them – it may also be of use to policymakers, social scientists, ethicists, and others who are seeking to understand the implications of organizations' potential exploitation of neuroprosthetic technologies and to establish the parameters for their use, in order to ensure that such technologies are employed in ways that are consistent with the rights of their human users and which benefit rather than harm the organizations deploying them and human society as a whole.

References

Ablett, Ruth, Shelly Park, Ehud Sharlin, Jörg Denzinger, and Frank Maurer. "A Robotic Colleague for Facilitating Collaborative Software Development." *Proceedings of Computer Supported Cooperative Work (CSCW 2006)*. ACM, 2006.

Adaptive Networks: Theory, Models and Applications, edited by Thilo Gross and Hiroki Sayama. Springer Berlin Heidelberg, 2009.

Advances in Neuromorphic Memristor Science and Applications, edited by Robert Kozma, Robinson E. Pino, and Giovanni E. Pazienza. Dordrecht: Springer Science+Business Media, 2012.

Ahlemann, Frederik, Eric Stettiner, Marcus Messerschmidt, Christine Legner, Kunal Mohan, and Daniel Schäfczuk. "People, adoption and introduction of EAM." In *Strategic Enterprise Architecture Management*, pp. 229-263. Springer Berlin Heidelberg, 2012.

Aier, Stephan. "The Role of Organizational Culture for Grounding, Management, Guidance and Effectiveness of Enterprise Architecture Principles." *Information Systems and E-Business Management* 12, no. 1 (2014): 43-70.

Alumur, Sibel, and Bahar Y. Kara. "Network hub location problems: The state of the art." *European journal of operational research* 190, no. 1 (2008): 1-21.

Amputation, Prosthetic Use, and Phantom Limb Pain: An Interdisciplinary Perspective, edited by Craig Murray. New York: Springer Science+Business Media, 2010.

Andersen, Peter A., and Janis F. Andersen. "Measures of Perceived Nonverbal Immediacy." In *The Sourcebook of Nonverbal Measures: Going Beyond Words*, edited by Valerie Manusov. Mahwah, NJ: Lawrence Erlbaum Associates, Inc., 2005.

Anderson, Walter Truett. "Augmentation, symbiosis, transcendence: technology and the future(s) of human identity." *Futures* 35, no. 5 (2003): 535-46.

Anderson, Michael L. "Embodied cognition: A field guide." *Artificial intelligence* 149, no. 1 (2003): 91-130.

Andrew, Alex M. "The decade of the brain: further thoughts." *Kybernetes* 26, no. 3 (1997): 255-264.

Ankarali, Z.E., Q.H. Abbasi, A.F. Demir, E. Serpedin, K. Qaraqe, and H. Arslan. "A Comparative Review on the Wireless Implantable Medical Devices Privacy and Security." In *2014 EAI 4th International Conference on Wireless Mobile Communication and Healthcare (Mobihealth)*, 246-49, 2014. doi:10.1109/MOBIHEALTH.2014.7015957.

ANSI/IEEE 1471-2000, IEEE Recommended Practice for Architectural Description for Software-Intensive Systems. IEEE Computer Society, 2000.

ArchiMate® 2.1 Specification. Berkshire: The Open Group, 2013.

Ariely, D., and G.S. Berns. "Neuromarketing: The Hope and Hype of Neuroimaging in Business." *Nature Reviews Neuroscience* 11, no. 4 (2010): 284-92.

Austerberry, David. *Digital Asset Management*. Second edition. Burlington, MA: Focal Press, 2013.

Ayaz, Hasan, Patricia A. Shewokis, Scott Bunce, Maria Schultheis, and Banu Onaral. "Assessment of Cognitive Neural Correlates for a Functional Near Infrared-Based Brain Computer Interface System." In *Foundations of Augmented Cognition. Neuroergonomics and Operational Neuroscience*, edited by Dylan D. Schmorrow, Ivy V. Estabrooke, and Marc Grootjen, 699-708. Lecture Notes in Computer Science 5638. Springer Berlin Heidelberg, 2009.

Baars, Bernard J. *In the Theater of Consciousness*. New York, NY: Oxford University Press, 1997.

Baddeley, Alan. "The episodic buffer: a new component of working memory?" *Trends in cognitive sciences* 4, no. 11 (2000): 417-23.

Baddeley, Alan. "Working memory: theories, models, and controversies." *Annual review of psychology* 63 (2012): 1-29.

Bainbridge, William Sims. *The Virtual Future*. London: Springer, 2011.

Băjenescu, Titu-Marius, and Marius I. Bâzu. *Reliability of Electronic Components: A Practical Guide to Electronic Systems Manufacturing*. Springer Berlin Heidelberg, 1999.

Baranyi, Péter, Adam Csapo, and Gyula Sallai. "Synergies Between CogInfoCom and Other Fields." In *Cognitive Infocommunications (CogInfoCom)*. Springer International Publishing, 2015.

Barfield, Woodrow. *Cyber-Humans: Our Future with Machines*. Springer Science+Business Media, 2015.

Barile, S., J. Pels, F. Polese, and M. Saviano. "An Introduction to the Viable Systems Approach and Its Contribution to Marketing." *Journal of Business Market Management* 5(2) (2012): 54-78.

Bean, Sally. "Re-Thinking Enterprise Architecture Using Systems and Complexity Approaches." *Journal of Enterprise Architecture* 6, no. 4 (2010).

Beer, Stafford. *Brain of the Firm*. Second edition. New York: John Wiley, 1981.

Bekey, G.A. *Autonomous Robots: From Biological Inspiration to Implementation and Control*. Cambridge, MA: MIT Press, 2005.

Bendle, Mervyn F. "Teleportation, cyborgs and the posthuman ideology." *Social Semiotics* 12, no. 1 (2002): 45-62.

Bergey, John, Stephen Blanchette, Jr., Paul Clements, Mike Gagliardi, John Klein, Rob Wojcik, and Bill Wood. "U.S. Army Workshop on Exploring Enterprise, System of Systems, System, and Software Architectures." Technical Report CMU/SEI-2009-TR-008 / ESC-TR-2009-008. Hanscom AFB, MA: Software Engineering Institute, 2009.

Berner, Georg. *Management in 20XX: What Will Be Important in the Future – A Holistic View*. Erlangen: Publicis Corporate Publishing, 2004.

Bhunia, Swarup, Abhishek Basak, Seetharam Narasimhan, and Maryam Sadat Hashemian. "Ultralow Power and Robust On-Chip Digital Signal Processing for Closed-Loop Neuro-Prosthesis." In *Implantable Bioelectronics: Devices, Materials, and Applications*, edited by Evgeny Katz. Weinheim: Wiley-VCH, 2014.

Birnbacher, Dieter. "Posthumanity, Transhumanism and Human Nature." In *Medical Enhancement and Posthumanity*, edited by Bert Gordijn and Ruth Chadwick, pp. 95-106. The International Library of Ethics, Law and Technology 2. Springer Netherlands, 2008.

Bischoff, Stefan, Stephan Aier, and Robert Winter. "Use It or Lose It? The Role of Pressure for Use and Utility of Enterprise Architecture Artifacts." In *2014 IEEE 16th Conference on Business Informatics*, vol. 2, pp. 133-140. IEEE, 2014.

Black, Michael J., Elie Bienenstock, John P. Donoghue, Mijail Serruya, Wei Wu, and Yun Gao. "Connecting brains with machines: the neural control of 2D cursor movement." In *Proceedings of the 1st International IEEE/EMBS Conference on Neural Engineering*, pp. 580-83, 2003.

Blank, S. Catrin, Sophie K. Scott, Kevin Murphy, Elizabeth Warburton, and Richard JS Wise. "Speech production: Wernicke, Broca and beyond." *Brain* 125, no. 8 (2002): 1829-38.

Bogue, Robert. "Brain-Computer Interfaces: Control by Thought." *Industrial Robot: An International Journal* 37, no. 2 (2010): 126-32.

Boly, Melanie, Anil K. Seth, Melanie Wilke, Paul Ingmundson, Bernard Baars, Steven Laureys, David B. Edelman, and Naotsugu Tsuchiya. "Consciousness in humans and non-human animals: recent advances and future directions." *Frontiers in Psychology* 4 (2013).

Bonaci, T., R. Calo, and H. Chizeck. "App Stores for the Brain." *IEEE Technology and Society Magazine,* 1932-4529/15 (2015): 32-39.

Borkar, Shekhar. "Designing reliable systems from unreliable components: the challenges of transistor variability and degradation." *Micro, IEEE* 25, no. 6 (2005): 10-16.

Bostrom, Nick. "Human Genetic Enhancements: A Transhumanist Perspective." In *Arguing About Bioethics*, edited by Stephen Holland, pp. 105-15. New York: Routledge, 2012.

Bostrom, Nick, and Anders Sandberg. "Cognitive Enhancement: Methods, Ethics, Regulatory Challenges." *Science and Engineering Ethics* 15, no. 3 (2009): 311-41.

Boucharas, Vasilis, Marlies van Steenbergen, Slinger Jansen, and Sjaak Brinkkemper. "The Contribution of Enterprise Architecture to the Achievement of Organizational Goals: A Review of the Evidence." In *Trends in Enterprise Architecture Research*, 1-15. Springer, 2010.

Braddon-Mitchell, David, and John Fitzpatrick. "Explanation and the Language of Thought." *Synthese* 83, no. 1 (April 1, 1990): 3-29.

Bradford, David L., and W. Warner Burke. *Reinventing Organization Development: New Approaches to Change in Organizations.* John Wiley & Sons, 2005.

Brain-Computer Interfaces: Principles and Practice, edited by Jonathan R. Wolpaw and Elizabeth Winter Wolpaw. New York: Oxford University Press, 2012.

Brandt, Thomas. *Vertigo: Its Multisensory Syndromes.* Springer Verlag London, 2003.

Breazeal, Cynthia. "Toward sociable robots." *Robotics and Autonomous Systems* 42 (2003): 167-75.

"Bridging the Bio-Electronic Divide." Defense Advanced Research Projects Agency, January 19, 2016. http://www.darpa.mil/news-events/2015-01-19. Accessed May 6, 2016.

Brunner, Peter, and Gerwin Schalk. "Brain-Computer Interaction." In *Foundations of Augmented Cognition. Neuroergonomics and Operational Neuroscience*, edited by Dylan D. Schmorrow, Ivy V. Estabrooke, and Marc Grootjen, 719-23. Lecture Notes in Computer Science 5638. Springer Berlin Heidelberg, 2009.

Brunner, P., L. Bianchi, C. Guger, F. Cincotti, and G. Schalk. "Current Trends in Hardware and Software for Brain-Computer Interfaces (BCIs)." *Journal of Neural Engineering* 8, no. 2 (2011): 25001.

Bublitz, J.C. "If Man's True Palace Is His Mind, What Is Its Adequate Protection? On a Right to Mental Self-Determination and Limits of Interventions into Other Minds." In *Technologies on the Stand: Legal and Ethical Questions in Neuroscience and Robotics*, edited by B. Van Den Berg and L. Klaming. Nijmegen: Wolf Legal Publishers, 2011.

Buckl, Sabine, Christian M. Schweda, and Florian Matthes. "A Situated Approach to Enterprise Architecture Management." In *2010 IEEE International Conference on Systems, Man and Cybernetics*, 587-92. IEEE, 2010.

Buckl, Sabine, Florian Matthes, and Christian M. Schweda. "A Viable System Perspective on Enterprise Architecture Management." In *IEEE International Conference on Systems, Man and Cybernetics, 2009. SMC 2009*, 1483-88. IEEE, 2009.

Butler, Jesse. *Rethinking Introspection*. Palgrave Macmillan UK, 2013.

C4ISR Architecture Framework Version 2.0. C4ISR Architecture Working Group (AWG), US Department of Defense, December 18, 1997. http://www.afcea.org/education/courses/archfwk2.pdf. Accessed December 4, 2016.

Cadle, James, Debra Paul, and Paul Turner. *Business Analysis Techniques: 72 Essential Tools for Success*. Swindon: British Informatics Society Limited, 2010.

Caetano, Artur, António Rito Silva, and José Tribolet. "A Role-Based Enterprise Architecture Framework." In *Proceedings of the 2009 ACM Symposium on Applied Computing*, pp. 253-58. ACM, 2009.

Cahill, Larry, and Michael T. Alkire. "Epinephrine enhancement of human memory consolidation: interaction with arousal at encoding." *Neurobiology of learning and memory* 79, no. 2 (2003): 194-98.

Callaghan, Vic. "Micro-Futures." Presentation at Creative-Science 2014, Shanghai, China, July 1, 2014.

Calverley, D.J. "Imagining a non-biological machine as a legal person." *AI & SOCIETY* 22, no. 4 (2008): 523-37.

Cameron, Oliver G. *Visceral Sensory Neuroscience: Interoception*. Oxford University Press, 2001.

Campbell, Lyle. *Historical Linguistics*. Third edition. The MIT Press, 2013.

Cane, Sheila, and Richard McCarthy. "Measuring the Impact of Enterprise Architecture." *Issues in Information Systems* 8, no. 2 (2007): 437-42.

Castronova, Edward. "Theory of the Avatar." CESifo Working Paper No. 863, February 2003. http://www.cesifo.de/pls/guestci/download/CESifo+Working+Papers+2003/CESifo+Working+Papers+February+2003/cesifo_wp863.pdf. Accessed January 25, 2016.

Cervera-Paz, Francisco Javier, and M. J. Manrique. "Auditory Brainstem Implants: Past, Present and Future Prospects." In *Operative Neuromodulation*, edited by Damianos E. Sakas and Brian A. Simpson, 437-42. Acta Neurochirurgica Supplements 97/2. Springer Vienna, 2007.

Chafe, Chris, and Sile O'Modhrain. "Musical muscle memory and the haptic display of performance nuance." In *Proceedings of the 1996 International Computer Music Conference*, pp. 1-4. Stanford: Stanford University, 1996.

Chan, Yolande E., and Blaize Horner Reich. "IT alignment: what have we learned?" *Journal of Information technology* 22, no. 4 (2007): 297-315.

Chang, Chia-Ke, Yu-Jung Li, and Chih-Cheng Lu. "RFID applied in recognition and identification for dental prostheses." In *Computerized Healthcare (ICCH), 2012 International Conference on*, pp. 43-45. IEEE, 2012.

Christen, Markus, and Sabine Müller. "Current status and future challenges of deep brain stimulation in Switzerland." *Swiss Medical Weekly* (2012): 142:w13570. doi:10.4414/smw.2012.13570.

"Chronic Pain and Spinal Cord Stimulation (SCS): Frequently Asked Questions." Boston Scientific, 2013. http://www.pae-eu.eu/wp-content/uploads/2015/03/NM-135814-AA-INTL-Spectra-Backgrounder_Final.pdf. Accessed December 8, 2016.

Church, George M., Yuan Gao, and Sriram Kosuri. "Next-generation digital information storage in DNA." *Science* 337, no. 6102 (2012): 1628.

Clancy, Frank. "At Military's Behest, Darpa Uses Neuroscience to Harness Brain Power." *Neurology Today* 6, no. 2 (2006): 4-8.

Clark, S.S., and K. Fu. "Recent Results in Computer Security for Medical Devices." In *Wireless Mobile Communication and Healthcare*, edited by K.S. Nikita, J.C. Lin, D.I. Fotiadis, and M.-T. Arredondo Waldmeyer, pp. 111-18. Lecture Notes of the Institute for Computer Sciences, Social Informatics and Telecommunications Engineering 83. Springer Berlin Heidelberg, 2012.

Clark, Andy. "Systematicity, structured representations and cognitive architecture: A reply to Fodor and Pylyshyn." In *Connectionism and the Philosophy of Mind*, pp. 198-218. Springer Netherlands, 1991.

Clark, Andy. *Natural-born cyborgs: Minds, Technologies, and the Future of Human Intelligence.* Oxford: Oxford University Press, 2004.

Clarke, Arthur C. "Hazards of Prophecy: The Failure of Imagination." In *Profiles of the Future: An Inquiry into the Limits of the Possible,* revised edition. Harper & Row, New York, 1973.

Clausen, J. "Conceptual and Ethical Issues with Brain–hardware Interfaces." *Current Opinion in Psychiatry* 24, no. 6 (2011): 495-501.

Claussen, Jens Christian, and Ulrich G. Hofmann. "Sleep, Neuroengineering and Dynamics." *Cognitive Neurodynamics* 6, no. 3 (May 27, 2012): 211-14.

A Clinical Guide to Transcranial Magnetic Stimulation, edited by Paul E. Holtzheimer and William McDonald. New York: Oxford University Press, 2014.

Clowes, Robert W. "The Cognitive Integration of E-Memory." *Review of Philosophy and Psychology* 4, no. 1 (January 26, 2013): 107-33.

"Cochlear Implant Quick Facts." American Speech-Language-Hearing Association. http://www.asha.org/public/hearing/Cochlear-Implant-Quick-Facts/. Accessed December 8, 2016.

"Cochlear Implants." National Institute on Deafness and Other Communication Disorders (NIDCD), May 3, 2016. https://www.nidcd.nih.gov/health/cochlear-implants. Accessed December 8, 2016.

Coeckelbergh, Mark. "From Killer Machines to Doctrines and Swarms, or Why Ethics of Military Robotics Is Not (Necessarily) About Robots." *Philosophy & Technology* 24, no. 3 (2011): 269-78.

Cognitive Psychology. Second edition. Edited by Nick Braisby and Angus Gellatly. Oxford: Oxford University Press, 2012.

Coker, Christopher. "Biotechnology and War: The New Challenge." *Australian Army Journal* vol. II, no. 1 (2004): 125-40.

Collins, Allison, and Norm Schultz. "A review of ethics for competitive intelligence activities." *Competitive Intelligence Review* 7, no. 2 (1996): 56-66.

Coma Science: Clinical and Ethical Implications, edited by Steven Laureys, Nicholas D. Schiff, and Adrian M. Owen. New York: Elsevier, 2009.

Comai, Alessandro. "Global code of ethics and competitive intelligence purposes: an ethical perspective on competitors." *Journal of Competitive Intelligence and Management* 1, no. 3, 2003.

Comas and Disorders of Consciousness, edited by Caroline Schnakers and Steven Laureys. Springer-Verlag London, 2012.

Communication in the Age of Virtual Reality, edited by Frank Biocca and Mark R. Levy. Hillsdale, NJ: Lawrence Erlbaum Associates, Publishers, 1995.

Computer Synthesized Speech Technologies: Tools for Aiding Impairment, edited by John Mullennix and Steven Stern. Hershey, PA: Medical Information Science Reference, 2010.

Conant, Roger C., and W. Ross Ashby. "Every Good Regulator of a System Must Be a Model of That System." *International journal of systems science* 1, no. 2 (1970): 89-97.

Content of Premarket Submissions for Management of Cybersecurity in Medical Devices: Guidance for Industry and Food and Drug Administration Staff. Silver Spring, MD: US Food and Drug Administration, 2014.

Converging Technologies for Improving Human Performance: Nanotechnology, Biotechnology, Information Technology and Cognitive Science, edited by William Sims Bainbridge. Dordrecht: Springer Science+Business Media, 2003.

Cornwall, Warren. "In Pursuit of the Perfect Power Suit." *Science* 350, issue 6258 (2015): 270-73.

Cory, Jr., Gerald A. "Language, Brain, and Neuron." In *Toward Consilience*, pp. 193-205. Springer US, 2000.

Cosgrove, G.R. "Session 6: Neuroscience, brain, and behavior V: Deep brain stimulation." Meeting of the President's Council on Bioethics. Washington, DC, June 24-25, 2004. https://bioethicsarchive.georgetown.edu/pcbe/transcripts/june04/session6.html. Accessed June 12, 2015.

Coughlin, Thomas M. *Digital Storage in Consumer Electronics: The Essential Guide*. Burlington, MA: Newnes, 2008.

Crane, Andrew. "In the company of spies: When competitive intelligence gathering becomes industrial espionage." *Business Horizons* 48, no. 3 (2005): 233-40.

Cummings, James J., and Jeremy N. Bailenson. "How immersive is enough? A meta-analysis of the effect of immersive technology on user presence." *Media Psychology* 19, no. 2 (2016): 272-309.

Curtis, H. *Biology*. Fourth edition. New York: Worth, 1983.

"Cybersecurity for Medical Devices and Hospital Networks: FDA Safety Communication." US Food and Drug Administration, June 13, 2013. http://www.fda.gov/MedicalDevices/Safety/AlertsandNotices/ucm356423.htm. Accessed May 3, 2016.

Cybersociety 2.0: Revisiting Computer-Mediated Communication and Community, edited by Steven G. Jones. Thousand Oaks: Sage Publications, 1998.

The Cyborg Experiments: The Extensions of the Body in the Media Age, edited by Joanna Zylinska. London: Continuum, 2002.

Cytowic, Richard E. *Synesthesia: A Union of the Senses*. Springer-Verlag New York, 1989.

Daft, Richard L. *Management*. Mason, OH: South-Western / Cengage Learning, 2011.

Daft, Richard L., Jonathan Murphy, and Hugh Willmott. *Organization Theory and Design*. Andover, Hampshire: Cengage Learning EMEA, 2010.

Daigle, K.R. "Manipulating the Mind: The Ethics of Cognitive Enhancement." Thesis, M.A. in Bioethics. Wake Forest University, 2010. https://wakespace.lib.wfu.edu/handle/10339/30407. Accessed May 8, 2016.

Dandamudi, Sivarama P. *Introduction to assembly language programming: from 8086 to Pentium processors*. Springer Science+Business Media New York, 1998.

Davies, Theresa Claire. *Audification of Ultrasound for Human Echolocation*. Dissertation, Ph.D. in Systems Design Engineering. Waterloo, Ontario: University of Waterloo, 2008.

Davis, Simon, Keith Nesbitt, and Eugene Nalivaiko. "A systematic review of cybersickness." In *Proceedings of the 2014 Conference on Interactive Entertainment*, pp. 1-9. ACM, 2014.

De Graaf, Maartje MA, and Somaya Ben Allouch. "Exploring influencing variables for the acceptance of social robots." *Robotics and Autonomous Systems* 61, no. 12 (2013): 1476-86.

De Melo-Martín, Inmaculada. "Genetically Modified Organisms (GMOs): Human Beings." In *Encyclopedia of Global Bioethics*, edited by Henk ten Have. Springer Science+Business Media Dordrecht. Version of March 13, 2015. doi: 10.1007/978-3-319-05544-2_210-1. Accessed January 21, 2016.

Deep Brain Stimulation for Parkinson's Disease, edited by Gordon H. Baltuch and Matthew B. Stern. Boca Raton: CRC Press, 2007.

Dellon, Brian, and Yoky Matsuoka. "Prosthetics, exoskeletons, and rehabilitation." *IEEE Robotics and Automation magazine* 14, no. 1 (2007): 30.

Denning, Tamara, Alan Borning, Batya Friedman, Brian T. Gill, Tadayoshi Kohno, and William H. Maisel. "Patients, pacemakers, and implantable defibrillators: Human values and security for wireless implantable medical devices." In *Proceedings of the SIGCHI Conference on Human Factors in Computing Systems*, 917-26. ACM, 2010.

Denning, Tamara, Yoky Matsuoka, and Tadayoshi Kohno. "Neurosecurity: Security and Privacy for Neural Devices." *Neurosurgical Focus* 27, no. 1 (2009): E7.

Digital Ecosystems: Society in the Digital Age, edited by Łukasz Jonak, Natalia Juchniewicz, and Renata Włoch. Warsaw: Digital Economy Lab, University of Warsaw, 2016.

DoD Architecture Framework Version 2.02. US Department of Defense, August 2010. http://dodcio.defense.gov/Portals/0/Documents/DODAF/DoDAF_v2-02_web.pdf. Accessed December 4, 2016.

Dormer, Kenneth J. "Implantable electronic otologic devices for hearing rehabilitation." In *Handbook of Neuroprosthetic Methods*, edited by Warren E. Finn and Peter G. LoPresti, pp. 237-60. Boca Raton: CRC Press, 2003.

Drummond, Katie. "Pentagon Preps Soldier Telepathy Push." *Wired*, May 14, 2009. https://www.wired.com/2009/05/pentagon-preps-soldier-telepathy-push/. Accessed December 5, 2016.

Dudai, Yadin. "The Neurobiology of Consolidations, Or, How Stable Is the Engram?" *Annual Review of Psychology* 55 (2004): 51-86.

Dumas II, Joseph D. *Computer Architecture: Fundamentals and Principles of Computer Design*. Boca Raton: CRC Press, 2006.

Edlinger, Günter, Cristiano Rizzo, and Christoph Guger. "Brain Computer Interface." In *Springer Handbook of Medical Technology*, edited by Rüdiger Kramme, Klaus-Peter Hoffmann, and Robert S. Pozos, 1003-17. Springer Berlin Heidelberg, 2011.

Electroreception, edited by Theodore H. Bullock, Carl D. Hopkins, Arthur N. Popper, and Richard R. Fay. New York: Springer Science+Business Media, 2005.

"Employee Tenure Summary." Washington, DC: US Department of Labor, Bureau of Labor Statistics, September 22, 2016. http://www.bls.gov/news.release/tenure.nro.htm. Accessed November 9, 2016.

Epinephrine in the Central Nervous System, edited by Jon M. Stolk, David C. U'Prichard, and Kjell Fuxe. Oxford University Press, 1988.

Ericsson, K. Anders, and Neil Charness. "Expert performance: Its structure and acquisition." *American Psychologist* 49, no. 8 (1994): 725-47.

Erikson, Jane. "Thought-Controlled Robotic Arm 'Makes a Big Negative a Whole Lot Better'." The University of Arizona Health Sciences, July 2013. http://medicine.arizona.edu/alumni/alumni-slide/thought-controlled-robotic-arm-%E2%80%98makes-big-negative-whole-lot-better%E2%80%99 Accessed December 5, 2016.

Evans, Dave. "The Internet of Everything: How More Relevant and Valuable Connections Will Change the World." Cisco Internet Solutions Business Group: Point of View, 2012. https://www.cisco.com/web/about/ac79/docs/innov/IoE.pdf. Accessed December 16, 2015.

Evans-Pughe, Christine. "Smarter Prosthetics: Researchers Are Working to Develop Artificial Limbs with Almost Life-like Levels of Functionality and Control." *Engineering and Technology* (June 2006): 32-36.

Evolutionary Linguistics, edited by April McMahon and Robert McMahon. New York: Cambridge University Press, 2013.

Fabbro, Franco, Salvatore M. Aglioti, Massimo Bergamasco, Andrea Clarici, and Jaak Panksepp. "Evolutionary aspects of self-and world consciousness in vertebrates." *Frontiers in human neuroscience* 9 (2015).

Fairchild, Alea M. *Technological Aspects of Virtual Organizations: Enabling the Intelligent Enterprise.* Springer Science+Business Media Dordrecht, 2004.

Fairclough, S.H. "Physiological Computing: Interfacing with the Human Nervous System." In *Sensing Emotions*, edited by J. Westerink, M. Krans, and M. Ouwerkerk, pp. 1-20. Philips Research Book Series 12. Springer Netherlands, 2010.

Falconer, Bruce. "Defense Research Agency Seeks to Create Supersoldiers." *Government Executive*, November 10, 2003. http://www.govexec.com/defense/2003/11/defense-research-agency-seeks-to-create-supersoldiers/15386/. Accessed May 22, 2016.

Farina, Dario, and Oskar Aszmann. "Bionic limbs: clinical reality and academic promises." *Science Translational Medicine* 6, no. 257 (2014): 257ps12.

Fayerman, Pamela. "Funding, doctors needed if brain stimulation surgery to expand in B.C." *Vancouver Sun*, 03.06.2013. http://www.vancouversun.com/health/Funding+doctors+needed+brain+stimulation+surgery+expand/8054047/story.html. Accessed December 8, 2016.

Federal Enterprise Architecture Framework, version 2. Washington, DC: Office of Management and Budget, January 29, 2013. https://www.whitehouse.gov/sites/default/files/omb/assets/egov_docs/fea_v2.pdf. Accessed November 18, 2016.

Fernandez-Lopez, Helena, José A. Afonso, J. H. Correia, and Ricardo Simoes. "The Need for Standardized Tests to Evaluate the Reliability of Data Transport in Wireless Medical Systems." In *Sensor Systems and Software*, edited by Francisco Martins, Luís Lopes, and Hervé Paulino, pp. 137-45. Lecture Notes of the Institute for Computer Sciences, Social Informatics and Telecommunications Engineering 102. Springer Berlin Heidelberg, 2012.

Ferrando, Francesca. "Posthumanism, Transhumanism, Antihumanism, Metahumanism, and New Materialisms: Differences and Relations." *Existenz: An International Journal in Philosophy, Religion, Politics, and the Arts* 8, no. 2 (Fall 2013): 26-32.

Fiehler, Katja, Immo Schütz, Tina Meller, and Lore Thaler. "Neural correlates of human echolocation of path direction during walking." *Multisensory research* 28, no. 1-2 (2015): 195-226.

"FIPA Device Ontology Specification." Foundation for Intelligent Physical Agents (FIPA), May 10, 2002. http://www.fipa.org/assets/XC00091D.pdf. Accessed February 9, 2015.

Fleischmann, Kenneth R. "Sociotechnical Interaction and Cyborg–Cyborg Interaction: Transforming the Scale and Convergence of HCI." *The Information Society* 25, no. 4 (2009): 227-35.

Fountas, Kostas N., and J. R. Smith. "A Novel Closed-Loop Stimulation System in the Control of Focal, Medically Refractory Epilepsy." In *Operative Neuromodulation*, edited by Damianos E. Sakas and Brian A. Simpson, 357-62. Acta Neurochirurgica Supplements 97/2. Springer Vienna, 2007.

Frewer, Lynn J., Ivo A. van der Lans, Arnout RH Fischer, Machiel J. Reinders, Davide Menozzi, Xiaoyong Zhang, Isabelle van den Berg, and Karin L. Zimmermann. "Public perceptions of agri-food applications of genetic modification – A systematic review and meta-analysis." *Trends in Food Science & Technology* 30, no. 2 (2013): 142-52.

Frewer Lynn, Jesper Lassen, B. Kettlitz, Joachim Scholderer, Volkert Beekman, and Knut G. Berdal. "Societal aspects of genetically modified foods." *Food and Chemical Toxicology* 42, no. 7 (2004): 1181-93.

Friedenberg, Jay. *Artificial Psychology: The Quest for What It Means to Be Human*. Philadelphia: Psychology Press, 2008.

Friedman, Batya, and Helen Nissenbaum. "Bias in Computer Systems." In *Human Values and the Design of Computer Technology*, edited by Batya Friedman, pp. 21-40. CSL Lecture Notes 72. Cambridge: Cambridge University Press, 1997.

Fritz, Robert. *Corporate Tides: The Inescapable Laws of Organizational Structure*. San Francisco: Berret-Koehler, 1996.

Fukuyama, Francis. *Our Posthuman Future: Consequences of the Biotechnology Revolution*. New York: Farrar, Straus, and Giroux, 2002.

The Future of Bioethics: International Dialogues, edited by Akira Akabayashi, Oxford: Oxford University Press, 2014.

Gallego, Juan Álvaro, Eduardo Rocon, Juan Manuel Belda-Lois, and José Luis Pons. "A neuro-prosthesis for tremor management through the control of muscle co-contraction." *Journal of neuroengineering and rehabilitation* 10, no. 1 (2013): 1.

Gammelgård, Magnus, Mårten Simonsson, and Åsa Lindström. "An IT Management Assessment Framework: Evaluating Enterprise Architecture Scenarios." *Information Systems and E-Business Management* 5, no. 4 (2007): 415-35.

Garg, Anant Bhaskar. "Embodied Cognition, Human Computer Interaction, and Application Areas." In *Computer Applications for Web, Human Computer Interaction, Signal and Image Processing, and Pattern Recognition*, pp. 369-74. Springer Berlin Heidelberg, 2012.

Gasson, M.N. "Human ICT Implants: From Restorative Application to Human Enhancement." In *Human ICT Implants: Technical, Legal and Ethical Considerations*, edited by Mark N. Gasson, Eleni Kosta, and Diana M. Bowman, pp. 11-28. Information Technology and Law Series 23. T. M. C. Asser Press, 2012.

Gasson, M.N. "ICT implants." In *The Future of Identity in the Information Society*, edited by S. Fischer-Hübner, P. Duquenoy, A. Zuccato, and L. Martucci, pp. 287-95. Springer US, 2008.

Gasson, M.N., Kosta, E., and Bowman, D.M. "Human ICT Implants: From Invasive to Pervasive." In *Human ICT Implants: Technical, Legal and Ethical Considerations*, edited by Mark N. Gasson, Eleni Kosta, and Diana M. Bowman, pp. 1-8. Information Technology and Law Series 23. T. M. C. Asser Press, 2012.

Gen, Mitsuo, Runwei Cheng, and Lin Lin. *Network Models and Optimization: Multiobjective Genetic Algorithm Approach*. Springer-Verlag London Limited, 2008.

Gene Therapy of the Central Nervous System: From Bench to Bedside, edited by Michael G. Kaplitt and Matthew J. During. Amsterdam: Elsevier, 2006.

Gill, Satinder P. "Socio-Ethics of Interaction with Intelligent Interactive Technologies." *AI & SOCIETY* 22, no. 3 (October 26, 2007): 283-300.

Giustozzi, Emilie Steele, and Betsy Van der Veer Martens. "The new competitive intelligence agents: 'Programming' competitive intelligence ethics into corporate cultures." *Webology* 8, no. 2 (2011): 1.

Gladden, Matthew E. "The Artificial Life-Form as Entrepreneur: Synthetic Organism-Enterprises and the Reconceptualization of Business." In *Proceedings of the Fourteenth International Conference on the Synthesis and Simulation of Living Systems*, edited by Hiroki Sayama, John Rieffel, Sebastian Risi, René Doursat and Hod Lipson, pp. 417-18. The MIT Press, 2014.

Gladden, Matthew E. "Cryptocurrency with a Conscience: Using Artificial Intelligence to Develop Money that Advances Human Ethical Values." *Annales: Ethics in Economic Life* 18, no. 4 (2015): 85-98.

Gladden, Matthew E. "Cybershells, Shapeshifting, and Neuroprosthetics: Video Games as Tools for Posthuman 'Body Schema (Re)Engineering'." Keynote presentation at the Ogólnopolska Konferencja Naukowa Dyskursy Gier Wideo, Facta Ficta / AGH, Kraków, June 6, 2015.

Gladden, Matthew E. "The Diffuse Intelligent Other: An Ontology of Nonlocalizable Robots as Moral and Legal Actors." In *Social Robots: Boundaries, Potential, Challenges*, edited by Marco Nørskov, pp. 177-98. Farnham: Ashgate, 2016.

Gladden, Matthew E. "Enterprise Architecture for Neurocybernetically Augmented Organizational Systems: The Impact of Posthuman Neuroprosthetics on the Creation of Strategic, Structural, Functional, Technological, and Sociocultural Alignment." Thesis project, MBA in Innovation and Data Analysis. Warsaw: Institute of Computer Science, Polish Academy of Sciences, 2016.

Gladden, Matthew E. "From Stand Alone Complexes to Memetic Warfare: Cultural Cybernetics and the Engineering of Posthuman Popular Culture." Presentation at the 50 Shades of Popular Culture International Conference. Facta Ficta / Uniwersytet Jagielloński, Kraków, February 19, 2016.

Gladden, Matthew E. *The Handbook of Information Security for Advanced Neuroprosthetics*. Indianapolis: Synthypnion Academic, 2015.

Gladden, Matthew E. "Implantable Computers and Information Security: A Managerial Perspective." In *Posthuman Management: Creating Effective Organizations in an Age of Social Robotics, Ubiquitous AI, Human Augmentation, and Virtual Worlds*. Second edition, pp. 285-300. Indianapolis: Defragmenter Media, 2016.

Gladden, Matthew E. "Information Security Concerns as a Catalyst for the Development of Implantable Cognitive Neuroprostheses." In *9th Annual EuroMed Academy of Business (EMAB) Conference: Innovation, Entrepreneurship and Digital Ecosystems (EUROMED 2016) Book of Proceedings*, edited by Demetris Vrontis, Yaakov Weber, and Evangelos Tsoukatos, pp. 891-904. Engomi: EuroMed Press, 2016.

Gladden, Matthew E. "Leveraging the Cross-Cultural Capacities of Artificial Agents as Leaders of Human Virtual Teams." *Proceedings of the 10th European Conference on Management Leadership and Governance*, edited by Visnja Grozdanić, pp. 428-35. Reading: Academic Conferences and Publishing International Limited, 2014.

Gladden, Matthew E. "Managerial Robotics: A Model of Sociality and Autonomy for Robots Managing Human Beings and Machines." *International Journal of Contemporary Management* 13, no. 3 (2014): 67-76.

Gladden, Matthew E. "Managing the Ethical Dimensions of Brain-Computer Interfaces in eHealth: An SDLC-based Approach." In *9th Annual EuroMed Academy of Business (EMAB) Conference: Innovation, Entrepreneurship and Digital Ecosystems (EUROMED 2016) Book of*

Proceedings, edited by Demetris Vrontis, Yaakov Weber, and Evangelos Tsoukatos, pp. 876-90. Engomi: EuroMed Press, 2016.

Gladden, Matthew E. "Neural Implants as Gateways to Digital-Physical Ecosystems and Posthuman Socioeconomic Interaction." In *Digital Ecosystems: Society in the Digital Age*, edited by Łukasz Jonak, Natalia Juchniewicz, and Renata Włoch, pp. 85-98. Warsaw: Digital Economy Lab, University of Warsaw, 2016.

Gladden, Matthew E. "Neuromarketing Applications of Neuroprosthetic Devices: An Assessment of Neural Implants' Capacities for Gathering Data and Influencing Behavior." In *9ᵗʰ Annual EuroMed Academy of Business (EMAB) Conference: Innovation, Entrepreneurship and Digital Ecosystems (EUROMED 2016) Book of Proceedings*, edited by Demetris Vrontis, Yaakov Weber, and Evangelos Tsoukatos, pp. 905-18. Engomi: EuroMed Press, 2016.

Gladden, Matthew E. "Organization Development and the Robotic-Cybernetic-Human Workforce: Humanistic Values for a Posthuman Future?" In *Posthuman Management: Creating Effective Organizations in an Age of Social Robotics, Ubiquitous AI, Human Augmentation, and Virtual Worlds*. Second edition, pp. 239-55. Indianapolis: Defragmenter Media, 2016.

Gladden, Matthew E. *Posthuman Management: Creating Effective Organizations in an Age of Social Robotics, Ubiquitous AI, Human Augmentation, and Virtual Worlds*. Second edition. Indianapolis: Defragmenter Media, 2016.

Gladden, Matthew E. *Sapient Circuits and Digitalized Flesh: The Organization as Locus of Technological Posthumanization*. Indianapolis: Defragmenter Media, 2016.

Gladden, Matthew E. "The Social Robot as 'Charismatic Leader': A Phenomenology of Human Submission to Nonhuman Power." In *Sociable Robots and the Future of Social Relations: Proceedings of Robo-Philosophy 2014*, edited by Johanna Seibt, Raul Hakli, and Marco Nørskov, pp. 329-39. Frontiers in Artificial Intelligence and Applications 273. IOS Press, 2014.

Gladden, Matthew E. "Utopias and Dystopias as Cybernetic Information Systems: Envisioning the Posthuman Neuropolity." *Creatio Fantastica* nr 3 (50) (2015).

Gockley, Rachel, Allison Bruce, Jodi Forlizzi, Marek Michalowski, Anne Mundell, Stephanie Rosenthal, Brennan Sellner, Reid Simmons, Kevin Snipes, Alan C. Schultz, and Jue Wang. "Designing Robots for Long-Term Social Interaction." In *2005 IEEE/RSJ International Conference on Intelligent Robots and Systems (IROS 2005)*, pp. 2199-2204. 2005.

Goldstein, E. Bruce. *Sensation and Perception*. Ninth edition. Belmont, CA: Wadsworth / CENGAGE Learning, 2014.

Goleman, Daniel. "What Makes a Leader?" *Harvard Business Review* 82 (1) (2004): 82-91.

Goodman, H. Maurice. *Basic Medical Endocrinology*. Fourth edition. Burlington, MA: Academic Press, 2009.

Graham, Stephen. "Imagining Urban Warfare: Urbanization and U.S. Military Technoscience." In *War, Citizenship, Territory*, edited by Deborah Cowen and Emily Gilbert. New York: Routledge, 2008.

Greve, Andrea, and Richard Henson. "What We Have Learned about Memory from Neuroimaging." In *The Wiley Handbook on the Cognitive Neuroscience of Memory*, edited by Donna Rose Addis, Morgan Barense, and Audrey Duarte, pp. 1-20. Malden, MA: John Wiley & Sons Ltd., 2015.

Grottke, M., H. Sun, R.M. Fricks, and K.S. Trivedi. "Ten fallacies of availability and reliability analysis." In *Service Availability*, pp. 187-206. Lecture Notes in Computer Science 5017. Springer Berlin Heidelberg, 2008.

Gunkel, David J. *The Machine Question: Critical Perspectives on AI, Robots, and Ethics*. Cambridge, MA: The MIT Press, 2012.

Gutnick, Tamar, Ruth A. Byrne, Binyamin Hochner, and Michael Kuba. "Octopus vulgaris uses visual information to determine the location of its arm." *Current Biology* 21, no. 6 (2011): 460-62.

Haki, Mohammad Kazem, Christine Legner, and Frederik Ahlemann. "Beyond EA Frameworks: Towards an Understanding of the Adoption of Enterprise Architecture Management." *ECIS 2012 Proceedings*, 2012.

Hameroff, Stuart, and Roger Penrose. "Consciousness in the universe: A review of the 'Orch OR' theory." *Physics of Life Reviews* 11, no. 1 (2014): 39-78.

Han, J.-H., S.A. Kushner, A.P. Yiu, H.-W. Hsiang, T. Buch, A. Waisman, B. Bontempi, R.L. Neve, P.W. Frankland, and S.A. Josselyn. "Selective Erasure of a Fear Memory." *Science* 323, no. 5920 (2009): 1492-96.

Hand, Eric. "Maverick scientist thinks he has discovered magnetic sixth sense in humans." *Science*, June 23, 2016. doi:10.1126/science.aaf5803. Accessed December 8, 2016.

Handbook of Cloud Computing, edited by Borko Furht and Armando Escalante. New York: Springer, 2010.

Handbook of Psychology, Volume 6: Developmental Psychology, edited by Richard M. Lerner, M. Ann Easterbrooks, and Jayanthi Mistry. Hoboken: John Wiley & Sons, Inc., 2003.

Hanson, R. "If uploads come first: The crack of a future dawn." *Extropy* 6, no. 2 (1994): 10-15.

Haraway, Donna. *Simians, Cyborgs, and Women: The Reinvention of Nature*. New York: Routledge, 1991.

Hargrove, Levi J., Ann M. Simon, Aaron J. Young, Robert D. Lipschutz, Suzanne B. Finucane, Douglas G. Smith, and Todd A. Kuiken. "Robotic leg control with EMG decoding in an amputee with nerve transfers." *New England Journal of Medicine* 369, no. 13 (2013): 1237-42.

Harman, Gilbert. *Thought*. Princeton: Princeton University Press, 1973.

Hatfield, B., A. Haufler, and J. Contreras-Vidal. "Brain Processes and Neurofeedback for Performance Enhancement of Precision Motor Behavior." In *Foundations of Augmented Cognition. Neuroergonomics and Operational Neuroscience*, edited by Dylan D. Schmorrow, Ivy V. Estabrooke, and Marc Grootjen, pp. 810-17. Lecture Notes in Computer Science 5638. Springer Berlin Heidelberg, 2009.

Haykin, Simon. *Neural Networks and Learning Machines*. Third edition. New York: Pearson Prentice Hall, 2009.

Hayles, N. Katherine. *How We Became Posthuman: Virtual Bodies in Cybernetics, Literature, and Informatics*. Chicago: University of Chicago Press, 1999.

Heim, Michael. *The Metaphysics of Virtual Reality*. New York: Oxford University Press, 1993.

Henderson, John C., and H. Venkatraman. "Strategic alignment: Leveraging information technology for transforming organizations." *IBM systems journal* 32, no. 1 (1993): 472-84.

Herbrechter, Stefan. *Posthumanism: A Critical Analysis*. London: Bloomsbury, 2013. [Kindle edition.]

Herzberg, Frederick. "One more time: How do you motivate employees." In *The Leader-Manager*, edited by John N. Williamson, pp. 433-48. New York: John Wiley & Sons, Inc., 1986.

Heylighen, Francis. "The Global Brain as a New Utopia." In *Renaissance der Utopie. Zukunftsfiguren des 21. Jahrhunderts*, edited by R. Maresch and F. Rötzer. Frankfurt: Suhrkamp, 2002.

Hochmair, Ingeborg. "Cochlear Implants: Facts." MED-EL, September 2013. http://www.medel.com/cochlear-implants-facts. Accessed December 8, 2016.

Hoffmann, Klaus-Peter, and Silvestro Micera. "Introduction to Neuroprosthetics." In *Springer Handbook of Medical Technology*, edited by Rüdiger Kramme, Klaus-Peter Hoffmann, and Robert S. Pozos, pp. 785-800. Springer Berlin Heidelberg, 2011.

Højsgaard, Hjalte. "Market-Driven Enterprise Architecture." *Journal of Enterprise Architecture* 7, no. 1 (2011), 28.

Hoogervorst, Jan. "Enterprise Architecture: Enabling Integration, Agility and Change." *International Journal of Cooperative Information Systems* 13, no. 03 (2004): 213-33.

Horling, Bryan, and Victor Lesser. "A Survey of Multi-Agent Organizational Paradigms." *The Knowledge Engineering Review* 19, no. 04 (2004): 281-316.

Howard, Robert. "The CEO as Organizational Architect: An Interview with Xerox's Paul Allaire." *Harvard Business Review* 70 (5) (1992): 106-21.

Hutchinson, Douglas T. "The quest for the bionic arm." *Journal of the American Academy of Orthopaedic Surgeons* 22, no. 6 (2014): 346-51.

Iacob, Maria-Eugenia, Lucas O. Meertens, Henk Jonkers, Dick A.C. Quartel, Lambert J.M. Nieuwenhuis, and M.J. Van Sinderen. "From Enterprise Architecture to Business Models and Back." *Software & Systems Modeling* 13, no. 3 (2012): 1059-83.

Implantable Biomedical Microsystems: Design Principles and Applications, edited by Swarup Bhunia, Steve Majerus, and Mohamad Sawan. Oxford: William Andrew, 2015.

Implantable Neuroprostheses for Restoring Function, edited by Kevin Kilgore. Cambridge: Woodhead Publishing, 2015.

Implantable Sensor Systems for Medical Applications, edited by Andreas Inmann and Diana Hodgins. Woodhead Publishing, 2013.

Information Storage and Management: Storing, Managing, and Protecting Digital Information in Classic, Virtualized, and Cloud Environments. Second edition. Edited by Somasundaram Gnanasundaram and Alok Shrivastava. Indianapolis: John Wiley & Sons, Inc., 2012.

ISO 15704:2000, Industrial automation systems – Requirements for enterprise-reference architectures and methodologies. ISO/TC 184/SC5. Geneva: ISO, 2000.

ISO 19439:2006, Enterprise integration – Framework for enterprise modelling. ISO/TC 184/SC5. Geneva: ISO, 2006.

ISO 27799:2016, Health informatics – Information security management in health using ISO/IEC 27002. ISO/TC 215. Geneva: ISO, 2016.

ISO/IEC 15288:2002, Systems engineering – System life cycle processes. ISO/IEC JTC 1/SC 7. Geneva: ISO, 2002.

ISO/IEC 7498-1:1994, Information technology – Open Systems Interconnection – Basic Reference Model: The Basic Model. ISO/IEC JTC 1. Geneva: ISO, 1994.

ISO/IEC/IEEE 42010:2010, Systems and software engineering – Architecture description. ISO/IEC JTC 1/SC 7. Geneva: ISO, 2011.

Johar, Swati. *Emotion, Affect and Personality in Speech: The Bias of Language and Paralanguage*. SpringerNature, 2016.

Josselyn, Sheena A. "Continuing the Search for the Engram: Examining the Mechanism of Fear Memories." *Journal of Psychiatry & Neuroscience : JPN* 35, no. 4 (2010): 221-28.

Jürgens, Uta, and Danko Nikolić. "Ideaesthesia: conceptual processes assign similar colours to similar shapes." *Translational Neuroscience* 3, no. 1 (2012): 22-27.

Kalat, James W. *Biological Psychology*. Ninth edition. Belmont, CA: Thomson Wadsworth, 2007.

Kanda, Takayuki, and Hiroshi Ishiguro. *Human-Robot Interaction in Social Robotics*. Boca Raton: CRC Press, 2013.

Kandjani, Hadi, Peter Bernus, and Lian Wen. "Enterprise Architecture Cybernetics for Complex Global Software Development: Reducing the Complexity of Global Software Development Using Extended Axiomatic Design Theory." In *2012 IEEE Seventh International Conference on Global Software Engineering*, pp. 169-73. IEEE, 2012.

Kaplan, Robert S., and David P. Norton. *The Strategy-Focused Organization: How Balanced Scorecard Companies Thrive in the New Business Environment*. Boston: Harvard Business School Press, 2001.

Katz, Gregory. "The hypothesis of a genetic protolanguage: An epistemological investigation." *Biosemiotics* 1, no. 1 (2008): 57-73.

Kazienko, Przemysław, Radosław Michalski, and Sebastian Palus. "Social Network Analysis as a Tool for Improving Enterprise Architecture." In *Agent and Multi-Agent Systems: Technologies and Applications*, edited by James O'Shea, Ngoc Thanh Nguyen, Keeley Crockett, Robert J. Howlett, and Lakhmi C. Jain, pp. 651-60. Lecture Notes in Computer Science 6682. Springer Berlin Heidelberg, 2011.

Keenan, James F. "Enhancing Prosthetics for Soldiers Returning from Combat with Disabilities." *ET Studies* 4, no. 1 (2013): 69-88.

Kelly, Kevin. "A Taxonomy of Minds." *The Technium*, February 15, 2007. http://kk.org/thetechnium/a-taxonomy-of-m/. Accessed January 25, 2016.

Kelly, Kevin. "The Landscape of Possible Intelligences." *The Technium*, September 10, 2008. http://kk.org/thetechnium/the-landscape-o/. Accessed January 25, 2016.

Kelly, Kevin. *Out of Control: The New Biology of Machines, Social Systems and the Economic World*. Basic Books, 1994.

Kerr, Paul K., John Rollins, and Catherine A. Theohary. "The Stuxnet Computer Worm: Harbinger of an Emerging Warfare Capability." Congressional Research Service, 2010.

Kim, Kwang Jin, Xiaobo Tan, Hyouk Ryeol Choi, and David Pugal. *Biomimetic Robotic Artificial Muscles*. Singapore: World Scientific Publishing Co. Pte. Ltd., 2013.

Kinicki, Angelo, and Brian Williams. *Management: A Practical Introduction*. Fifth edition. New York: McGraw Hill, 2010.

KleinOsowski, A., Ethan H. Cannon, Phil Oldiges, and Larry Wissel. "Circuit design and modeling for soft errors." *IBM Journal of Research and Development* 52, no. 3 (2008): 255-63.

Koene, Randal A. "Embracing Competitive Balance: The Case for Substrate-Independent Minds and Whole Brain Emulation." In *Singularity Hypotheses*, edited by Amnon H. Eden, James H. Moor, Johnny H. Søraker, and Eric Steinhart, pp. 241-67. The Frontiers Collection. Springer Berlin Heidelberg, 2012.

Kohno, T., T. Denning, and Y. Matsuoka. "Security and Privacy for Neural Devices." *Neurosurgical Focus* 27 (2009): 1-4.

Koltko-Rivera, Mark E. "The potential societal impact of virtual reality." *Advances in virtual environments technology: Musings on design, evaluation, and applications* 9 (2005).

Koops, B.-J., and R. Leenes. "Cheating with Implants: Implications of the Hidden Information Advantage of Bionic Ears and Eyes." In *Human ICT Implants: Technical, Legal and Ethical Considerations*, edited by Mark N. Gasson, Eleni Kosta, and Diana M. Bowman, pp. 113-34. Information Technology and Law Series 23. T. M. C. Asser Press, 2012.

Kosta, E., and D.M. Bowman, "Implanting Implications: Data Protection Challenges Arising from the Use of Human ICT Implants." In *Human ICT Implants: Technical, Legal and Ethical Considerations*, edited by Mark N. Gasson, Eleni Kosta, and Diana M. Bowman, pp. 97-112. Information Technology and Law Series 23. T. M. C. Asser Press, 2012.

Kostov, Aleksander, and Mark Polak. "Parallel man-machine training in development of EEG-based cursor control." *IEEE Transactions on Rehabilitation Engineering* 8, no. 2 (2000): 203-05.

Kourany, J.A. "Human Enhancement: Making the Debate More Productive." *Erkenntnis* 79, no. 5 (2013): 981-98.

Kowalewska, Agata. "Symbionts and Parasites – Digital Ecosystems." In *Digital Ecosystems: Society in the Digital Age*, edited by Łukasz Jonak, Natalia Juchniewicz, and Renata Włoch, pp. 73-84. Warsaw: Digital Economy Lab, University of Warsaw, 2016.

Kraemer, Felicitas. "Me, Myself and My Brain Implant: Deep Brain Stimulation Raises Questions of Personal Authenticity and Alienation." *Neuroethics* 6, no. 3 (May 12, 2011): 483-97. doi:10.1007/s12152-011-9115-7.

Krishnan, Armin. "Enhanced Warfighters as Private Military Contractors." In *Super Soldiers: The Ethical, Legal and Social Implications*, edited by Jai Galliott and Mianna Lotz. London: Routledge, 2015.

Krishnan, Armin. "From Psyops to Neurowar: What Are the Dangers?" ISAC-ISSS 2014 Annual Conference on Security Studies. University of Texas, Austin, Texas, November 16, 2014. http://web.isanet.org/Web/Conferences/ISSS%20Austin%202014/Archive/b137347c-6281-466d-b9e7-ef7e0e5d363c.pdf. Accessed May 8, 2016.

Kruchten, Philippe B. "The 4+1 view model of architecture." *IEEE software* 12, no. 6 (1995): 42-50.

Kshirsagar, Sumedha, Chris Joslin, Won-Sook Lee, and Nadia Magnenat-Thalmann. "Personalized Face and Speech Communication over the Internet." In *Proceedings of IEEE Virtual Reality 2001*, edited by Haruo Takemura and Kiyoshi Kiyokawa, pp. 37-44. Los Alamitos, CA: IEEE, 2001.

Kusek, Kristen. "The $3 Million Suit: Wyss Institute Wins DARPA Grant to Further Develop its Soft Exosuit." *Harvard Gazette*, September 11, 2014. http://news.harvard.edu/gazette/story/2014/09/the-3-million-suit/. Accessed December 5, 2016.

Kyllonen, Patrick C., and Raymond E. Christal. "Reasoning ability is (little more than) working-memory capacity?!" *Intelligence* 14, no. 4 (1990): 389-433.

LaFleur, Karl, Kaitlin Cassady, Alexander Doud, Kaleb Shades, Eitan Rogin, and Bin He. "Quadcopter control in three-dimensional space using a noninvasive motor imagery-based brain-computer interface." *Journal of neural engineering* 10, no. 4 (2013): 046003.

Lamm, Ehud, and Ron Unger. *Biological Computation*. Boca Raton: CRC Press, 2011.

Land, Martin Op 't, Erik Proper, Maarten Waage, Jeroen Cloo, and Claudia Steghuis. "Positioning Enterprise Architecture." In *Enterprise Architecture*, pp. 25-47. The Enterprise Engineering Series. Springer Berlin Heidelberg, 2009.

Lebedev, M. "Brain-Machine Interfaces: An Overview." *Translational Neuroscience* 5, no. 1 (March 28, 2014): 99-110.

Leder, Felix, Tillmann Werner, and Peter Martini. "Proactive Botnet Countermeasures: An Offensive Approach." In *The Virtual Battlefield: Perspectives on Cyber Warfare*, volume 3, edited by Christian Czosseck and Kenneth Geers, pp. 211-25. IOS Press, 2009.

Leed, Maren. *Offensive Cyber Capabilities at the Operational Level: The Way Ahead*. Washington, DC: Center for Strategic and International Studies, 2013.

Li, S., F. Hu, and G. Li, "Advances and Challenges in Body Area Network." In *Applied Informatics and Communication*, edited by J. Zhan, pp. 58-65. Communications in Computer and Information Science 22. Springer Berlin Heidelberg, 2011.

Lilley, Stephen. *Transhumanism and Society: The Social Debate over Human Enhancement*. Springer Science & Business Media, 2013.

Lin, James C. "Hearing microwaves: The microwave auditory phenomenon." *IEEE Antennas and Propagation Magazine* 43, no. 6 (2001): 166-68.

Lind, Jürgen. "Issues in agent-oriented software engineering." In *Agent-Oriented Software Engineering*, pp. 45-58. Springer Berlin Heidelberg, 2001.

Lindström, Åsa, Pontus Johnson, Erik Johansson, Mathias Ekstedt, and Mårten Simonsson. "A Survey on CIO Concerns – Do Enterprise Architecture Frameworks Support Them?" *Information Systems Frontiers* 8, no. 2 (2006).

Ling, Geoffrey SF, Peter Rhee, and James M. Ecklund. "Surgical innovations arising from the Iraq and Afghanistan wars." *Annual review of medicine* 61 (2010): 457-68.

Linsenmeier, Robert A. "Retinal Bioengineering." In *Neural Engineering*, edited by Bin He, pp. 421-84. Bioelectric Engineering. Springer US, 2005.

Liu, Hanjun, and Manwa L. Ng. "Electrolarynx in voice rehabilitation," *Auris Nasus Larynx* 34, no. 3 (2007): 327-32.

Liu, Kecheng, Lily Sun, Dian Jambari, Vaughan Michell, and Sam Chong. "A Design of Business-Technology Alignment Consulting Framework." In *Advanced Information Systems Engineering*, edited by Haralambos Mouratidis and Colette Rolland, pp. 422-35. Lecture Notes in Computer Science 6741. Springer Berlin Heidelberg, 2011.

Lloyd, David. "Biological Time Is Fractal: Early Events Reverberate over a Life Time." *Journal of Biosciences* 33, no. 1 (March 1, 2008): 9-19.

LoBello, Lucia, and Emanuele Toscano. "An adaptive approach to topology management in large and dense real-time wireless sensor networks." *IEEE Transactions on Industrial Informatics* 5, no. 3 (2009): 314-24.

Logan, Lynne Romeiser. "Rehabilitation techniques to maximize spasticity management." *Topics in stroke rehabilitation* 18, no. 3 (2011): 203-11.

Lohn, Andrew J., Patrick R. Mickel, James B. Aimone, Erik P. Debenedictis, and Matthew J. Marinella. "Memristors as Synapses in Artificial Neural Networks: Biomimicry Beyond Weight Change." In *Cybersecurity Systems for Human Cognition Augmentation*, edited by Robinson E. Pino, Alexander Kott, and Michael Shevenell, pp. 135-50. Springer International Publishing, 2014.

Longuet-Higgins, H.C. "Holographic Model of Temporal Recall." *Nature* 217, no. 5123 (1968): 104.

Lorence, Daniel, Anusha Sivaramakrishnan, and Michael Richards. "Transaction-Neutral Implanted Data Collection Interface as EMR Driver: A Model for Emerging Distributed Medical Technologies." *Journal of Medical Systems* 34, no. 4 (March 20, 2009): 609-17.

Luber, B., C. Fisher, P.S. Appelbaum, M. Ploesser, and S.H. Lisanby. "Non-invasive brain stimulation in the detection of deception: Scientific challenges and ethical consequences." *Behavioral Sciences and the Law* 27, no. 2 (2009): 191-208.

Lune, Howard. *Understanding Organizations*, Cambridge: Polity Press, 2010.

Lyon, David. "Beyond Cyberspace: Digital Dreams and Social Bodies." In *Education and Society*, third edition, edited by Joseph Zajda, pp. 221-38. Albert Park: James Nicholas Publishers, 2001.

Ma, Jianhua, Kim-Kwang Raymond Choo, Hui-huang Hsu, Qun Jin, William Liu, Kevin Wang, Yufeng Wang, and Xiaokang Zhou. "Perspectives on Cyber Science and Technology for Cyberization and Cyber-Enabled Worlds." In *2016 IEEE 14th International Conference on Dependable, Autonomic and Secure Computing, 14th International Conference on Pervasive Intelligence and Computing, 2nd International Conference on Big Data Intelligence and Computing and Cyber Science and Technology Congress (DASC/PiCom/DataCom/CyberSciTech)*, pp. 1-9. IEEE, 2016.

Ma, Wei Ji, Masud Husain, and Paul M. Bays. "Changing concepts of working memory." *Nature neuroscience* 17, no. 3 (2014): 347-56.

MacLennan, Elzavita, and Jean-Paul Van Belle. "Factors Affecting the Organizational Adoption of Service-Oriented Architecture (SOA)." *Information Systems and E-Business Management* 12, no. 1 (January 5, 2013): 71-100.

MacVittie, Kevin, Jan Halámek, Lenka Halámková, Mark Southcott, William D. Jemison, Robert Lobel, and Evgeny Katz. "From 'cyborg' lobsters to a pacemaker powered by implantable biofuel cells." *Energy & Environmental Science* 6, no. 1 (2013): 81-86.

Magnetic Stimulation in Clinical Neurophysiology. Second edition. Edited by Mark Hallett and Sudhansu Chokroverty. Philadelphia: Elsevier Butterworth Heinemann, 2005.

Magoulas, Thanos, Aida Hadzic, Ted Saarikko, and Kalevi Pessi. "Alignment in Enterprise Architecture: A Comparative Analysis of Four Architectural Approaches." *Electronic Journal Information Systems Evaluation* 15, no. 1 (2012).

Maguire, Gerald Q., and Ellen M. McGee. "Implantable brain chips? Time for debate." *Hastings Center Report* 29, no. 1 (1999): 7-13.

Maitra, Amit K. "Offensive cyber-weapons: technical, legal, and strategic aspects." *Environment Systems and Decisions* 35, no. 1 (2015): 169-82.

Mak, Stephen. "Ethical Values for E-Society: Information, Security and Privacy." In *Ethics and Policy of Biometrics*, edited by Ajay Kumar and David Zhang, 96-101. Lecture Notes in Computer Science 6005. Springer Berlin Heidelberg, 2010.

Mangun, George R. *The Neuroscience of Attention: Attentional Control and Selection*. Oxford University Press, 2012.

Martin, Richard, and Edward Robertson. "A Comparison of Frameworks for Enterprise Architecture Modeling." In *Conceptual Modeling - ER 2003*, edited by Il-Yeol Song, Stephen W. Liddle, Tok-Wang Ling, and Peter Scheuermann. Lecture Notes in Computer Science 2813. Springer Berlin Heidelberg, 2003.

Mascarenhas, S., R. Prada, A. Paiva, and G.J. Hofstede. "Social Importance Dynamics: A Model for Culturally-Adaptive Agents." In *Intelligent Virtual Agents*, pp. 325-38. Lecture Notes in Computer Science no. 8108. Springer Berlin Heidelberg, 2013.

McCullagh, P., G. Lightbody, J. Zygierewicz, and W.G. Kernohan, "Ethical Challenges Associated with the Development and Deployment of Brain Computer Interface Technology." *Neuroethics* 7, no. 2 (July 28, 2013): 109-22.

McCulloh, Ian A., Helen L. Armstrong, and Anthony N. Johnson. *Social Network Analysis with Applications*. Hoboken, NJ: John Wiley & Sons, Inc., 2013.

McGaugh, James L., and Benno Roozendaal. "Role of adrenal stress hormones in forming lasting memories in the brain." *Current opinion in neurobiology* 12, no. 2 (2002): 205-10.

McGee, E.M. "Bioelectronics and Implanted Devices." In *Medical Enhancement and Posthumanity*, edited by Bert Gordijn and Ruth Chadwick, pp. 207-24. The International Library of Ethics, Law and Technology 2. Springer Netherlands, 2008.

McGrath, Michael J., and Cliodhna Ní Scanaill. "Regulations and Standards: Considerations for Sensor Technologies." In *Sensor Technologies*, pp. 115-35. Apress, 2013.

McIntosh, Daniel. "The Transhuman Security Dilemma." *Journal of Evolution and Technology* 21, no. 2 (2010): 32-48.

Mehlman, Maxwell J. *Transhumanist Dreams and Dystopian Nightmares: The Promise and Peril of Genetic Engineering*. Baltimore: The Johns Hopkins University Press, 2012.

Mercanzini, André, and Philippe Renaud. *Microfabricated Cortical Neuroprostheses*. Boca Raton: CRC Press, 2010.

Merkel, R., G. Boer, J. Fegert, T. Galert, D. Hartmann, B. Nuttin, and S. Rosahl. "Central Neural Prostheses." In *Intervening in the Brain: Changing Psyche and Society*, pp. 117-60. Ethics of Science and Technology Assessment 29. Springer Berlin Heidelberg, 2007.

Mezzanotte, Sr., Dominic M., and Josh Dehlinger. "Enterprise Architecture: A Framework Based on Human Behavior Using the Theory of Structuration." In *Software Engineering Research, Management and Applications 2012*, edited by Roger Lee, pp. 65-79. Studies in Computational Intelligence 430. Springer Berlin Heidelberg, 2012.

Miah, Andy. "A Critical History of Posthumanism." In *Medical Enhancement and Posthumanity*, edited by Bert Gordijn and Ruth Chadwick, pp. 71-94. The International Library of Ethics, Law and Technology 2. Springer Netherlands, 2008.

Miller, Kai J., and Jeffrey G. Ojemann. "A Simple, Spectral-Change Based, Electrocorticographic Brain–Computer Interface." In *Brain-Computer Interfaces*, edited by Bernhard Graimann, Gert Pfurtscheller, and Brendan Allison, pp. 241-58. The Frontiers Collection. Springer Berlin Heidelberg, 2009.

Miller, Jr., Gerald Alva. "Conclusion: Beyond the Human: Ontogenesis, Technology, and the Posthuman in Kubrick and Clarke's 2001." In *Exploring the Limits of the Human through Science Fiction*, pp. 163-90. American Literature Readings in the 21st Century. Palgrave Macmillan US, 2012.

Min Neo, Hui, and Romain Fonsegrives. "For these 'cyborgs,' keys are so yesterday." AFP / Yahoo! Tech, September 4, 2015. https://www.yahoo.com/tech/cyborgs-keys-yesterday-114442441.html. Accessed December 8, 2016.

"Mind over Mouth? Study Could Lead to Communicating via Thoughts." UCI News, August 13, 2008. https://news.uci.edu/briefs/mind-over-mouth-study-could-lead-to-communicating-via-thoughts/. Accessed December 5, 2016.

Mitcheson, Paul D. "Energy harvesting for human wearable and implantable bio-sensors." In *Engineering in Medicine and Biology Society (EMBC), 2010 Annual International Conference of the IEEE*, pp. 3432-36. IEEE, 2010.

Mizraji, Eduardo, Andrés Pomi, and Juan C. Valle-Lisboa. "Dynamic Searching in the Brain." *Cognitive Neurodynamics* 3, no. 4 (June 3, 2009): 401-14.

Møller, Aage R. *Sensory Systems: Anatomy and Physiology*. Third edition. Aage R Møller, 2014.

Moravec, Hans. *Mind Children: The Future of Robot and Human Intelligence*. Cambridge: Harvard University Press, 1990.

Moreno, Jonathan. "DARPA On Your Mind." *Cerebrum* vol. 6 issue 4 (2004): 92-100.

Moxon, David. *Memory*. Oxford: Heinemann Educational Publishers, 2000.

Mubin, Omar, Christoph Bartneck, Loe Feijs, Hanneke Hooft van Huysduynen, Jun Hu, and Jerry Muelver. "Improving speech recognition with the robot interaction language." *Disruptive Science and Technology* 1, no. 2 (2012): 79-88.

Mubin, Omar, Joshua Henderson, and Christoph Bartneck. "Talk ROILA to your Robot." In *Proceedings of the 15th Conference on International conference on multimodal interaction*, pp. 317-18. ACM, 2013.

Mueller, Scott. *Upgrading and Repairing PCs, 20th Edition*. Indianapolis: Que, 2012.

Mulhauser, Gregory R. "On the end of a quantum mechanical romance." *Psyche* 2, no. 5 (1995).

Murphy, Robin. *Introduction to AI Robotics*. Cambridge, MA: The MIT Press, 2000.

Muscolino, Joseph E. *Kinesiology: The Skeletal System and Muscle Function*. Third edition. St. Louis: Elsevier, 2017.

Nadler, David, and Michael Tushman. *Competing by Design: The Power of Organizational Architecture*. Oxford University Press, 1997. [Kindle edition.]

Nakakawa, Agnes, Patrick van Bommel, and H. A. Erik Proper. "Quality Enhancement in Creating Enterprise Architecture: Relevance of Academic Models in Practice." In *Advances in Enterprise Engineering II*, edited by Erik Proper, Frank Harmsen, and Jan L. G. Dietz, pp. 109-33. Lecture Notes in Business Information Processing 28. Springer Berlin Heidelberg, 2009.

The Nature of Time: Geometry, Physics and Perception, edited by Rosolino Buccheri, Metod Saniga, and William Mark Stuckey. Springer Science+Business Media Dordrecht, 2003.

Nayar, Pramod K. *An Introduction to New Media and Cybercultures*. Chichester: John Wiley & Sons Ltd., 2010.

Negoescu, R. "Conscience and Consciousness in Biomedical Engineering Science and Practice." In *International Conference on Advancements of Medicine and Health Care through Technology*, edited by Simona Vlad, Radu V. Ciupa, and Anca I. Nicu, pp. 209-14. IFMBE Proceedings 26. Springer Berlin Heidelberg, 2009.

Neubauer, André, Jürgen Freudenberger, and Volker Kühn. *Coding Theory: Algorithms, Architectures and Applications*. Chichester: John Wiley & Sons Ltd., 2007.

Neuper, Christa, and Gert Pfurtscheller. "Neurofeedback Training for BCI Control." In *Brain-Computer Interfaces*, edited by Bernhard Graimann, Gert Pfurtscheller, and Brendan Allison, pp. 65-78. The Frontiers Collection. Springer Berlin Heidelberg, 2009.

Neuroimaging and Memory, edited by Jonathan K. Foster. Hove, East Sussex: Psychology Press Ltd., 1999.

The Neuroscience of Sleep, edited by Robert Stickgold and Matthew P. Walker. London: Elsevier, 2009.

Neuweiler, Gerhard. "Evolutionary aspects of bat echolocation." *Journal of Comparative Physiology A* 189, no. 4 (2003): 245-256.

Niku, Saeed B. *Introduction to Robotics: Analysis, Control, Applications*. Second edition. Hoboken: John Wiley & Sons, Inc., 2011.

NIST Special Publication 1800-1b: Securing Electronic Health Records on Mobile Devices: Approach, Architecture, and Security Characteristics. Leah Kauffman, editor-in-chief. Gaithersburg, MD: National Institute of Standards & Technology, 2015.

NIST Special Publication 800-100: Information Security Handbook: A Guide for Managers. Edited by P. Bowen, J. Hash, and M. Wilson. Gaithersburg, MD: National Institute of Standards & Technology, 2006.

NIST Special Publication 800-53, Revision 4: Security and Privacy Controls for Federal Information Systems and Organizations. Joint Task Force Transformation Initiative. Gaithersburg, MD: National Institute of Standards & Technology, 2013.

Niven, Jeremy E. "Invertebrate neurobiology: Visual direction of arm movements in an octopus." *Current Biology* 21, no. 6 (2011): R217-R218.

Noran, Ovidiu. "A Mapping of Individual Architecture Frameworks (GRAI, PERA, C4ISR, CI-MOSA, ZACHMAN, ARIS) onto GERAM." In *Handbook on Enterprise Architecture*, edited by Peter Bernus, Laszlo Nemes, and Günter Schmidt, pp. 65-210. International Handbooks on Information Systems. Springer Berlin Heidelberg, 2003.

Nouvel, Pascal. "A Scale and a Paradigmatic Framework for Human Enhancement." In *Inquiring into Human Enhancement*, edited by Simone Bateman, Jean Gayon, Sylvie Allouche, Jérôme Goffette, and Michela Marzano, pp. 103-18. Palgrave Macmillan UK, 2015.

Novaković, Branko, Dubravko Majetić, Josip Kasać, and Danko Brezak. "Artificial Intelligence and Biorobotics: Is an Artificial Human Being our Destiny?" In *Annals of DAAAM for 2009 & Proceedings of the 20th International DAAAM Symposium "Intelligent Manufacturing & Automation: Focus on Theory, Practice and Education,"* edited by Branko Katalinic, pp. 121-22. Vienna: DAAAM International, 2009.

Null, Linda, and Julia Lobur. *The Essentials of Computer Organization and Architecture*. Second edition. Sudbury, MA: Jones and Bartlett Publishers, 2006.

Obaid, M., I. Damian, F. Kistler, B. Endrass, J. Wagner, and E. André. "Cultural Behaviors of Virtual Agents in an Augmented Reality Environment." In *Intelligent Virtual Agents*, pp. 412-18. Lecture Notes in Computer Science no. 7502. Springer Berlin Heidelberg, 2012.

Ochsner, Beate, Markus Spöhrer, and Robert Stock. "Human, non-human, and beyond: cochlear implants in socio-technological environments." *NanoEthics* 9, no. 3 (2015): 237-50.

Okun, Michael S. "Parkinson's Disease: Guide to Deep Brain Stimulation Therapy." Second edition. National Parkinson Foundation, 2014. http://www.parkinson.org/sites/default/files/Guide_to_DBS_Stimulation_Therapy.pdf. Accessed December 8, 2016.

Olson, Eric T. "Personal Identity." *The Stanford Encyclopedia of Philosophy* (Fall 2015 Edition), edited by Edward N. Zalta. http://plato.stanford.edu/archives/fall2015/entries/identity-personal/. Accessed January 17, 2016.

Osterwalder, Alexander, and Yves Pigneur. *Business Model Generation: A Handbook for Visionaries, Game Changers, and Challengers*. John Wiley & Sons, 2010.

The Oxford Handbook of Philosophy of Perception, edited by Mohan Matthen. Oxford: Oxford University Press, 2015.

Panno, Joseph. *Gene Therapy: Treating Disease by Repairing Genes*. New York: Facts on File, 2005.

Panoulas, Konstantinos J., Leontios J. Hadjileontiadis, and Stavros M. Panas. "Brain-Computer Interface (BCI): Types, Processing Perspectives and Applications." In *Multimedia Services in Intelligent Environments*, edited by George A. Tsihrintzis and Lakhmi C. Jain, pp. 299-321. Smart Innovation, Systems and Technologies 3. Springer Berlin Heidelberg, 2010.

Park, M.C., M.A. Goldman, T.W. Belknap, and G.M. Friehs. "The Future of Neural Interface Technology." In *Textbook of Stereotactic and Functional Neurosurgery*, edited by A.M. Lozano, P.L. Gildenberg, and R.R. Tasker, pp. 3185-3200. Heidelberg/Berlin: Springer, 2009.

Parker, Donn B. "Toward a New Framework for Information Security." In *The Computer Security Handbook*, fourth edition, edited by Seymour Bosworth and M. E. Kabay. John Wiley & Sons, 2002.

Patel, Rajeev, Robert J. Torres, and Peter Rosset. "Genetic engineering in agriculture and corporate engineering in public debate: risk, public relations, and public debate over genetically modified crops." *International journal of occupational and environmental health* 11, no. 4 (2005).

Patil, P.G., and D.A. Turner. "The Development of Brain-Machine Interface Neuroprosthetic Devices." In *Neurotherapeutics* 5, no. 1 (January 1, 2008): 137-46.

Patoine, Brenda. "Progress Report 2010: Deep Brain Stimulation – The 2010 Progress Report on Brain Research." The Dana Foundation, January 2010. http://www.dana.org/Publications/ReportDetails.aspx?id=44348. Accessed December 8, 2016.

Payr, S., and Trappl, R. "Agents across Cultures." In *Intelligent Virtual Agents*, pp. 320-24. Lecture Notes in Computer Science 2792. Springer Berlin Heidelberg, 2003.

Pazzaglia, Mariella, and Marco Molinari. "The embodiment of assistive devices – from wheelchair to exoskeleton." *Physics of Life Reviews* 16 (2016): 163-75.

Pearce, David. "The Biointelligence Explosion." In *Singularity Hypotheses*, edited by A.H. Eden, J.H. Moor, J.H. Søraker, and E. Steinhart, pp. 199-238. The Frontiers Collection. Berlin/Heidelberg: Springer, 2012.

Peigneux, Philippe. "Neuroimaging Studies of Sleep and Memory in Humans." In *Sleep, Neuronal Plasticity and Brain Function*, edited by Peter Meerlo, Ruth M. Benca, and Ted Abel, pp. 239-68. Springer Berlin Heidelberg, 2015.

Pérez Ríos, José. "Systems Thinking, Organisational Cybernetics and the Viable System Model." In *Design and Diagnosis for Sustainable Organizations*, pp. 1-64. Springer Berlin Heidelberg, 2012.

Peterson, David J. *The Art of Language Invention: From Horse-Lords to Dark Elves, The Words Behind World-Building*. New York: Penguin Books, 2015.

Pino, Robinson E., and Alexander Kott. "Neuromorphic Computing for Cognitive Augmentation in Cyber Defense." In *Cybersecurity Systems for Human Cognition Augmentation*, edited by Robinson E. Pino, Alexander Kott, and Michael Shevenell, pp. 19-46. Springer International Publishing, 2014.

Polcar, Jiri, and Petr Horejsi. "Knowledge Acquisition and Cyber Sickness: A Comparison of VR Devices in Virtual Tours." *MM Science Journal* (June 2015): 613-16.

Pollatsek, Alexander. "The Role of Sound in Silent Reading." In *The Oxford Handbook of Reading*, edited by Alexander Pollatsek and Rebecca Treiman, pp. 185-201. New York: Oxford University Press, 2015.

Postmarket Management of Cybersecurity in Medical Devices: Draft Guidance for Industry and Food and Drug Administration Staff. Silver Spring, MD: US Food and Drug Administration, 2016.

Prestes, E., J.L. Carbonera, S. Rama Fiorini, V.A.M. Jorge, M. Abel, R. Madhavan, A. Locoro, et al. "Towards a Core Ontology for Robotics and Automation." *Robotics and Autonomous Systems* 61, no. 11 (November 2013): 1193-1204.

Pribram, K.H. "Prolegomenon for a Holonomic Brain Theory." In *Synergetics of Cognition*, edited by Hermann Haken and Michael Stadler, pp. 150-84. Springer Series in Synergetics 45. Springer Berlin Heidelberg, 1990.

Pribram, K.H., and S.D. Meade. "Conscious Awareness: Processing in the Synaptodendritic Web – The Correlation of Neuron Density with Brain Size." *New Ideas in Psychology* 17, no. 3 (December 1, 1999): 205-14.

Primer on the Autonomic Nervous System. Third edition. Edited by David Robertson, Italo Biaggioni, Geoffrey Burnstock, Phillip A. Low, and Julian F.R. Paton. London: Academic Press, 2012.

Principe, José C., and Dennis J. McFarland. "BMI/BCI Modeling and Signal Processing." In *Brain-Computer Interfaces*, pp. 47-64. Springer Netherlands, 2008.

"Products and Procedures." Medtronic, 2016. http://professional.medtronic.com/pt/neuro/dbs-md/prod/index.htm. Accessed December 8, 2016.

294 • Neuroprosthetic Supersystems Architecture

"Prosthetics: Sponsor." Johns Hopkins Applied Physics Laboratory. http://www.jhuapl.edu/prosthetics/program/sponsor.asp. Accessed December 5, 2016.

Proudfoot, Diane. "Software Immortals: Science or Faith?" In *Singularity Hypotheses*, edited by Amnon H. Eden, James H. Moor, Johnny H. Søraker, and Eric Steinhart, pp. 367-92. The Frontiers Collection. Springer Berlin Heidelberg, 2012.

Putnam, Hilary. *Reason, Truth and History*. Cambridge: Cambridge University Press, 1981.

Radvansky, Gabriel A. *Human Memory*. Second edition. New York: Routledge, 2016.

Ramirez, S., X. Liu, P.-A. Lin, J. Suh, M. Pignatelli, R.L. Redondo, T.J. Ryan, and S. Tonegawa. "Creating a False Memory in the Hippocampus." *Science* 341, no. 6144 (2013): 387-91.

Rao, R.P.N., A. Stocco, M. Bryan, D. Sarma, T.M. Youngquist, J. Wu, and C.S. Prat. "A direct brain-to-brain interface in humans." *PLoS ONE* 9, no. 11 (2014).

Rao, Umesh Hodeghatta, and Umesha Nayak. *The InfoSec Handbook*. New York: Apress, 2014.

Regalado, Antonio. "Engineering the perfect baby." *MIT Technology Review* 118, no. 3 (2015): 27-33.

Rehm, M., André, E., and Nakano, Y. "Some Pitfalls for Developing Enculturated Conversational Agents." In *Human-Computer Interaction: Ambient, Ubiquitous and Intelligent Interaction*, pp. 340-48. Lecture Notes in Computer Science 5612. Springer Berlin Heidelberg, 2009.

Rehm, M., Y. Nakano, E. André, T. Nishida, N. Bee, B. Endrass, M. Wissner, A.A. Lipi, and H. Huang, "From observation to simulation: generating culture-specific behavior for interactive systems." *AI & SOCIETY* vol. 24, no. 3 (2009): 267-80.

Reynolds, Dwight W., Christina M. Murray, and Robin E. Germany. "Device Therapy for Remote Patient Management." In *Electrical Diseases of the Heart*, edited by Ihor Gussak, Charles Antzelevitch, Arthur A. M. Wilde, Paul A. Friedman, Michael J. Ackerman, and Win-Kuang Shen, pp. 809-25. Springer London, 2008.

Robertazzi, Thomas G. *Networks and Grids: Technology and Theory*. Springer-Verlag New York, 2007.

Robinett, W. "The consequences of fully understanding the brain." In *Converging Technologies for Improving Human Performance: Nanotechnology, Biotechnology, Information Technology and Cognitive Science*, edited by M.C. Roco and W.S. Bainbridge, pp. 166-70. National Science Foundation, 2002.

Roden, David. *Posthuman Life: Philosophy at the Edge of the Human*. Abingdon: Routledge, 2014.

Rodriguez, Ana Maria. *Autism Spectrum Disorders*. Minneapolis: Twenty-First Century Books, 2011.

Rohloff, Michael. "Framework and Reference for Architecture Design." In *AMCIS 2008 Proceedings*, 2008. http://citeseerx.ist.psu.edu/viewdoc/download?doi=10.1.1.231.8261&rep=rep1&type=pdf. Accessed October 21, 2015.

Rookes, Paul, and Jane Willson. *Perception: Theory, Development and Organisation*. London: Routledge, 2000.

Roosendaal, Arnold. "Implants and Human Rights, in Particular Bodily Integrity." In *Human ICT Implants: Technical, Legal and Ethical Considerations*, edited by Mark N. Gasson, Eleni Kosta, and Diana M. Bowman, pp. 81-96. Information Technology and Law Series 23. T. M. C. Asser Press, 2012.

Rosenbaum, David A. *Human Motor Control*. Second edition. Burlington, MA: Elsevier, 2010.

Rotter, Pawel, Barbara Daskala, and Ramon Compañó. "Passive Human ICT Implants: Risks and Possible Solutions." In *Human ICT Implants: Technical, Legal and Ethical Considerations*, edited by Mark N. Gasson, Eleni Kosta, and Diana M. Bowman, pp. 55-62. Information Technology and Law Series 23. T. M. C. Asser Press, 2012.

Rowlands, Mark. *Can Animals Be Moral?* Oxford: Oxford University Press, 2012.

Royakkers, Lambèr, and Rinie van Est. "A literature review on new robotics: automation from love to war." *International journal of social robotics* 7, no. 5 (2015): 549-70.

Rozhok, Andrii. *Orientation and Navigation in Vertebrates.* Springer-Verlag Berlin Heidelberg, 2008.

Rugg, Michael D., Jeffrey D. Johnson, and Melina R. Uncapher. "Encoding and Retrieval in Episodic Memory: Insights from fMRI." In *The Wiley Handbook on the Cognitive Neuroscience of Memory*, edited by Donna Rose Addis, Morgan Barense, and Audrey Duarte, pp. 84-107. Malden, MA: John Wiley & Sons, 2015.

Rutherford, Andrew, Gerasimos Markopoulos, Davide Bruno, and Mirjam Brady-Van den Bos. "Long-Term Memory: Encoding to Retrieval." In *Cognitive Psychology*, second edition, edited by Nick Braisby and Angus Gellatly, pp. 229-65. Oxford: Oxford University Press, 2012.

Rutten, W. L. C., T. G. Ruardij, E. Marani, and B. H. Roelofsen. "Neural Networks on Chemically Patterned Electrode Arrays: Towards a Cultured Probe." In *Operative Neuromodulation*, edited by Damianos E. Sakas and Brian A. Simpson, pp. 547-54. Acta Neurochirurgica Supplements 97/2. Springer Vienna, 2007.

Rynes, Sara L., Barry Gerhart, and Kathleen A. Minette. "The importance of pay in employee motivation: Discrepancies between what people say and what they do." *Human Resource Management* 43, no. 4 (2004): 381-94.

Saha, S.K. *Introduction to Robotics.* New Delhi: Tata McGraw-Hill Publishing Company Limited, 2008.

Sakas, Damianos E., I. G. Panourias, B. A. Simpson, and E. S. Krames. "An Introduction to Operative Neuromodulation and Functional Neuroprosthetics, the New Frontiers of Clinical Neuroscience and Biotechnology." In *Operative Neuromodulation*, edited by Damianos E. Sakas and Brian A. Simpson, pp. 2-10. Acta Neurochirurgica Supplements 97/1. Springer Vienna, 2007.

Salvini, Pericle, Cecilia Laschi, and Paolo Dario. "From robotic tele-operation to tele-presence through natural interfaces." In *The First IEEE/RAS-EMBS International Conference on Biomedical Robotics and Biomechatronics, 2006. BioRob 2006*, pp. 408-13. IEEE, 2006.

Sanchez, Justin C. *Neuroprosthetics: Principles and Applications.* Boca Raton: CRC Press, 2016.

Sandberg, Anders. "Ethics of brain emulations." *Journal of Experimental & Theoretical Artificial Intelligence* 26, no. 3 (2014): 439-57.

Sandor, Christian, Martin Fuchs, Alvaro Cassinelli, Hao Li, Richard Newcombe, Goshiro Yamamoto, and Steven Feiner. "Breaking the Barriers to True Augmented Reality." arXiv preprint, *arXiv:1512.05471 [cs.HC]*, December 17, 2015. http://arxiv.org/abs/1512.05471. Accessed January 25, 2016.

Sasse, Martina Angela, Sacha Brostoff, and Dirk Weirich. "Transforming the 'weakest link' – a human/computer interaction approach to usable and effective security." *BT technology journal* 19, no. 3 (2001): 122-31.

Sayood, Khalid. *Introduction to Data Compression.* Waltham, MA: Morgan Kaufmann, 2012.

Sayrafian-Pour, K., W.-B. Yang, J. Hagedorn, J. Terrill, K. Yekeh Yazdandoost, and K. Hamaguchi. "Channel Models for Medical Implant Communication." *International Journal of Wireless Information Networks* 17, no. 3-4 (December 9, 2010): 105-12.

Schermer, Maartje. "The Mind and the Machine. On the Conceptual and Moral Implications of Brain-Machine Interaction." *NanoEthics* 3, no. 3 (December 1, 2009): 217-30.

Schwartz, Bennett L. *Memory: Foundations and Applications.* Second edition. Thousand Oaks, CA: SAGE Publications, Inc., 2014.

"Security Risk Assessment Framework for Medical Devices." Washington, DC: Medical Device Privacy Consortium, 2014.

Self-Organizing Networks (SON): Self-Planning, Self-Optimization and Self-Healing for GSM, UMTS, and LTE, edited by Juan Ramiro and Khalid Hamied. Chichester: John Wiley & Sons Ltd., 2012.

Sensory Evolution on the Threshold: Adaptations in Secondarily Aquatic Vertebrates, edited by J.G.M. Thewissen and Sirpa Nummela. Berkeley: University of California Press, 2008.

Settle-Murphy, Nancy M. *Leading Effective Virtual Teams: Overcoming Time and Distance to Achieve Exceptional Results.* Boca Raton: CRC Press, 2012.

Shahinpoor, Mohsen, Kwang J. Kim, and Mehran Mojarrad. *Artificial Muscles: Applications of Advanced Polymeric Nanocomposites.* Boca Raton: Taylor & Francis, 2007.

Shekhar, Sandhya. *Managing the Reality of Virtual Organizations.* Springer India, 2016.

Sherwood, Lauralee. *Fundamentals of Human Physiology.* Fourth edition. Belmont, CA: Brooks/Cole, 2012.

Shoniregun, Charles A., Kudakwashe Dube, and Fredrick Mtenzi. "Introduction to E-Healthcare Information Security." In *Electronic Healthcare Information Security,* pp. 1-27. Advances in Information Security 53. Springer US, 2010.

Siegel, Allan, and Hreday Sapru. *Essential Neuroscience.* Baltimore: Lippincott Williams & Wilkins, 2006.

Siewert, Charles. "Consciousness and Intentionality." *The Stanford Encyclopedia of Philosophy* (Fall 2011 Edition), edited by Edward N. Zalta. http://plato.stanford.edu/archives/fall2011/entries/consciousness-intentionality/.

Siewiorek, Daniel, and Robert Swarz. *Reliable Computer Systems: Design and Evaluation.* Second edition. Burlington: Digital Press, 1992.

Sloman, Aaron. "Some Requirements for Human-like Robots: Why the recent over-emphasis on embodiment has held up progress." In *Creating brain-like intelligence,* pp. 248-77. Springer Berlin Heidelberg, 2009.

Smith, C.U.M., *Biology of Sensory Systems.* Second edition. Chichester: John Wiley & Sons Ltd., 2008.

Smolensky, Paul. "The constituent structure of connectionist mental states: A reply to Fodor and Pylyshyn." *The Southern Journal of Philosophy* 26, no. S1 (1988): 137-61.

Snider, Greg S. "Cortical Computing with Memristive Nanodevices." *SciDAC Review* 10 (2008): 58-65.

Snyder, Allan W., Elaine Mulcahy, Janet L. Taylor, D. John Mitchell, Perminder Sachdev, and Simon C. Gandevia. "Savant-like skills exposed in normal people by suppressing the left fronto-temporal lobe." *Journal of integrative neuroscience* 2, no. 02 (2003): 149-58.

Snyder, Allan. "Explaining and inducing savant skills: privileged access to lower level, less-processed information." *Philosophical Transactions of the Royal Society of London B: Biological Sciences* 364, no. 1522 (2009): 1399-1405.

Social Robots and the Future of Social Relations, edited by Johanna Seibt, Raul Hakli, and Marco Nørskov. Amsterdam: IOS Press, 2014.

Social Robots from a Human Perspective, edited by Jane Vincent, Sakari Taipale, Bartolomeo Sapio, Giuseppe Lugano, and Leopoldina Fortunati. Springer International Publishing, 2015.

Social Robots: Boundaries, Potential, Challenges, edited by Marco Nørskov. Farnham: Ashgate Publishing, 2016.

Sohl-Dickstein, Jascha, Santani Teng, Benjamin M. Gaub, Chris C. Rodgers, Crystal Li, Michael R. DeWeese, and Nicol S. Harper. "A device for human ultrasonic echolocation." *IEEE Transactions on Biomedical Engineering* 62, no. 6 (2015): 1526-34.

Sosinsky, Barrie. *Networking Bible.* Indianapolis: Wiley Publishing Inc., 2009.

Soussou, Walid V., and Theodore W. Berger. "Cognitive and Emotional Neuroprostheses." In *Brain-Computer Interfaces,* pp. 109-23. Springer Netherlands, 2008.

"Spinal Cord Stimulation." American Association of Neurological Surgeons, October 2008. http://www.aans.org/Patient%20Information/Conditions%20and%20Treatments/Spinal%20Cord%20Stimulation.aspx. Accessed December 8, 2016.

Spohrer, Jim. "NBICS (Nano-Bio-Info-Cogno-Socio) Convergence to Improve Human Performance: Opportunities and Challenges." In *Converging Technologies for Improving Human Performance: Nanotechnology, Biotechnology, Information Technology and Cognitive Science,* edited by M.C. Roco and W.S. Bainbridge, pp. 101-17. Arlington, VA: National Science Foundation, 2002.

Srinivasan, G. R. "Modeling the cosmic-ray-induced soft-error rate in integrated circuits: an overview." *IBM Journal of Research and Development* 40, no. 1 (1996): 77-89.

Stahl, B. C. "Responsible Computers? A Case for Ascribing Quasi-Responsibility to Computers Independent of Personhood or Agency." *Ethics and Information Technology* 8, no. 4 (2006): 205-13.

Stallings, William. *Cryptography and Network Security: Principles and Practice.* Seventh edition. Harlow: Pearson Education Limited, 2017.

Starr, Cecie, and Beverly McMillan. *Human Biology.* Eleventh edition. Boston: Cengage Learning, 2016.

Stelzer, Dirk. "Enterprise Architecture Principles: Literature Review and Research Directions." In *Service-Oriented Computing. ICSOC/ServiceWave 2009 Workshops,* pp. 12-21. Springer, 2010.

Stieglitz, Thomas. "Restoration of Neurological Functions by Neuroprosthetic Technologies: Future Prospects and Trends towards Micro-, Nano-, and Biohybrid Systems." In *Operative Neuromodulation,* edited by Damianos E. Sakas, Brian A. Simpson, and Elliot S. Krames, pp. 435-42. Acta Neurochirurgica Supplements 97/1. Springer Vienna, 2007.

Streib, James T. *Guide to Assembly Language: A Concise Introduction.* Springer-Verlag London Limited, 2011.

"Studies Find Disparities in Use of Deep Brain Stimulation." Parkinson's Disease Foundation, January 29, 2014. http://www.pdf.org/en/science_news/release/pr_1391019029. Accessed December 8, 2016.

Sundberg, Hakan P. "Building the Enterprise Architecture: A Bottom-Up Evolution?" In *Advances in Information Systems Development,* edited by Wita Wojtkowski, W. Gregory Wojtkowski, Jože Zupancic, Gabor Magyar, and Gabor Knapp, 287. Springer US, 2007.

"Surgeons Publish Study on Auditory Brainstem Implant Procedure." *The Hearing Review,* May 22, 2015. http://www.hearingreview.com/2015/05/surgeons-publish-study-auditory-brainstem-implant-procedure/. Accessed December 8, 2016.

Synaesthesia: Theoretical, artistic and scientific foundations, edited by María José de Córdoba Serrano, Dina Riccò, and Sean A. Day. Granada: International Foundation Artecittà Publishing, 2014.

Szoldra, P. "The government's top scientists have a plan to make military cyborgs." *Tech Insider*, January 22, 2016. http://www.techinsider.io/darpa-neural-interface-2016-1. Accessed May 6, 2016.

Talan, J. "DARPA: On the Hunt for Neuroprosthetics to Enhance Memory." *Neurology Today* 14, no. 20 (2014): 8-10.

Tang, Hao, Yun Fu, Jilin Tu, Mark Hasegawa-Johnson, and Thomas S. Huang. "Humanoid audio–visual avatar with emotive text-to-speech synthesis." *IEEE Transactions on multimedia* 10, no. 6 (2008): 969-81.

Targeted Muscle Reinnervation: A Neural Interface for Artificial Limbs, edited by Todd A. Kuiken, Aimee E. Schultz Feuser, and Ann K. Barlow. Boca Raton: CRC Press, 2014.

Tarín, C., L. Traver, P. Martí, and N. Cardona. "Wireless Communication Systems from the Perspective of Implantable Sensor Networks for Neural Signal Monitoring." In *Wireless Technology*, edited by S. Powell and J.P. Shim, pp. 177-201. Lecture Notes in Electrical Engineering 44. Springer US, 2009.

Taylor, N. R., and J. G. Taylor. "The Neural Networks for Language in the Brain: Creating LAD." In *Computational Models for Neuroscience*, edited by Robert Hecht-Nielsen and Thomas McKenna, pp. 245-65. Springer London, 2003.

Taylor, Annette Kujawski, "Hyperthymesia." In *Encyclopedia of Human Memory*, edited by Annette Kujawski Taylor, pp. 547-50. Santa Barbara: Greenwood, 2013.

Taylor, Dawn M. "Functional Electrical Stimulation and Rehabilitation Applications of BCIs." In *Brain-Computer Interfaces*, pp. 81-94. Springer Netherlands, 2008.

Taylor, Dawn M., Stephen I. Helms Tillery, and Andrew B. Schwartz. "Direct cortical control of 3D neuroprosthetic devices." *Science* 296, no. 5574 (2002): 1829-32.

Teng, Santani, and David Whitney. "The acuity of echolocation: spatial resolution in the sighted compared to expert performance." *Journal of visual impairment & blindness* 105, no. 1 (2011): 20-32.

Tennison, Michael N., and Jonathan D. Moreno. "Neuroscience, Ethics, and National Security: The State of the Art." *PLOS Biology*, March 20, 2012. http://dx.doi.org/10.1371/journal.pbio.1001289. Accessed December 5, 2016.

Thanos, Solon, P. Heiduschka, and T. Stupp. "Implantable Visual Prostheses." In *Operative Neuromodulation*, edited by Damianos E. Sakas and Brian A. Simpson, pp. 465-72. Acta Neurochirurgica Supplements 97/2. Springer Vienna, 2007.

Thornton, Stephanie. *Understanding Human Development: Biological, Social and Psychological Processes from Conception to Adult Life*. New York: Palgrave Macmillan, 2008.

Thorpe, Julie, Paul C. van Oorschot, and Anil Somayaji. "Pass-thoughts: authenticating with our minds." In *Proceedings of the 2005 Workshop on New Security Paradigms*, pp. 45-56. ACM, 2005.

TOGAF® Version 9.1. Berkshire: The Open Group, 2011.

Tomas, David. "Feedback and Cybernetics: Reimaging the Body in the Age of the Cyborg." In *Cyberspace, Cyberbodies, Cyberpunk: Cultures of Technological Embodiment*, edited by Mike Featherstone and Roger Burrows, pp. 21-43. London: SAGE Publications, 1995.

Transcranial Magnetic Stimulation in Clinical Psychiatry, edited by Mark S. George and Robert H. Belmaker. Washington, DC: American Psychiatric Publishing, Inc., 2007.

Treffert, Darold A. "Accidental Genius." *Scientific American* 311, no. 2 (2014): 52-57.

Treffert, Darold A. "The savant syndrome: an extraordinary condition. A synopsis: past, present, future." *Philosophical Transactions of the Royal Society of London B: Biological Sciences* 364, no. 1522 (2009): 1351-57.

Troyk, Philip R., and Stuart F. Cogan. "Sensory Neural Prostheses." In *Neural Engineering*, edited by Bin He, pp. 1-48. Bioelectric Engineering. Springer US, 2005.

Turner, Patrick, John Gøtze, and Peter Bernus. "Architecting the Firm – Coherency and Consistency in Managing the Enterprise." In *On the Move to Meaningful Internet Systems: OTM 2009 Workshops*, edited by Robert Meersman, Pilar Herrero, and Tharam Dillon, pp. 162-71. Lecture Notes in Computer Science 5872. Springer Berlin Heidelberg, 2009.

Upper Motor Neurone Syndrome and Spasticity: Clinical Management and Neurophysiology, second edition, edited by Michael P. Barnes and Garth R. Johnson. Cambridge: Cambridge University Press, 2008.

Van den Berg, Bibi. "Pieces of Me: On Identity and Information and Communications Technology Implants." In *Human ICT Implants: Technical, Legal and Ethical Considerations*, edited by Mark N. Gasson, Eleni Kosta, and Diana M. Bowman, pp. 159-73. Information Technology and Law Series 23. T. M. C. Asser Press, 2012.

Van der Raadt, Bas and Hans van Vliet. "Assessing the Efficiency of the Enterprise Architecture Function." In *Advances in Enterprise Engineering II*, edited by Erik Proper, Frank Harmsen, and Jan L. G. Dietz, 63. Lecture Notes in Business Information Processing 28. Springer Berlin Heidelberg, 2009.

Van der Torre, Leendert, Marc M. Lankhorst, Hugo ter Doest, Jan T. P. Campschroer, and Farhad Arbab. "Landscape Maps for Enterprise Architectures." In *Advanced Information Systems Engineering*, edited by Eric Dubois and Klaus Pohl. Lecture Notes in Computer Science 4001. Springer Berlin Heidelberg, 2006.

Van Drongelen, Wim, Hyong C. Lee, and Kurt E. Hecox. "Seizure Prediction in Epilepsy." In *Neural Engineering*, edited by Bin He, pp. 389-419. Bioelectric Engineering. Springer US, 2005.

Vänni, Kimmo J., and Annina K. Korpela. "Role of Social Robotics in Supporting Employees and Advancing Productivity." In *Social Robotics*, pp. 674-83. Springer International Publishing, 2015.

Versace, Massimiliano, and Ben Chandler. "The Brain of a New Machine." *IEEE spectrum* 47, no. 12 (2010): 30-37.

Vinciarelli, A., M. Pantic, D. Heylen, C. Pelachaud, I. Poggi, F. D'Errico, and M. Schröder. "Bridging the Gap between Social Animal and Unsocial Machine: A survey of Social Signal Processing." *IEEE Transactions on Affective Computing* 3:1 (January-March 2012): 69-87.

Viola, M. V., and Aristides A. Patrinos. "A Neuroprosthesis for Restoring Sight." In *Operative Neuromodulation*, edited by Damianos E. Sakas and Brian A. Simpson, pp. 481-86. Acta Neurochirurgica Supplements 97/2. Springer Vienna, 2007.

Virtual Organizations: Systems and Practices, edited by Luis M. Camarinha-Matos, Hamideh Afsarmanesh, and Martin Ollus. Boston: Springer Science+Business Media, 2005.

Wallach, Wendell, and Colin Allen. *Moral machines: Teaching robots right from wrong*. Oxford University Press, 2008.

Warwick, K. "The Cyborg Revolution." *Nanoethics* 8 (2014): 263-73.

Warwick, K., and M. Gasson. "Implantable Computing." In *Digital Human Modeling*, edited by Y. Cai, pp. 1-16. Lecture Notes in Computer Science 4650. Berlin/Heidelberg: Springer, 2008.

Weiland, James D., Wentai Liu, and Mark S. Humayun. "Retinal Prosthesis." *Annual Review of Biomedical Engineering* 7, no. 1 (2005): 361-401.

Weinberger, Sharon. "Pentagon to Merge Next-Gen Binoculars with Soldiers' Brains." *Wired*, May 1, 2007. https://www.wired.com/2007/05/binoculars/. Accessed December 5, 2016.

Weiss, Simon, and Robert Winter. "Development of Measurement Items for the Institutionalization of Enterprise Architecture Management in Organizations." In *Trends in Enterprise Architecture Research and Practice-Driven Research on Enterprise Transformation*, edited by Stephan Aier, Mathias Ekstedt, Florian Matthes, Erik Proper, and Jorge L. Sanz, pp. 268-83. Lecture Notes in Business Information Processing 131. Springer Berlin Heidelberg, 2012.

Wells, M.J. *Octopus: Physiology and Behaviour of an Advanced Invertebrate*. Springer Netherlands, 1978.

Westlake, Philip R. "The possibilities of neural holographic processes within the brain." *Biological Cybernetics* 7, no. 4 (1970): 129-53.

White, Stephen E. "Brave new world: Neurowarfare and the limits of international humanitarian law." *Cornell International Law Journal* 41 (2008): 177.

Widge, A.S., C.T. Moritz, and Y. Matsuoka. "Direct Neural Control of Anatomically Correct Robotic Hands." In *Brain-Computer Interfaces*, edited by D.S. Tan and A. Nijholt, pp. 105-19. Human-Computer Interaction Series. London: Springer, 2010.

Wiener, Norbert. *Cybernetics: Or Control and Communication in the Animal and the Machine*, second edition. Cambridge, MA: The MIT Press, 1961. [Quid Pro ebook edition for Kindle, 2015.]

Wilkinson, Jeff, and Scott Hareland. "A cautionary tale of soft errors induced by SRAM packaging materials." *IEEE Transactions on Device and Materials Reliability* 5, no. 3 (2005): 428-33.

Williams, Theodore J., and Hong Li. "PERA and GERAM – Enterprise Reference Architectures in Enterprise Integration." In *Information Infrastructure Systems for Manufacturing II*, edited by John J. Mills and Fumihiko Kimura, pp. 3-30. IFIP – The International Federation for Information Processing 16. Springer US, 1999.

Wilson, Margaret. "Six views of embodied cognition." *Psychonomic bulletin & review* 9, no. 4 (2002): 625-36.

Wise, Kensall D., Amir M. Sodagar, Ying Yao, Mayurachat Ning Gulari, Gayatri E. Perlin, and Khalil Najafi. "Microelectrodes, microelectronics, and implantable neural microsystems." *Proceedings of the IEEE* 96, no. 7 (2008): 1184-1202.

Wise, Kensall D., D. J. Anderson, J. F. Hetke, D. R. Kipke, and K. Najafi. "Wireless implantable microsystems: high-density electronic interfaces to the nervous system." *Proceedings of the IEEE* 92, no. 1 (2004): 76-97.

Woisetschläger, David M. "Consumer Perceptions of Automated Driving Technologies: An Examination of Use Cases and Branding Strategies." In *Autonomous Driving*, pp. 687-706. Springer Berlin Heidelberg, 2016.

Wolf-Meyer, Matthew. "Fantasies of extremes: Sports, war and the science of sleep." *BioSocieties* 4, no. 2 (2009): 257-71.

Wooldridge, M., and N. R. Jennings. "Intelligent agents: Theory and practice." *The Knowledge Engineering Review*, 10(2) (1995): 115-52.

Yampolskiy, Roman V. "The Universe of Minds." arXiv preprint, *arXiv:1410.0369 [cs.AI]*, October 1, 2014. http://arxiv.org/abs/1410.0369. Accessed January 25, 2016.

Yonck, Richard. "Toward a standard metric of machine intelligence." *World Future Review* 4, no. 2 (2012): 61-70.

Zebda, Abdelkader, S. Cosnier, J.-P. Alcaraz, M. Holzinger, A. Le Goff, C. Gondran, F. Boucher, F. Giroud, K. Gorgy, H. Lamraoui, and P. Cinquin. "Single glucose biofuel cells implanted in rats power electronic devices." *Scientific Reports* 3, article 1516 (2013).

Zhao, QiBin, LiQing Zhang, and Andrzej Cichocki. "EEG-Based Asynchronous BCI Control of a Car in 3D Virtual Reality Environments." *Chinese Science Bulletin* 54, no. 1 (January 11, 2009): 78-87.

Zofi, Yael. *A Manager's Guide to Virtual Teams.* New York: American Management Association, 2012.

Zullo, Letizia, German Sumbre, Claudio Agnisola, Tamar Flash, and Binyamin Hochner. "Non-somatotopic organization of the higher motor centers in octopus." *Current Biology* 19, no. 19 (2009): 1632-36.

Index

About the Author

Matthew E. Gladden is a management consultant and researcher whose work focuses on the organizational implications of emerging technologies such as those relating to artificial intelligence, social robotics, virtual reality, neuroprosthetic enhancement, and artificial life. He lectures internationally on the relationship of such posthumanizing technologies to organizational life, and his research has been published in journals such as the *International Journal of Contemporary Management, Annals of Computer Science and Information Systems, Informatyka Ekonomiczna / Business Informatics, Creatio Fantastica,* and *Annales: Ethics in Economic Life,* as well as by IOS Press, Ashgate Publishing, the Digital Economy Lab of the University of Warsaw, and the MIT Press. His books include *Sapient Circuits and Digitalized Flesh: The Organization as Locus of Technological Posthumanization* (2016); *Posthuman Management: Creating Effective Organizations in an Age of Social Robotics, Ubiquitous AI, Human Augmentation, and Virtual Worlds* (second edition, 2016); and *The Handbook of Information Security for Advanced Neuroprosthetics* (second edition, 2017).

He is the founder and CEO of consulting firms NeuraXenetica LLC and Cognitive Firewall LLC. He previously served as Administrator of the Department of Psychology at Georgetown University and Associate Director of the Woodstock Theological Center and has also taught philosophical ethics and worked in computer game design. He is a member of ISACA, ISSA, and the Academy of Management and its divisions for Managerial and Organizational Cognition, Technology and Innovation Management, and Business Policy and Strategy. He completed his MBA in Innovation and Data Analysis at the Institute of Computer Science of the Polish Academy of Sciences and holds certificates in Advanced Business Management and Nonprofit Management from Georgetown University and a BA in Philosophy from Wabash College.